STUDIES
IN THE HISTORY
OF MEDIAEVAL SCIENCE

Studies
in the History
of Mediaeval Science

CHARLES HOMER HASKINS

FREDERICK UNGAR PUBLISHING CO.
NEW YORK

Copyright 1924, 1927 by
Harvard University Press

Copyright renewed 1955 by
Clare A. Haskins

Republished 1960 by arrangement
with the author's estate

PRINTED IN THE UNITED STATES OF AMERICA

Library of Congress Catalog Card No. 60-53135

TO

J. F. J.
G. L. H.
F. J. T.

PREFACE

THE history of European science in the Middle Ages is twofold. On the one hand it is concerned with the recovery and assimilation of the science of antiquity, little known at first and only gradually brought into the West, to some extent as enlarged by the Arabs, in the course of the twelfth and thirteenth centuries; while on the other hand, it has to take account of the advance of knowledge by the processes of observation and experiment in western Europe. The first phase deals primarily with translation from the Arabic and the Greek, in Spain, Sicily, North Africa, and the East, as a preliminary to the full assimilation of these successive increments of ancient learning and the Arabic additions thereto. The second, more obscure, has to trace the extension of knowledge by such means as the observation of plants and animals, especially dogs, hawks, and horses, the actual treatment of disease, geographical exploration, and the growth of the experimental habit. On both these sides a consecutive and comprehensive history still remains to be written, while at many points monographic investigation is entirely lacking.

Toward the materials for such a history the present volume is meant to offer a contribution. It is limited to the twelfth and early thirteenth centuries, the period of scientific revival, and to certain specific topics worked out primarily from the manuscript sources. After a survey of the place of Spain in the introduction of Arabic science into Europe, the pioneers of the new learning are studied in the person of Adelard of Bath, tutor of King Henry II, that extraordinary traveller in distant lands and student and translator of the mathematics, astronomy, astrology, and philosophy of his time, and in his immediate and little known succes-

vii

sors, Hermann of Carinthia and Hugh of Santalla. The astronomy of the computists and of the Platonists of Chartres is then noted as the background for the reception of the Aristotelian physics and the Ptolemaic astronomy in the course of the twelfth century, and the coming of the new astronomy and mathematics is illustrated in detail by the series of scholars who brought them to England. Some account is given of the modest part of Syria in the transmission of Arabic knowledge. The Greek phase of the mediaeval renaissance is then examined, and illustrated in detail by a study of the Sicilian translators which brings into fresh relief the significance of Sicily as a centre of diffusion for Greek mathematics, astronomy, and philosophy. A parallel movement is traced in northern Italy in the person of Latins resident at Constantinople, who brought to the West something of the stored up knowledge and superstition of the Byzantine capital. Then the court of the Emperor Frederick II is presented on its scientific side as the meeting-point of these Greek and Arabic currents, and as a fruitful centre of inquiry and experiment, as seen particularly in the writings of the emperor's adviser, Michael Scot, and in Frederick's own treatise on falcons, a highly characteristic product of this extraordinary mind. Other studies deal with the introduction of the abacus into the English exchequer, with Syrian astronomy and western falconry, and with a list of textbooks which sums up the curriculum of the close of the twelfth century. Of the ancient authors upon whom mediaeval learning depended, special attention is given to Aristotle, Ptolemy, and their influence, without neglecting Plato, Euclid, and the Greek physicians.

This series of studies was planned and in large measure written before the appearance of Lynn Thorndike's *History of Magic and Experimental Science* (New York, 1923), largely even before the publication of Pierre Duhem's *Le système du monde de Platon à*

Copernic (Paris, 1913–17); but much use has been made of
both. While the present volume traverses portions of the larger
field covered by each of these more ambitious works, its point of
view remains distinct. Thorndike's chief interest is in magic
taken in the broadest sense; Duhem is primarily concerned with
tracing cosmological ideas; whereas the present volume ap-
proaches mediaeval science from the point of view of the general
history of culture in the Middle Ages, and thus touches other
phases of science as well as philosophy, classical learning, and
even institutions in their relations to science. It is designed in
the first instance as a contribution to the history of the mediaeval
renaissance and the influence of eastern culture upon the West.

While the effort is to advance knowledge at critical points
rather than to tell a continuous story, the chapters have been
grouped so as to bring out the general connection, while three
general chapters (I, VIII, XII) sum up the present state of our
information on the Spanish translators from the Arabic, on Greek
studies, and on the court of Frederick II. Certain of these chapters are new, I, II (largely), III, V, VII, X (largely); others which
have appeared in various journals in the course of the past fifteen
years have been carefully revised, and in most instances extended
as the result of further investigation. Each chapter is based, in
part at least, upon unprinted sources and brings to light a certain
amount of material not previously known. Most of this research
has been performed on the spot, but it has been greatly facilitated
by photographic reproductions. These photographs of manuscripts were made possible by grants from the Woodbury Lowery
fund of Harvard University; they are available for the use of other
investigators in the Harvard Library.

The list would be long of the many scholars who have aided my
researches, but for special help in relation to manuscripts I must
mention particularly His Eminence Cardinal Ehrle and Mon-

signori G. Mercati and A. Pelzer at the Vatican; Mgr. L. Gra-
matica at the Ambrosian; Comm. I. Giorgi of the Biblioteca
Casanatense; the late Professor Eduardo de Hinojosa of Madrid;
Señor E. Hurtebise of the Archives of the Crown of Aragon at
Barcelona; MM. Henri Omont and Lucien Auvray at the Bib-
liothèque Nationale; Mr. J. A. Herbert at the British Museum;
Dr. H. H. E. Craster at the Bodleian; Mr. R. Livingstone of
Corpus Christi College, Oxford; the Provost of Eton; Professor
J. S. Reid at Cambridge; and Professor Clemens Baeumker of
Munich; besides many other librarians and scholars from Lisbon
to Vienna and from Edinburgh to Palermo. For other forms of
suggestion and assistance I am especially indebted to my master,
M. Charles-V. Langlois, who was good enough to review certain
chapters in the *Journal des savants* in 1919; Dr. Reginald Lane
Poole, to whom no student of the twelfth century ever turns in
vain; Mr. C. C. J. Webb, and Dr. Charles Singer; Dr. A. Birkenmajer, of Cracow; Dr. F. Liebermann, of Berlin; Professor J. L.
Heiberg of Copenhagen; the late Professor H. Suter of Zurich;
Professor R. Sabbadini of Milan; Professors D. E. Smith of Columbia, D. P. Lockwood of Haverford, L. C. Karpinski of Michigan, and Lynn Thorndike of Western Reserve; and to my
colleagues Messrs. Maurice De Wulf, E. K. Rand, George Sarton,
E. C. Streeter, and H. A. Wolfson. Mr. George W. Robinson,
Secretary of the Graduate School of Arts and Sciences of Harvard
University, has rendered invaluable assistance in correcting the
proofs and has prepared the index of proper names.

This book is dedicated to three friends who in my early years,
one as teacher, two as fellow-students and colleagues, contributed most to the formation of my ideals of scholarship.

CAMBRIDGE, April, 1924.

PREFACE TO THE SECOND EDITION

THE necessity of reprinting has given opportunity not only to correct certain printers' errors in the text, but also to make the following additions and corrections, which are the result partly of further research and partly of suggestions made by reviewers and others.

December, 1926.

ADDITIONS AND CORRECTIONS

CHAPTER I

PAGE

4. Mention should be made of the work of those scholars who ascribe to the Arabs of Spain an important influence on European music. See J. Ribera, *La música de los cantigas* (Madrid, 1922); and H. G. Farmer, "Clues for the Arabian Influence on European Musical Theory," in *Journal of the Royal Asiatic Society*, 1925, pp. 61–80; as well as his further studies in Arabic musical MSS.

5. On the relative influence of translations from the Arabic and translations from the Greek, cf. my paper on "Arabic Science in Western Europe," in *Isis*, vii. 478–485 (1925).

11, line 11. Read: 'Plato's dated versions.'

15, n. 46. The Alfonso who translated Averroës is placed in the fourteenth century by Birkenmajer, *Vermischte Untersuchungen* (Münster, 1922), pp. 17, 212.

CHAPTER II

20. F. Bliemetzrieder announces a considerable volume on Adelard of Bath for publication at Graz early in 1927.

26. There is also a fragment of the *Questiones naturales* at Leyden, MS. Voss. Lat. 4° 84, which Sandys (*History of Classical Scholarship*,[3] i. 655) mistook for an abridgment of the *Causae* ascribed to Quintilian.

CHAPTER III

55, 56. As Professor Thorndike has pointed out (*American Historical Review*, xxx. 346), the text here gives the impression of taking somewhat too literally the rhetorical phrases of Hermann's preface.

CHAPTER IV

78, n. 47. On the problem of the two geomancies, see Birkenmajer, in *Philosophisches Jahrbuch*, xxxviii. 282. Munich, Cod. lat. 3216, f. 18, should be added to the list of copies of Hugo.

79, notes 49, 50. MS. Lat. 1461 should be 4161, in which we read at f. 71 v 'Hugonis Sazeliensis translatio.'

Chapter V

87. The library of the Michelsberg at Bamberg had, between 1112 and 1123, a 'Liber Sarracenus de mathematica' and two Greek books on the same subject: G. Becker, *Catalogi bibliothecarum antiqui* (Bonn, 1885), p. 192, nos. 79, 114, 115.

88. Of course there were other sources of Platonism, such as Apuleius, Augustine, and Boethius.

89, line 14. Bernard of Chartres should be mentioned as well as Bernard Silvester; I did not mean to imply their identity. Cf. my paper on "An Italian Master Bernard," in the volume of studies dedicated to Reginald Lane Poole (Oxford, 1927).

91, n. 56. I now have a photograph of this MS. on which one of my students is at work.

95. Another such collection in B. N., MS. lat. 7412, ff. 80–89 v (saec. xii), is entitled *Questiones physice* ('Primis mensibus' . . .).

99. The library of Whitby, *ca.* 1180, had a 'Liber Mamnonis': Becker, *Catalogi*, p. 227, no. 47.

111. Professor Powicke (*E. H. R.*, xl. 422) corrects my date for Simon of Bredon as too early.

n. 163 end. Add MS. Vat. lat. 4075, copied in 1411.

Chapter VI

122. J. Ruska, *Arabische Alchimisten*, i (Heidelberg, 1924), has now thrown grave doubt upon the authenticity of the Latin Morienus as well as its ascription to Robert of Chester.

Chapter VII

133. For a Graeco-Arabic medical MS. from southern Italy, see below, under p. 184.

134. The Cesena MS., which also has the synonyms, has in the concluding date 1107, 26 January 'feria iiii.'

135. A treatise *De modo medendi* found in several MSS. is ascribed to Stephen of Antioch by R. Ganszyniec in *Archiv für Geschichte der Medizin*, xiv. 110–113 (1923).

136. Another copy of the *Experimentarius* is at Magdalene College, Cambridge, Pepys MS. 911, according to Sandys, *History of Classical Scholarship*,[3] i. 536. On Bernard Silvester, see above, p. 89.

139. Philip was archdeacon of Tripoli in 1267, according to Birkenmajer, *Philosophisches Jahrbuch*, xxxviii. 282.

CHAPTER VIII

148. The Italian journeys of John of Salisbury have now been reduced to six by R. L. Poole, *The Early Correspondence of John of Salisbury* (British Academy, 1924).

151 (cf. pp. 181, 233). On Boethius as a source of the mediaeval method of literal translation, see E. K. Rand, in *Jahrbücher für classische Philologie*, Suppl. xxvi. 429 ff. (1901).

154, n. 45. See also Heiberg, "Wie die Schriften des Alterthums an uns gelangt sind," in *Scientia*, xxxix. 81–88, 153–162 (1926).

CHAPTER IX

170, 188. On learning at the court of Henry II, see my paper in the *Essays Presented to Thomas Frederick Tout* (Manchester, 1925), pp. 71–77.

179. Other MSS. of the Latin version of Proclus are Vat. lat. 11600, ff. 137–139 v (saec. xv); and B.N., MS. lat. 15453, f. 412, a copy which Birkenmajer would date 1243, well before William of Moerbeke (*Philosophisches Jahrbuch*, xxxviii. 282).

184. An eleventh-century Greek medical MS. from southern Italy with Arabic glosses (ff. 113 v–115 v, 116, 118 v) is in B. N., Suppl. grec, MS. 1297. The provenance appears from a Latin note in a Beneventan hand on f. 89.

CHAPTER X

195, 197. The fragment of the Latin version of the *libellus* of Chrysolanus I have completed in *Byzantion*, ii. 234–236 (1926), from MS. 233 of Prague and a MS. at the Biblioteca Nazionale Centrale at Florence (Conv. soppr. I. iv. 21).

196. It appears from the new material noted under p. 218 that the visit of Henry of Grado to Constantinople falls *ca.* 1158. Cf. P. F. Kehr, *Italia pontificia*, vii, 2, p. 64, no. 121.

207, n. 57. Delete the reference to the Psalter.

208, n. 93. The account of the Latin versions of Galen in Donald Campbell, *Arabian Medicine* (London, 1926), follows Diels closely. According to the Vatican catalogue Burgundio is the author of the translation of the *De iuvamento membrorum* in Ms. Pal. lat. 1099, f. 1, but the MS. is silent on this point.

n. 94. The version of 1185 is also so dated in the Cesena MS. xxv. 2.

214. On the theological writings of Hugo Eterianus I am sorry to have overlooked the careful study of R. Lechat, in *Mélanges Charles Moeller* (Louvain, 1914), i. 485–507.

216. The *Oneirocriticon* of Achmet has now been edited by Drexl (Leipzig, 1925), who finds Leo's translation useful for the text. There is another MS. of Leo in the Riccardian, MS. 859, ff. 1–51 v (saec. xv).

218–221. Two further translations of Paschal the Roman, a *Dialogue with a Jew* of 1158 (1163?) and a life of the Virgin by Epiphanius, I have described in *Byzantion*, ii. 231–234 (1926). There is another undated text of the *Liber thesauri* in MS. Vat. lat. 4436, f. 1; and a sixth MS. of the Latin *Kiranides* at the Laurentian, MS. Ashburnham 1520 (1443), f. 1 (saec. xiv), with the date 1169.

221. The translation of Albumasar is ascribed to Stephen of Messina in 1262: Steinschneider, *E. U.*, no. 114.

222. The Leipzig dissertation of F. A. Krah (1924) throws doubt on the connection of Zacharias with Constantinople. See *Mitteilungen zur Geschichte der Medizin*, xxiii. 61 (1924).

CHAPTER XI

223–241. F. Bliemetzrieder, in *Philosophisches Jahrbuch*, xxxviii. 230–249 (1925), makes an elaborate argument for Burgundio the Pisan as the author of the version of the *Posterior Analytics* preserved at Toledo, but he brings forward no fresh evidence. Burgundio throws no light on the question in the preface of 1173 which lists earlier translations from the Greek (*infra*, p. 232, n. 37), for he omits both *Analytics* from his enumeration of the versions made by Boethius, and says nothing of his own translations or those of James of Venice.

239. For 'Egidio Colonna' read 'Egidio Romano.' Cf. Mandonnet, in *Revue des sciences philosophiques*, iv. 480–499 (1910); and Dyroff, in *Philosophisches Jahrbuch*, xxxviii. 18–25 (1925).

CHAPTER XII

242. On the different estimates of Frederick, see K. Hampe, *Kaiser Friedrich II in der Auffassung der Nachwelt* (Stuttgart, 1925). W. Cohn, *Das Zeitalter der Hohenstaufen in Sizilien* (Breslau, 1925), adds nothing on intellectual matters. For interesting sidelights, see A. Haseloff, *Die Bauten der Hohenstaufen in Unteritalien*, i (Leipzig, 1920).

250. On grammar and rhetoric at the University of Naples, see now my paper on "Magister Gualterius Esculanus" in the *Mélanges Ferdinand Lot* (Paris, 1925), pp. 245–257.

257. 'Pulvis domini F. imperatoris,' also in the Laurentian MS. xii. 27, f. 48.

263. Respecting the mission to Norway, Dr. Birkenmajer (*Philosophisches Jahrbuch*, xxxviii. 283) calls attention to earlier passages in Thomas of Cantimpré and Albertus Magnus.

269. Bartholomew of Messina also translated from the Greek for Manfred Hippocrates *De natura puerorum*: Vatican, MS. Vat. lat. 2382, ff. 95–98 v; Vat. lat. 2417, ff. 257 v–263). Another MS. of the version of Hierocles is also at the Vatican, MS. Reg. lat. 1301.

CHAPTER XIII

273, n. 8. Subsequent inspection of the MS. entirely confirms the reading 'MCC etc.'

277. It is suggested by Monsignore G. Mercati (*Miscellanea Ehrle*, v. 121) that the copy of the *De animalibus* and the *Abbreviatio Avicenne* now in the Chigi library at the Vatican (MS. E. 251) may have been copied for Frederick II. For another thirteenth-century version of Arabic abridgments of the *De animalibus*, see Monsignore A. Pelzer's memoir on "Pierre Gallego" in the *Miscellanea Ehrle*, i. 407–456.

279. The version of Averroës's commentary on the *De anima* is explicitly ascribed to Michael Scot in B. N., MS. lat. 14385, ff. 133–160 v.

280 f. Of the treatise on alchemy ascribed to Scot I have ready for publication in *Isis* a study based on comparison of the Palermo and Oxford MSS. For Catania on p. 281 we should read Sarzana.

283. The Jew Andrew who helped Scot with his translations is very probably to be identified with a Master Andrew, canon of Palencia, whom in 1225 (when Scot was probably at the Curia) the Pope praises for his knowledge of Arabic, Hebrew, Chaldee, and Latin, as well as the seven liberal arts. See the bull of Honorius III analyzed in Pressutti, *Regesta*, no. 5445, and printed in full in my paper on "Michael Scot in Spain" in the *Homenaje A. Bonilla y San Martín* (Madrid, 1925).

283, n. 69. On the scholastic prologues of this type, see Miss Hope Allen, in *Romanic Review*, viii. 434–462 (1917).

292, 296. Monsignore Mercati informs me that instead of 'Montepulciano' the text should read 'Bulicame,' near Viterbo. Cf. Dante, *Inferno*, xiv. 79.

CHAPTER XIV

302. On MS. E., see A. Restori, in *Revue des langues romanes*, xxxix. 289–293 (1896).

318, n. 122. Dr. Birkenmajer points out that the *Liber de animalibus* is more probably the translation of Aristotle by Theodore Gaza.

319. Cf. below under p. 351.

CHAPTER XV

329. The enigmatical William ℞ appears as 'Ego Guillelmus ℞' in a computistic fragment posterior to 1072 in B. N., MS. lat. 14069, f. 15 v (saec. xii).

CHAPTER XVI

338. Another early MS. of Nimrod, resembling the Venetian, is B. N., MS. lat. 14754, ff. 203–232 v (saec. xii).

342, 345. 'Alexander' has been placed in Syria *ca.* 800 by Cumont, in *Revue archéologique*, fifth series, iii. 17 f. (1916), a conclusion which supports the date here suggested for Nimrod.

CHAPTER XVII

346. An early treatise on hawks of a certain Grimaldus 'baiuli et comitis sacri palatii ad Karulum regem' is preserved at Poitiers, MS. 184, ff. 70–73.

349. Another MS. of an Italian version at Rome, Biblioteca Vittorio Emmanuele, MS. 506, f. 74 (saec. xv).

351. On the translations of Daniel of Cremona, see C. Frati, "Re Enzo e un' antica versione francese di due trattati di falconeria," in *Miscellanea Tassoniana* (Modena, 1908), pp. 61–81.

354. The treatise of Peter de l'Astor is printed from a Bologna codex by A. Restori, *Revue des langues romanes*, xxxix. 294–301 (1896).

355. In the Barberini MS. 12 at the Vatican, ff. 94–95 (saec. xiv), there is a fragment 'De avibus rapacibus marescallia' beginning, 'Ad ancipitris tesgam . . .'

CHAPTER XVIII

367, 368. For the use of the *Physics* and *Metaphysics* by David of Dinant, whose writings were condemned at Paris in 1210, see G. Théry, *David de Dinant* (Kain, 1925), pp. 72–83.

371, n. 72. For 'saec. xiii' read 'saec. xii exeuntis'; and add 'B. N., MS. lat. 7647 (saec. xiii ineuntis).'

374. On Honein ben Ishak see *Isis*, vi. 282–292, viii. 685–724.

CONTENTS

THE SCIENCE OF THE ARABS

TRANSLATIONS FROM THE GREEK

THE COURT OF FREDERICK II

OTHER STUDIES

LIST OF ABBREVIATIONS

B. E. C. . . . . . . *Bibliothèque de l'Ecole des Chartes.* Paris, 1839– .

B. M. . . . . *Bibliotheca Mathematica.* Stockholm, 1884– .
(Third series unless otherwise noted.)

B. N. . . . . Bibliothèque Nationale, Paris.

B. Z. . . . . *Byzantinische Zeitschrift.* Leipzig, 1892– .

Beiträge . . . . . . *Beiträge zur Geschichte der Philosophie des Mittelalters,* ed. Cl. Baeumker and others. Münster, 1891– .

Bullettino . . . . . *Bullettino della bibliografia e della storia delle scienze mathematiche e fisiche,* ed. B. Boncompagni. Rome, 1868–87.

Cantor . . . . . . Moritz Cantor, *Vorlesungen über Geschichte der Mathematik.* Leipzig, i (third edition), 1907, ii (second edition), 1900.

Duhem . . . . . . Pierre Duhem, *Le système du monde de Platon à Copernic.* Paris, 1913–17.

E. H. R. . . . . . . *English Historical Review.* London, 1886– .

Jourdain . . . . . A. Jourdain, *Recherches critiques sur l'âge et l'origine des traductions latines d'Aristote.* Second edition, Paris, 1843.

Krumbacher . . . . Karl Krumbacher, *Geschichte der byzantinischen Litteratur.* Second edition, Munich, 1897.

M. G. H. . . . . . *Monumenta Germaniae Historica.* Hanover, etc., 1826– .

Steinschneider, *E. U.* Moritz Steinschneider, *Die europäischen Uebersetzungen aus dem Arabischen,* in *Sitzungsberichte* of the Vienna Academy, phil.-hist. Kl., cxlix, no. 4, cli, no. 1 (1904–05).

Steinschneider, *H. U.* Moritz Steinschneider, *Die hebräischen Uebersetzungen des Mittelalters.* Berlin, 1893.

Suter . . . . H. Suter, *Die Mathematiker und Astronomen der Araber und ihre Werke* (= *Abhandlungen zur Geschichte der mathematischen Wissenschaften,* x, xiv, 1900, 1902).

Thorndike . . . . Lynn Thorndike, *A History of Magic and Experimental Science.* New York, 1922.

Wüstenfeld . . . . F. Wüstenfeld, *Die Uebersetzungen arabischer Werke in das Lateinische,* in *Abhandlungen* of the Göttingen Academy, xxii (1877).

Z. M. Ph. . . . . . *Zeitschrift für Mathematik und Physik.* Leipzig, 1856– .

STUDIES
IN THE HISTORY
OF MEDIAEVAL SCIENCE

STUDIES IN THE HISTORY OF MEDIAEVAL SCIENCE

CHAPTER I

TRANSLATORS FROM THE ARABIC IN SPAIN[1]

THE recovery of ancient science and philosophy in the twelfth and thirteenth centuries marks an epoch in the history of European intelligence. "The introduction of Arabic texts into the studies of the West," says Renan, "divides the history of science and philosophy in the Middle Ages into two perfectly distinct periods. In the first the human mind has, to satisfy its curiosity, only the meagre fragments of the Roman schools heaped together in the compilations of Martianus Capella, Bede, Isidore, and certain technical treatises whose wide circulation saved them from oblivion. In the second period ancient science comes back once more to the West, but this time more fully, in the Arabic commentaries or the original works of Greek science for which the Romans had substituted compends "[2] — Hippocrates and Galen, the entire body of Aristotle's writings, the mathematics and astronomy of the Arabs. The full recovery of this ancient learning, supplemented by what the Arabs had gained from the Orient and from their own observation, constitutes the scientific renaissance of the Middle Ages.

[1] Read before the American Philosophical Society, 19 April 1923, but not heretofore published.

[2] Renan, *Averroès* (Paris, 1869), p. 200. The standard accounts of the translations from the Arabic are: F. Wüstenfeld, "Die Uebersetzungen arabischer Werke in das Lateinische," in *Abhandlungen* of the Göttingen Academy, xxii (1877); M. Steinschneider, *Die hebräischen Uebersetzungen des Mittelalters* (Berlin, 1893); idem, *Die arabischen Uebersetzungen aus dem Griechischen* (Leipzig, 1897) — a factitious collection from *Centralblatt für Bibliothekswesen*, Beihefte v and xii; Virchow's *Archiv*, cxxiv; *Zeitschrift für Mathematik und Physik*, xxxi; and *Zeitschrift der deutschen morgenländischen Gesellschaft*, l; idem, "Die europäischen Übersetzungen aus dem Arabischen," in *Sitzungsberichte* of the Vienna Academy, phil.-hist. Klasse, cxlix, cli (1904-1905). See also his *Introduction to the Arabic Literature of the Jews* (London, 1901); and his many special articles.

The most important channel by which the new learning reached western Europe ran through the Spanish peninsula. "Spain," says W. P. Ker,[3] "from the Rock in the South, which is a pillar of Hercules, to the Pass in the North, which is Roncesvalles, is full of the visions of stories." It has its romance of commerce, from the 'corded bales' of the Tyrian trader to the silver fleets of the Indies; of discovery and conquest, as personified in Columbus and the conquistadores; of crusading and knight errantry in the Cid and Don Quixote. It has also its romance of scholarship, of adventure in new paths of learning and even in forbidden bypaths. In consequence of the Saracen conquest, the Peninsula became for the greater portion of the Middle Ages a part of the Mohammedan East, heir to its learning and its science, to its magic and astrology, and the principal means of their introduction into western Europe. When, in the twelfth century, the Latin world began to absorb this oriental lore, the pioneers of the new learning turned chiefly to Spain, where one after another sought the key to knowledge in the mathematics and astronomy, the astrology and medicine and philosophy which were there stored up; and throughout the twelfth and thirteenth centuries Spain remained the land of mystery, of the unknown yet knowable, for inquiring minds beyond the Pyrenees. The great adventure of the European scholar lay in the Peninsula.

Spain, of course, was not the only route by which Arabic science reached the West. Already in the eleventh century Constantine the African was at work in Africa or the East at his more or less trustworthy paraphrases of medical writers, and one of these versions, the *Regalis dispositio* of Ali-ben-Abbas, was subsequently improved and completed by Stephen of Pisa at Antioch.[4] Adelard of Bath can be followed in Syria more surely than in

---

[3] *Two Essays* (Glasgow, 1918), p. 23.

[4] Infra, Chapter VII. On Constantine, cf. Thorndike, i, c. 32. Constantine's biographer, Petrus Diaconus, tells us (Migne, clxxiii. 1050) that he himself translated the lapidary of 'Evax rex Arabum'; but Petrus is a shaky authority (cf. E. Caspar, *Petrus Diaconus und die Monte Cassineser Fälschungen*, Berlin, 1909), and the problem of the origin of the Latin lapidaries is highly complicated. See Steinschneider, *H.U.*, pp. 956 f. and his references; J. Ruska, *Das Steinbuch des Aristoteles* (Heidelberg, 1912); Thorndike, i, c. 34; Caspar, pp. 28 f.

Spain.[5] North Africa was apparently the source of the new arithmetic of Leonard of Pisa.[6] Some Arabic material, like Achmet's
Dream-book, came via the Byzantine Empire.[7] A more important
intermediary was Sicily, where the Arabs had ruled from 902 to
1091, and where the Mohammedan population remained a considerable element after the Norman conquest. Here about the
middle of the twelfth century Edrisi wrote his great compendium
of Arabic geography, and Eugene of Palermo translated the *Optics*
of Ptolemy from the Arabic. In the next century the hospitality
of Frederick II to Arab learning is well known. Michael Scot's
later years were spent at his court, and Jewish translations of
Averroës were dedicated to him. While these examples show the
influence of Spain, the emperor's relations extended to many other
parts of Islam.[8] On the side of astronomy and astrology translation from the Arabic went on under Frederick's son and successor,
Manfred, and still later under Charles of Anjou. Moreover, there
is a considerable amount of material from the Arabic of unknown
origin, some of which, like the alchemical treatises, was modified
and enlarged before it reached its current Latin form, and in all
this it is impossible to fix the relative part played by Spain and by
other countries. There was also, as we shall see,[9] a large body of
science and philosophy derived directly from the Greek. Nevertheless, the broad fact remains that the Arabs of Spain were the
principal source of the new learning for western Europe.

The science of mediaeval Spain was, of course, an importation
from the Mohammedan East. It was not specifically Arab, save
for the Arab power of absorbing rapidly the older culture of the
Byzantine Empire, Egypt, Syria, and the lands beyond. Fundamentally it was chiefly Greek, either by way of direct translation
or through the intermediary of Syriac and perhaps Hebrew versions of Aristotle, Ptolemy, Euclid, Hippocrates, and the rest, but
developed in many fruitful ways by elements from the farther
East and by a certain amount of specific observation and dis-

---

[5] Infra, Chapter II.

[6] Cantor, ii, c. 41; S. Günther, *Geschichte der Mathematik* (Leipzig, 1908), i, c. 15.

[7] Infra, Chapter X, n. 137.

[8] Infra, Chapter XII.   [9] Infra, Chapters VIII–XI.

covery under the caliphs. The men of science were from all parts of Islam, few indeed being Arabs, but they shared the speech and culture which gave the several caliphates their common civilization.

The Spanish element in this Saracen culture awaits clearer definition. The current books are likely either to reproduce the highly colored reports of Moorish writers, such as the conventional account of Cordova with its 600 mosques and its library of 600,000 volumes, or to deal with generalities concerning Saracen learning and science which have little that is distinctively Spanish.[10] Spain clearly participated, but what did she contribute? Nothing significant in the way of translation into Arabic from the older literature which was the source of Arabic science,[11] for this was to be found only in the East, and reached Spain only in the Arabic versions. Something, undoubtedly, in the discussion and elaboration of this material on Spanish soil. Yet when we examine the lists of Arabic writers on medicine, mathematics, astronomy, and cognate subjects, the number of those who wrote in Spain is not large, and most of these are known to us only from the general phrases of the Arabic cataloguers.[12] The list includes the philoso-

[10] A critical account of the libraries of Mohammedan Spain is lacking. J. Ribera, *Bibliófilos y bibliotecas en la España musulmana* (Saragossa, 1896), is a sketch without references.

[11] The only exception I know is the MS. of Dioscorides said to have been brought from Constantinople in the tenth century as a present from the emperor. See Steinschneider's citations, in Virchow's *Archiv*, cxxiv. 482.

[12] F. Wüstenfeld, *Geschichte arabischer Aerzte und Naturforscher* (Göttingen, 1840); L. Leclerc, *Histoire de la médicine arabe* (Paris, 1876); Suter, particularly nos. 84–87, 90–92, 100, 107, 109–111, 128, 134–136, 159–161, 163, 168–170, 176, 179, 182, 188–190, 194–197, 200–202, 208–213, 219–227, 234–247, 249, 252, 255–259, 264 f., 267, 269, 272, 274 f., 277, 279–282, 284–286, 289 f., 294–296, 301–304, 308, 311 f., 315, 321–323, 325–327, 329–332, 334 f., 339, 342, 350, 355, 373, 379, 384, 388, 390 f., 402, 407–410, 420, 444; Brockelmann, *Geschichte der arabischen Litteratur* (Weimar, 1898), i; and the *Encyclopaedia of Islam*, passim. The best sketch of Arabic astronomy and astrology is that of Nallino, in Hastings' *Cyclopaedia of Religion and Ethics*, xii, pp. 88–101 (1922). No help can be gained from Spanish works such as Eduardo García del Real, *Historia de la medicina en España* (Madrid, 1921); or Norbert Font y Sagué, *Historia de les ciencies naturals á Catalunya* (Barcelona, 1908). A. Bonilla y San Martin, *Historia de la filosofía española*, i (Madrid, 1911) is useful, as is the brief account, with bibliography, in A. Ballesteros y Beretta, *Historia de España*, ii (Barcelona, 1920). See also J. A. Sanchez Pérez, *Biografías de*

phers Avempace of Saragossa,[13] Abubacer (ibn Tofail), and Averroës; the astronomers al-Bitrogi and ibn Aflah, who joined them
in criticising, apparently on the basis of Greek sources, Ptolemy's
theory of planetary motion; their predecessor Maslama, who introduced the astronomy of the East into Spain and adapted the
tables of al-Khwarizmi to the meridian of Cordova;[14] and al-
Zarkali (Arzachel), observer and designer of instruments, who
determined more accurately the angle of the ecliptic, discussed the
precession of the equinoxes, and composed the canons which accompanied the standard tables of Toledo.[15] To these should be
added some physicians of note, like the family of Avenzoar[16] and
the surgeon Abul-Kasim, one or two writers on agriculture, and an
occasional geographer like al-Bekri, ibn Jubair, and Benjamin of
Tudela. Benjamin suggests the Jewish element, which prospered
greatly under the western caliphs and held an important position
in the intellectual life of the age. Spain produced Avicebron (ibn
Gabirol) and the most eminent among mediaeval Jewish philosophers, Maimonides, who, however, removed early in life to the
East; and Spanish Jews coöperated with Moslem scientists so
that the share of each cannot easily be distinguished.[17] Among
the Moslems the outstanding mind would seem to have been Averroës, yet it has been remarked of him that his influence was far
less in Islam than in western Christendom. At the same time,
Spain seems to have possessed the principal writers of the Mohammedan East and versions of the Greek works from which they
drew, and it was in transmitting to western Europe the fulness of

matemáticos árabes que florecieron en España (Madrid, 1921); and his edition of the
Algebra of Abenbéder (Madrid, 1916); as well as the sketch of David Eugene Smith,
History of Mathematics (Boston, 1923), i. 205–211. M. Menendez y Pelayo, "Inven-
tario bibliográfico de la ciencia española," in his *Ciencia española* (Madrid, 1888),
iii. 127–445, is useful mainly for the later period.

[13] I have not seen the articles of Asin, in the *Revista de Aragón*, 1900–01.

[14] H. Suter, *Die astronomischen Tafeln des Muhammed ibn Musa al-Khwarizmi in
der Bearbeitung des Maslama ibn Ahmed al-Madjriti*, published by the Royal Danish
Academy, Copenhagen, 1914. See Chapter II, no. 3.

[15] Steinschneider, "Etudes sur Zarkali," in *Bullettino*, xiv, xvi–xviii, xx (1881–
87); and, for the astronomers in general, Suter, and Duhem, ii.

[16] G. Colin, *Avenzoar* (Paris, 1911).

[17] See below, n. 57.

eastern learning that the Peninsula seems chiefly to have served the advancement of knowledge.

Down to the twelfth century the share of Christian Spain in the diffusion of Saracen learning seems to have been small; indeed the oldest catalogues of its monastic and cathedral libraries are confined to the Latin tradition of the earlier Middle Ages, and with the exception of one noteworthy manuscript they show no trace of Mohammedan science until far into the twelfth century.[18] Nevertheless, it is important to note that the most learned man of the tenth century, Gerbert of Aurillac, the future Silvester II, certainly visited the county of Barcelona in his youth and studied mathematics there under Atto, bishop of Vich. There is no certain evidence that he penetrated farther into the Peninsula; but later, in 984, we find him sending to Miro Bonusfilius, bishop of Gerona, for the treatise of a certain Joseph the Wise on multiplication and division, and asking Lupitus of Barcelona, likewise unknown to us, for a *liber de astrologia* which Lupitus has translated.[19] This latter work, at least, was obviously translated from

[18] For early Spanish libraries, see in general R. Beer, *Handschriftenschätze Spaniens* (Vienna, 1894); and, for MSS. in the Visigothic hand, the list in C. U. Clark, *Collectanea Hispanica* (Paris, 1920). Further references are in R. Foulché-Delbosc and L. Barrau-Dihigo, *Manuel de l'hispanisant*, i (New York, 1920). The best study of a particular library is that of Beer, "Die Handschriften des Klosters Santa Maria de Ripoll," in Vienna *Sitzungsberichte*, clv, 3, clviii, 2 (1907, 1908). See also Delisle, "MSS. de l'abbaye de Sillos," in his *Mélanges de paléographie*, pp. 53–116 (cf. Férotin, *Histoire de l'abbaye de Sillos*, Paris, 1897); Denifle's catalogue of the Tortosa MSS., *Revue des bibliothèques*, vi. 1–61 (1896); and the scattered notices in Villanueva, *Viage literario*. The uncatalogued MSS. of the provincial library of Tarragona I examined on the spot in 1913.

The only clear example of Arabic influence yet pointed out is MS. Ripoll 225, of the tenth century, to which we shall return below (note 21). Two interesting manuals of technology edited by J. M. Burnam, who suggests their derivation from Ripoll, show no Arabic influence. See his "Recipes from Codex Matritensis A16 (ahora 19)," in *University of Cincinnati Studies*, viii, 1 (1912); *A Classical Technology* (Boston, 1920); and cf. *Bulletin Hispanique*, xxii. 229–233 (1920). So a Ripoll MS. of 1056 now in the Vatican (Reg. Lat. 123) contains only the older Latin astronomy. See Pijoán, in *Trabajos* of the *Escuela española de arqueología é historia en Roma*, i. 1–10 (1912); Saxl, in Heidelberg *Sitzungsberichte*, 1915, no. 5, pp. 45–59.

[19] Richer, *Historiae*, iii, ch. 43; *Lettres de Gerbert*, ed. Havet, nos. 17, 24, 25. Cf. M. M. Büdinger, *Ueber Gerberts wissenschaftliche und politische Stellung* (Marburg, 1851), pp. 7–25; Beer, in Vienna *Sitzungsberichte*, clv, 3, pp. 46–59; Manitius, *Geschichte der lateinischen Litteratur des Mittelalters*, ii. 729–742 (1923). For Joseph the Wise, cf. Suter, no. 182.

the Arabic; it has been conjecturally identified with a treatise on the astrolabe, very possibly with the source of a treatise on this subject which Bubnov ascribed hesitatingly and on no very conclusive grounds to Gerbert himself.[20] Whoever the author, he worked from Arabic sources, as is seen by the Arabic terms which he takes over, and it so happens that a fragment of his work which was unknown to Bubnov still exists in a codex of the tenth century among the manuscripts of Santa Maria de Ripoll at Barcelona.[21] Apart, however, from this doubtful work, it seems now agreed that there is no direct influence of Arabian mathematics visible in Gerbert's writings.[22] Throughout the eleventh century Arabic influence is limited to the technical terms of the astrolabe and the names of stars, with the possible addition of the astrology of Alchandrinus.[23]

In general, the lure of Spain began to act only in the twelfth century, and the active impulse toward the spread of Arabic learning came from beyond the Pyrenees and from men of diverse origins. The chief names are Adelard of Bath, Plato of Tivoli, Robert of Chester, Hermann of Carinthia, with his pupil Rudolf of Bruges, and Gerard of Cremona, while in Spain itself we have Dominicus Gondisalvi, Hugh of Santalla, and a group of Jewish

---

[20] *Gerberti Opera Mathematica* (Berlin, 1899), pp. 109 ff. The discovery of evidence from the tenth century (see the following note and Thorndike, i, ch. 30) requires a reopening of the question.

[21] Archives of the Crown of Aragon, MS. Ripoll 225, 105 folios; cf. Beer, *loc. cit.*, pp. 57–59. The MS., which I examined in 1913, is in some confusion and needs to be collated with the several early treatises on the astrolabe (cf. Bubnov, pp. cv–cviii). It begins in the middle of the work ascribed to Gerbert: [super]ponitur tabule . . . (Bubnov, p. 123, l. 5). Then, f. 1 v, 'De mensura astrolabii. Philosophi quorum sagaci studio . . .' F. 7 v, 'De mensura volvelli.' F. 9 v–10, table of stars with Arabic names. F. 24 v, 'Incipit astrolabii sententie. Quicumque vult scire certas horas noctium et dierum . . .' F. 25 v, 'Explicit prologus. Incipit de nominibus laborum laboratorum in ipsa tabula. In primis Almucantarat . . .' F. 30 v, 'Incipiunt capitula orologii regis Ptolomei. Quomodo scias altitudinem solis . . . .' F. 35, 'Incipiunt regule de quarta parte astrolabii . . . .' F. 39, a new treatise; cf. Beer, p. 59.

[22] Bubnov, *Gerberti Opera Mathematica*; Cantor, *Vorlesungen*, i. ch. 39.

[23] Bubnov, pp. 124 ff., 370–375; Thorndike, i, ch. 30. An Arabic-Latin glossary of the eleventh century has been edited by C. F. Seybold (Tübingen, 1900); cf. E. Böhmer, in *Romanische Studien*, i. 221–230 (1871); Götz, *Corpus glossariorum Latinorum*, i. 188 f.

scholars, Petrus Alphonsi, John of Seville, Savasorda, and Abraham ben Ezra. Much in their biography and relations with one another is still obscure. Their work was at first confined to no single place, but translation was carried on at Barcelona, Tarazona, Segovia, Leon, Pamplona, as well as beyond the Pyrenees at Toulouse, Béziers, Narbonne, and Marseilles. Later, however, the chief centre became Toledo. An exact date for this new movement cannot be fixed, now that criticism has removed the year 1116 from an early title of Plato of Tivoli,[24] but the astronomical tables of Adelard are dated 1126, and this whole group of translators, save Gerard of Cremona, can be placed within the second quarter of the twelfth century. They owed much to ecclesiastical patronage, especially to Raymond, archbishop of Toledo, and his contemporary Michael, bishop of Tarazona. Besides a large amount of astrology, inevitable in an age which regarded astrology as merely applied astronomy and a study of great practical utility, their attention was given mainly to astronomy and mathematics.

Adelard of Bath, to begin with the earliest of this group, can be connected with Spain by indirect evidence only. He was a translator of mathematical and astronomical works from the Arabic, but, as he speaks specifically of sojourns in Syria and southern Italy, his knowledge of both the learning and the language of the Saracens may well have been gained outside of the Peninsula. Nevertheless the astronomical tables which he turned into Latin in 1126 were, in this form, the work of the Spanish astronomer, Maslama, and based upon the meridian of Cordova, and it is quite unlikely that Adelard found these elsewhere than in Spain. Where his other versions, such as the translation of Euclid's *Elements*, were made, it is impossible to say, but it is clear that he must be viewed in a European rather than a Spanish perspective.[25] He is also interesting as the first of a long line of Englishmen who played an important part in this whole movement and whose writings serve as an index of the absorption of the new learning in the North.[26]

[24] See below, n. 29.   [26] See Chapter VI.
[25] See the following chapter.

Plato of Tivoli, whose biography is known only from his translations,[27] was until recently supposed to have made a mathematical translation as early as 1116, his *Liber embadorum* of Savasorda being dated 15 Safar in year 510 of the Hegira.[28] I showed, however, in 1911 [29] that this date did not correspond with the position of the sun and planets as therein described, which requires an emendation of the text to 540 (DXL from which the L has been lost), thus bringing us down to 13 August 1145. The *Liber embadorum*, interesting for the introduction of Arabic trigonometry and mensuration into the West, and for its apparent influence on the geometry of Leonard of Pisa, is hence the latest of Plato's versions. The others, mostly dated at Barcelona between 1134 [30] and 1138, include the astronomy of al-Battani, which Plato preferred to the longer *Almagest* of Ptolemy,[31] and a certain number of miscellaneous astrological treatises, among them Ptolemy's own *Quadripartitum* (1138). He had the help of the Jew Savasorda (Abraham ben Chija) and was also in relations with John David, to whom we shall come later.

Hermann of Carinthia and Robert of Chester constitute a sort of literary partnership working at various places in northern Spain and southern France.[32] Hermann appears first, translating a work of Arabic astrology in 1138, and by 1141 the two are together in the region of the Ebro, where Peter of Cluny found them and engaged them, along with Master Peter of Toledo and his

[27] B. Boncompagni, "Delle versioni fatte da Platone Tiburtino traduttore del secolo duodecimo," in *Atti dell' Accademia Pontificia dei Lincei*, iv. 249–286 (1851); Wüstenfeld, pp. 39–44; Steinschneider, *E. U.*, no. 98; Thorndike, ii. 119.

[28] M. Curtze, *Der "Liber embadorum" des Savasorda in der Uebersetzung des Plato von Tivoli* (*Abhandlungen zur Geschichte der mathematischen Wissenschaften*, xii), Leipzig, 1902.

[29] *Romanic Review*, ii. 2; *E. H. R.*, xxvi. 491. The astronomical facts were verified by my colleague, the late Professor R. W. Willson.

[30] To the evidence for the year 1134 as the date of the version of Hali, *De electionibus*, should be added MS. 10063 of the Biblioteca Nacional, f. 32; and MS. 5–5–14 of the Biblioteca Colombina at Seville.

[31] C. A. Nallino, *Al-Battani sive Albatenii Opus astronomicum*, in *Pubblicazioni del R. Osservatorio di Brera in Milano*, xl (1904). To the Latin MSS. there enumerated (p. li) should be added MS. 5–1–21 of the Biblioteca Colombina, ca. 1200.

[32] For a critical study of Hermann and his writings, see Chapter III, below; for Robert, Chapter VI.

own secretary, on a Latin version of the Koran. From the next few years we have a number of works in the name of Hermann or Robert, with frequent dedications by one to the other, which together cover a wide range of mathematical, astronomical, astrological, and philosophical studies. Of outstanding importance among them are Hermann's version of Ptolemy's *Planisphere*, otherwise unknown, and his *De essentiis*, as well as lost mathematical works; and Robert's astronomical tables, his version of the alchemy of Morienus, one of the earliest of such works to reach the West, and his highly significant translation of the *Algebra* of al-Khwarizmi for the Latin world. An astronomical treatise of Hermann's pupil, Rudolf of Bruges, belongs to the same group.

Hugh of Santalla is likewise connected with the north of Spain, of which he was apparently a native.[33] His patron was Michael, bishop of Tarazona in Aragon from 1119 to 1151, and his work was probably done there or in the neighborhood, as we find him mentioning a library at Roda or Rueda. His numerous translations have to do with astrology, geomancy, and various forms of divination, including the *Centiloquium* and several Arabic authors.

While it thus appears that the work of translation was early active at several places in northern Spain, Toledo soon became the most important centre. Reconquered by the Christians in 1085, the seat of the primate and soon the residence of the king of Castile, the historic city on the Tagus was the natural place of exchange for Christian and Saracen learning. "At this ancient centre of scientific teaching were to be found a wealth of Arabic books and a number of masters of the two tongues, and with the help of these Mozarabs and resident Jews there arose a regular school for the translation of Arabic-Latin science which drew from all lands those who thirsted for knowledge, and left the signature of Toledo on many of the most famous versions of Arabic learning." [34] Of a formal school the sources tell us very little, but the succession of translators is clear for more than a century, beginning about 1135 and continuing until the time of Alfonso X

[33] See Chapter IV, below.
[34] V. Rose, "Ptolemäus und die Schule von Toledo," in *Hermes*, viii. 327 (1874).

(1252–84). The first initiative seems to have been due to Archbishop Raymond, 1125 to 1151,[35] as seen in the dedications of the two Toletan translators of this period, Dominicus Gondisalvi, or Gundissalinus, and a converted Jew named John. So far as we can judge from these, the archbishop's interests were chiefly philosophical. Gundissalinus, archdeacon of Segovia, is the author of several translations and adaptations of Arabic and Jewish philosophy: the *Metaphysics* and other works of Avicenna, the *Fons vitae* of Avicebron (ibn Gabirol), the classification of the sciences of al-Farabi, the philosophy of Algazel (al-Gazzali).[36] At least at the outset his ignorance of Arabic put him in close dependence on John, who gave him the Spanish word which the archdeacon then turned into Latin,[37] so that there is little evidence of direct translation by Gondisalvi.[38] John son of David (Avendehut) is an enigmatical personage who still needs investigation. He is usually identified with a John of Spain, of Seville, or of Luna, who meets us between 1135 and 1153 as a voluminous translator and compiler from the Arabic.[39] The score of works ascribed to him are

[35] B. Gams, *Kirchengeschichte von Spanien*, iii, 1, pp. 20–23, 37; Jaffé-Löwenfeld, *Regesta*, no. 7231.

[36] Jourdain, pp. 107–113; Wüstenfeld, pp. 38 f.; Bonilla, *Filosofia española*, i. 316–359; Ueberweg-Baumgartner, *Grundriss der Geschichte der Philosophie*[10] (Berlin, 1915), ii. 405 f., 412, 414-416, 153*; Correns, *Dominicus Gundisalvi de Unitate*, in *Beiträge*, i, no. 1 (1891); Baeumker, *Avencebrolis Fons Vitae, ibid.*, i, nos. 2–4 (1892); Bülow, *Des Dominicus Gundissalinus Schrift Von der Unsterblichkeit der Seele, ibid.*, ii, no. 3 (1897); Baur, *Gundissalinus De divisione philosophiae, ibid.*, iv, nos. 2–3 (1903); Baeumker, *Alfarabi, Ueber den Ursprung der Wissenschaften, ibid.*, xix, no. 3 (1916); Furiani, *Des Dominicus Gundissalinus Abhandlung de anima, ibid.*, xxiv, nos. 2–4 (in press); Thorndike, ii. 78–82.

[37] Preface to version of Avicenna's *De anima* in Jourdain, p. 449; Correns, pp. 32 f.; and Bonilla, i. 447. I have verified the text from MS. Bodley 463, f. 139.

[38] Steinschneider, *E. U.*, no. 49.

[39] The best list is in Steinschneider, *E. U.*, no. 68. See also Wüstenfeld, pp. 25–38; Bonilla, i. 319–323; *B. M.*, vi. 114 (1905), ix. 2; Thorndike, ii. 73–78, including his appendix on "Some Mediaeval Johns," pp. 94–98. Thorndike calls attention to a brief tract at St. John's College, Oxford, MS. 188, f. 99 v, which has the following reference: 'Scire oportet vos, karissimi lectores, quod debetis aliquos annos scire super quod cursus planetarum valeatis ordinare vel per quos possitis ordinatos cursus in libro quem ego Johannis Yspalensis interpres existens rogatu et ope duorum Angligenarum, Gauconis scilicet et Willelmi, de arabico in latinum transtuli.' In MS. 10053 of the Biblioteca Nacional, which contains several of John's treatises, we have however (f. 86 v): 'Scire debes, karissime lector, quia oportebit te aliquos annos

chiefly astrological—Albumasar, Omar, Thebit, Messahala, Hali, as well as the *Centiloquium* attributed to Ptolemy—in the forms which became widely current in western Europe; but to these should be added the astronomical manual of al-Fargani,[40] an interesting treatise of al-Khwarizmi on arithmetic, and a popular version of the medical portion of the *Secretum secretorum*.[41] He was also in relations with the translators who worked outside of Toledo, for translations are dedicated to him by Plato of Tivoli and Rudolf of Bruges.[42]

The latter half of the twelfth century saw the most industrious and prolific of all these translators from the Arabic, Gerard of Cremona.[43] Fortunately we have a brief biographical note and list of his works, drawn up by his pupils in imitation of the catalogue of Galen's writings and affixed to Gerard's version of Galen's *Tegni*, lest the translator's light be hidden under a bushel and others receive credit for work which he left anonymous. From this we learn that, a scholar from his youth and master of the content of Latin learning, he was drawn to Toledo by love of

scire super quos cursus planetarum valeas ordinare vel per quos possis ordinatos cursus in libro quem ego Johannes Ispanus interpres existens de arabico in latinum transtuli.'

[40] Nallino, *Al-Battani*, p. lvii, dates the version of al-Fargani 11 March 1135, and the *Centiloquium* 17 March 1136.

[41] With a dedication to 'T., queen of Spain.' See Foerster, *De Aristotelis quae feruntur Secretis secretorum* (Kiel, 1888); R. Steele's edition of Roger Bacon's *Secretum secretorum*, pp. xvi–xviii; Thorndike, ii. 269 f.

[42] Steinschneider, *E. U.*, nos. 98 *i*, 104 *b*.

[43] The standard monograph is Boncompagni, "Della vita e delle opere di Gherardo cremonese," in *Atti dell' Accademia pontificia dei Lincei*, iv. 387–493 (1851). The contemporary list of his translations here first edited will also be found in Wüstenfeld, p. 57; it is edited with special reference to the medical works by Sudhoff, in *Archiv für die Geschichte der Medizin*, viii. 73–82 (1914). Cf. Thorndike, ii. 87 ff.; *B. M.*, vi. 239–248 (1905). The best critical list of his translations is in Steinschneider, *E. U.*, no. 46. I have noted the following further manuscripts (numbers of Gerard's treatises as in Steinschneider): 10 (34) St. Mark's, vi, 37, "secundum translationem Gerardi"; 21 (45), Madrid, 1407, f. 69 v; 22 (46), Biblioteca Colombina 5-5-21; 33 (27), Vatican, Vat. lat. 3096, dated Toledo 1140 or 1143 (?); 39 (5), Madrid, 10010, f. 1 v–13; 42 (24), Madrid, 10006; 44 (11), Madrid, 10010, f. 69; 46 (62), Madrid, 1193, Escorial i. f. 8; 57 (18), Colombina 7-6-2, f. 141 v; 74 (20), Madrid 10010, ff. 84 v–86 v; 75 (28), Escorial, ii. O. 10. f. 84 v; 76 (29), Madrid, 10053, f. 1 (fragment); 84(68), University of Bologna, Lat. 449 (760), inc. 'Si quis partem,' and in an Italian version at Florence, Biblioteca Nazionale, II. 1. 372.

Ptolemy's *Almagest*, which he could not find among the Latins. There he discovered a multitude of Arabic books in every field, and, pitying the poverty of the Latins, learned Arabic in order to translate them. His version of the *Almagest* bears the date of 1175.[44] Before his death, which came at Toledo in 1187 at the age of seventy-three, he had turned into Latin the seventy-one Arabic works of this catalogue, beside perhaps a score of others. Three of these are logical, the *Posterior Analytics* of Aristotle [45] with the commentaries of Themistius and al-Farabi; several are mathematical, including Euclid's *Elements*, the *Spherics* of Theodosius, a tract of Archimedes, and various treatises on geometry, algebra, and optics. The catalogue of works on astronomy and astrology is considerable, as is also the list of the scientific writings of Aristotle, but the longest list of all is medical, Galen and Hippocrates and the rest, who were chiefly known in these versions throughout the later Middle Ages.[46] Indeed, more of Arabic science in general passed into western Europe at the hands of Gerard of Cremona than in any other way. Where Gerard's versions have been tested, they have been found closely literal and reasonably accurate; but we are told that he had the assistance of a Mozarab named Galippus, so that we cannot say how far the versions were his own. Both Gerard and Galippus lectured on astronomy in the hearing of Daniel of Morley, an Englishman who had left Paris in disgust to hear the wiser philosophers of the world at Toledo, whence he returned home with a store of precious manuscripts.[47]

After Gerard of Cremona, Roger Bacon lists Alfred the Englishman, Michael Scot, and Hermann the German as the principal translators from the Arabic,[48] all of whom worked in Spain in the earlier thirteenth century. Alfred was a philosopher, concerned especially with expounding the natural philosophy of Aristotle, although he was also known for his version of two pseudo-Aris-

---

[44] Infra, Chapter V, n. 139.   [45] Infra, Chapter XI.

[16] The medical translations of Mark, canon of Toledo, belong apparently to the same period. See Rose, in *Hermes*, viii. 338, who gives one of his prefaces; and Steinschneider, *E. U.*, no. 81; Diels, in Berlin *Abhandlungen*, 1905, pp. 86 f. Alfonso of Toledo, translator of a tract of Averroës (Steinschneider, *E. U.*, no. 12) has not been dated.

[47] Infra, Chapter VI, n. 39.   [48] *Opus tertium*, ed. Brewer, p. 91.

totelian treatises.[49] Michael Scot first appears at Toledo in 1217 as the translator of al-Bitrogi *On the Sphere*, and by 1220 he had made the standard Latin version of Aristotle *On Animals*, not to mention his share in the transmission of the commentaries of Averroës on Aristotle and his own important works on astrology.[50] Hermann the German, toward the middle of the century, was likewise concerned with Aristotle and Averroës, particularly the *Ethics, Poetics,* and *Rhetoric* and the commentaries thereon.[51] Lesser writers of the same period concerned themselves with astrology and medicine.[52]

The thirteenth century is an age of royal patrons of learning,[53] and it is fitting that the culmination of the Christian science of Spain should come in the reign of Alfonso the Wise, king of Castile from 1252 to 1284. This is no place to discuss the many-sided intellectual activity of this prince, a glory of which Spanish historians are justly proud.[54] On the side of science he shone in astronomy and astrology, as seen in the Alfonsine tables, in a collection of treatises on astronomical instruments, and in a group of works on astrology. These were not original, save for a certain amount of specific observation, but were based on well known Arabic works, some of them already translated into Latin in the

---

[49] Infra, Chapter VI, n. 47.   [50] Infra, Chapter XIII.

[51] The versions are dated 1240–44, one perhaps in 1256: Jourdain, pp. 135 ff.; Steinschneider, *E. U.*, no. 51; Luquet, in *Revue de l'histoire des religions*, xliv. 407–422 (1901); C. Marchesi, *L'etica Nicomachea nella tradizione latina medievale* (Messina, 1904); Bonilla, *Filosofia española*, i. 368–371; Grabmann, *Aristotelesueber-setzungen* (*Beiträge*, xvii, no. 5), especially pp. 208 ff.; A. Pelzer, in *Revue néo-scolastique*, xxiii. 323 ff. (1921). Hermann's *Summa Alexandrinorum* is also at Seville, MS. Colombina 7-4-22.

[52] E. g., Salio (Steinschneider, *E. U.*, no. 107; Thorndike, ii. 221); and Stephen of Saragossa (*E. U.*, no. 113). Rufinus of Alessandria (*ibid.*, no. 105; Rose, in *Hermes*, viii. 337) belongs to this period and not to 1168 if his master in Arabic was a Dominican; indeed his ophthalmological version is specifically dated at Murcia in 1271 in MS. Bern 216, f. 42 v.

[53] Cf. what is said of Frederick II in Chapter XII, infra.

[54] On Alfonso's astronomical work, see A. Wegener, "Die astronomischen Werke Alfons X," in *B. M.*, vi. 129–185 (1905); Dreyer, in *Monthly Notices of the Royal Astronomical Society*, lxxx. 243–267 (1920); on his translators, Steinschneider, *E. U.*, nos. 4, 9, 40, 55, 60, 61, 69, 87, 93, 97, 108; on his influence, Duhem, ii. 259–266, and passim. For the reign in general, see the forthcoming book of A. Ballesteros y Beretta; and cf. his *Sevilla en el siglo XIII* (Madrid, 1913), ch. 11. Wegener dates the *Instruments* 1256 ff.; the *Tables* ca. 1270; the astrological collection 1276–79.

preceding century. These were, however, elaborated, reconciled, systematized, regrouped, and often rewritten at Alfonso's command. The account of an astronomical congress at his court has been shown to be a legend, as was probably also his so-called astronomical college. He had the aid of two Jewish scholars, Isaac ibn Sid and Jehuda ben Moses Cohen, as well as of certain Christian translators like Egidio of Parma. Of the results of their labor the Alfonsine *Tables* are the most famous, although the current texts do not represent their original form of ca. 1270. Seventy-five mediaeval manuscripts and thirteen early editions are known. The *Libros de saber*, describing the various instruments of astronomy, on the other hand, seem to have lain neglected until the Castilian text was printed in five volumes by the Spanish Academy in 1863 and following years.[55] The unpublished astrological collection has still to be specially studied, as also the magical book of the enigmatical Picatrix which is assigned to this reign.[56]

Jews, both orthodox and converted, play a large part in the work of translation in Spain and southern France.[57] Sometimes

[55] *Libros del saber de astronomía del Rey D. Alfonso X de Castilla*, ed. Manuel Rico y Sinobas, Madrid, 1863–67. On the MSS., cf. Tallgren, in *Neuphilologische Mitteilungen*, 1908, p. 110.

[56] On Picatrix, see Thorndike, ii, ch. 66; H. Ritter, in *Bibliothek Warburg* (Leipzig, 1923), pp. 94–124. See also the *Lapidario del Rey D. Alfonso X*, ed. J. F. Montana in facsimile (Madrid, 1881); and cf. Steinschneider, in *Zeitschrift der deutschen morgenländischen Gesellschaft*, xlix. 266–270 (1895); F. Boll, *Sphära* (Leipzig, 1903), pp. 430–434.

[57] See, in general, J. Amador de los Rios, *Historia de los Judíos de España* (Madrid, 1875–76), i. cc. 3, 5, 7; and the check-list of Spanish-Jewish writers, with references and bibliography, in Joseph Jacobs, *An Inquiry into the Sources of the History of the Jews in Spain* (New York, 1894); Graetz, *Geschichte der Juden*, v, vi; Steinschneider, *H. U.*, passim; "The Arabic Literature of the Jews," in *Jewish Quarterly Review*, ix–xiii, and separately (London, 1901); and for mathematics the Spanish section of his articles on "Die Mathematik bei den Juden," in *B. M.*, 2, ix. 47–50, 97–104, x. 33–42, 77–83, 109–114, xi. 13–18 (1895–97). Steinschneider also has special articles on Savasorda and Abraham ibn Ezra in *Z. M. Ph.*, xii. 1–44 (1867), and in *Abhandlungen zur Geschichte der Mathematik*, iii. 57–128 (1880). On the diffusion of Abraham ibn Ezra among the Latins, see A. Birkenmajer, *Bibljoteka Ryszarda de Fournival* (Cracow, 1922), pp. 35–42, 50 f.; D. E. Smith and J. Ginsburg, "Rabbi ben Ezra and the Hindu-Arabic Problem," in *American Mathematical Monthly*, xxv. 99–108 (1918). On the methods of these translators, see Renan, *Averroès*, pp. 202–204; Nallino, *al-Battani*, pp. xxx f.; and for John of Seville, see also Dyroff, in Boll, *Sphära*, p. 484; for Gerard of Cremona, O. Bardenhewer, *Die pseudo-aristotelische Schrift Ueber das reine Gute* (Freiburg, 1882), pp. 148 f., 192 ff.

they are themselves the authors or translators, as in the case of Petrus Alphonsi, John of Seville, Abraham ibn Ezra, and the astronomers of Alfonso X just mentioned. Sometimes they act as interpreters for Christian translators who receive the chief credit, as, for example, Savasorda for Plato of Tivoli and a certain Andrew or Abuteus for Michael Scot. Apparently such interpreting frequently took the form of translating from Arabic into the current Spanish idiom, which the Christian translator then turned into Latin. This fact helps to explain the inaccuracies of many of the versions, although in general they are slavishly literal, even to carrying over the Arabic article. We must also bear in mind that there was a large amount of translation from Arabic into Hebrew and then later into Latin, as any one can verify by turning to Steinschneider's great volume on Hebrew translations.

In this process of translation and transmission accident and convenience played a large part. No general survey of the material was made, and the early translators groped somewhat blindly in the mass of works suddenly disclosed to them. Brief works were often taken first because they were brief and the fundamental treatises were long and difficult; commentators were often preferred to the subject of the commentary. Moreover, the translators worked in different places, so that they might easily duplicate one another's work, and the earliest or most accurate version was not always the most popular. Much was translated to which the modern world is indifferent, something was lost which we should willingly recover, yet the sum total is highly significant. From Spain came the *Metaphysics* and natural science of Aristotle and his Arabic commentators in the form which was to transform European thought in the thirteenth century. The Spanish translators made most of the current versions of Galen and Hippocrates and of the Arab physicians like Avicenna. Out of Spain came the new Euclid, the new algebra, and treatises on perspective and optics. Spain was the home of astronomical tables and astronomical observation from the days of Maslama and Zarkali to those of Alfonso the Wise, and the meridian of Toledo was long the standard of computation for the West, while we must also note the current compends of astronomy, like al-Fargani, as well as the generally received version of Ptolemy's *Almagest*, for the love of which Gerard of Cremona made the long journey to Toledo. The great body of eastern astrology came through Spain, as did something of eastern alchemy.

By the close of the thirteenth century Arabic science had been transmitted to western Europe and absorbed, and Spain's work as an intermediary was done. Meanwhile the Peninsula had gained a European reputation as the centre of the black art, and the familiar associations of Toledo, Cordova, Seville, and Salamanca were now with demons and necromancers.[58] Spain became the scene of visions and prophecies, of mystifications like Virgil of Cordova, of legends like the university of demonology at Toledo connected with the magic cave of Hercules. Association with Spain was enough to condemn even a learned Pope like Gerbert to the role of a magician who had sold his soul to the devil, and to make of poor Michael Scot

> A wizard, of such dreaded fame,
> That when, in Salamanca's cave,
> Him listed his magic wand to wave,
> The bells would ring in Notre-Dame!

In the mediaeval mind the science of magic lay close to the magic of science.

[58] On Spain as the home of magic see particularly Rose, in *Hermes*, viii. 343 f.; H. Grauert, "Meister Johann von Toledo," in Munich *Sitzungsberichte*, phil.-hist. Classe, 1901, pp. 111–325; J. Wood Brown, *Michael Scot*, chs. 9, 10; F. Picavet, *Gerbert*, ch. 6; Thorndike, passim; S. M. Waxman, "Chapters on Magic in Spanish Literature," in *Revue hispanique*, xxxviii. 325–463 (1916).

CHAPTER II

ADELARD OF BATH

ADELARD of Bath, the pioneer student of Arabic science and philosophy in the twelfth century, and "the greatest name in English science before Robert Grossetete and Roger Bacon," [1] still remains in many ways a dim and shadowy figure in the history of European learning. The older writers upon literary history give lists of works attributed to him, but they tell us nothing of his life beyond the fact that he lived under Henry I and travelled in various distant lands; [2] and while more recent studies have made clearer his place in the history of mediaeval philosophy,[3] his work as a whole has yet to be examined, and many fundamental facts in his biography still elude us.[4] Except for a bare mention in the Pipe Roll of 1130 Adelard is known only from his own writings, which consist in part of translations and in part of independent treatises, and a list of these is the necessary point of departure for any further study.

1. *De eodem et diverso.* Edited, with commentary, from the unique MS., B. N., Lat. 2389, by Willner, in *Beiträge*, iv. no. 1.[5] Besides the evidence of the dedicatory letter and the title, Adelard's authorship is established by the following passage in his *Astrolabe*:

[1] Wright, *Biographia Britannica literaria* (London, 1846), ii. 94.

[2] Tanner, *Bibliotheca Britannico-Hibernica* (London, 1748), p. 55, reproduces Leland's account, with notes drawn from Bale, Pits, Oudin, and his own reading.

[3] Jourdain, pp. 97–99, 258–277, 452 ff.; Hauréau, *Histoire de la philosophie scolastique*, i. 348–361; Willner, *Des Adelard von Bath Traktat De eodem et diverso*, in *Beiträge*, iv. no. 1 (Münster, 1903); De Wulf, *Histoire de la philosophie médiévale* (Louvain, 1912), pp. 217–219; Ueberweg-Baumgartner, *Grundriss* [10] (Berlin, 1915), ii. 310–317.

[4] The best of the earlier accounts is that of Wright (ii. 94–101), supplemented by Boncompagni in *Bullettino*, xiv. 1–90 (1881). I took up the problem first in 1911, with results here revised and supplemented (*E. H. R.*, xxvi. 491-498; xxviii. 515 f.; xxxvii. 398 f.). Thorndike has a good but by no means a final chapter (ii, ch. 36). The notice in the *Dictionary of National Biography* is superficial; that of Dom Berlière in Baudrillart's *Dictionnaire*, i. 522 f., is useful.

[5] Extracts in Jourdain, pp. 260–273, 452–454.

Sunt et alię metiendi corpora demonstrationes, sed quoniam in eo libro quem de eodem et diverso scripsimus dictę sunt magisque geometricę quam astrolabicę dici possunt, eas preterimus.[6]

The *De eodem*, of which the scene is laid near Tours while the author is still *iuvenis*, is one of Adelard's earliest works. He has already travelled widely and feels called to explain his wandering life to his nephew; but there is no intimation that he has gone farther than southern Italy and Sicily,[7] and he shows the influence of Greek rather than of Arabic learning. There are a few Greek but no Arabic words. More definite evidence respecting the date is afforded by the dedication to William, bishop of Syracuse, who is last found in 1115 and whose successor is in office in the following year.[8] The date of his accession is more difficult to determine, in the scarcity of Syracusan documents from this period: he is first mentioned as bishop at the Lateran Council of March, 1112,[9] but as he there represented the whole body of Sicilian prelates, he had doubtless been in office for some time, perhaps since 1104, when Pirro places the death of his predecessor. Furthermore, Adelard speaks of having played the *cithara* before the queen in the course of his musical studies in France the preceding year,[10] and as there was no queen of France between the death of Philip I in 1108 and the marriage of Louis VI in 1115,[11] the treatise, unless the bishop of Syracuse was still alive in 1116, would not be later than 1109. It is possible, but not probable,

---

[6] McClean MS. 165, f. 84; Arundel MS. 377, f. 70. The demonstrations will be found on pp. 29–31 of the edition of the *De eodem*.

[7] P. 33: 'Et ego certe, cum a Salerno veniens in Grecia maiore quendam philosophum grecum, qui pre ceteris artem medicine naturasque rerum disserebat, sententiis pretemptarem.' Cf. p. 32: 'Quod enim gallica studia nesciunt, transalpina reserabunt; quod apud Latinos non addisces, Grecia facunda docebit.' There is nothing here to justify the usual interpretation (Jourdain, p. 97; Wright, p. 95) that Adelard visited Greece. Much for his purposes was to be found in souther. Italy and Sicily; see Chapter IX, below.

[8] Pirro, *Sicilia sacra* (1733), i. 620, ii. 799; Garufi, *I documenti inediti dell' epoca normanna*, pp. 10, 14; Caspar, *Roger II*, pp. 488, 491, nos. 25, 33; Chalandon, *Histoire de la domination normande*, i. 364.

[9] 'Wilihelmus Siracusanus legatus pro omnibus Siculis': *Constitutiones et Acta Publica* (*M. G. H.*), i. 572.

[10] P. 25: 'Cum preterito anno in eadem musica gallicis studiis totus sudares [Philosophy is addressing Adelard] adessetque in serotino tempore magister artis una cum discipulis, cum eorum regineque rogatu citharam tangeres.'

[11] It is possible, but not likely, that the title may have been here given to Bertrada after Philip's death; nor, between 1108 and 1115, could either of Philip's daughters have been meant.

that the reference is to the queen of England,[12] but Matilda is not found on the French side of the Channel after 1109.[13]

2. *Regule abaci*, dedicated 'H. suo.' Edited by Boncompagni in *Bullettino*, xiv. 1–134. This evidently belongs to the earlier part of Adelard's life, for its authorities are Boethius and Gerbert,[14] and it shows no trace of Arabic influence.

3. *Ezich Elkauresmi per Athelardum bathoniensem ex arabico sumptus*, a translation of the important astronomical tables of Mohammed ben Musa al-Khwarizmi, as revised by Maslama at Cordova.[15] Bodleian, MS. Auct. F. 1. 9 (Bernard, no. 4137), ff. 99 v–159 v, a fine manuscript of the twelfth century; Chartres, MS. 214, ff. 41–102, likewise of the twelfth century; Bibliothèque Mazarine, MS. 3642, ff. 82–87, incomplete; Madrid, Biblioteca Nacional, MS. 10016, f. 8, as revised by Robert of Chester;[16] Oxford, Corpus Christi College, MS. 283, ff. 113–144, incomplete, with tables as far as p. 167 in Suter's edition, mixed in with some material of Petrus Alphonsi.[17] Trigonometrical portions were edited by A. A. Björnbo from the first two MSS. in the *Festskrift til H. G. Zeuthen* (Copenhagen, 1909), pp. 1–17; the whole is published, with commentary, from Björnbo's papers with the use of the first four MSS., by H. Suter, *Die astronomischen Tafeln des Muhammed ibn Mūsā al-Khwārizmī in der Bearbeitung des Maslama ibn Ahmed al-Madjrīlī und der latein. Uebersetzung des Athelhard von Bath*, in the *Selsk. Skrifter* of the Copenhagen Academy, 1914. In the Corpus

[12] This is Thorndike's theory (ii. 44 f.); the suggestion was made to me by R. L. Poole in a letter of 1910, coupled with the possibility that the 'Gallica studia' were not necessarily in France (see, however, the usage in notes 7 and 37), but I have not been convinced by it, nor would it apparently affect the date.

[13] Haskins, *Norman Institutions*, p. 310; W. Farrer, *Itinerary of King Henry I* (1919)

[14] See, however, Bubnov, *Opera Gerberti*, p. 215 n.

[15] The Mazarine MS. has 'Liber ezich iafaris elkauresmy,' which led Wüstenfeld (p. 21) to ascribe the tables to abu Maʻashar Jaʻafar. See, however, Steinschneider, *H. U.*, pp. 568–570; Nallino, "Al-Huwarizmi," in *Atti dei Lincei*, fifth series, ii. 11. That Maslama's edition was used by Adelard is seen from the mention of Cordova in the tables and the use of the era of the Hegira in place of that of Yezdegerd. The mention of the Spanish era is also noteworthy. The treatise begins: 'Liber iste septem planetarum atque draconis statum continet. . . .'

[16] On this MS., which I discovered in 1909 and studied in 1913, see infra, Chapter VI, n. 32.

[17] This MS., of the twelfth century, unknown to Björnbo and Suter, I found in 1914 but have studied only from photographs. See infra, Chapter VI, n. 11. The Latin months are used in the tables, which differ in some other respects from those in Suter.

MS. (f. 142 v) Petrus Anfulsus calls himself 'translator huius libri,' but his description (ff. 142 v–144) follows Adelard's and relates only to the concordance of calendars (ff. 113–114), so that we may have merely the confusion of two treatises by a copyist. In these calendars, one of which coincides with that in the *Liber ysagogarum* (no. 4 below), the basal year is 1115. Adelard formally asserts his own authorship in his *Astrolabe*:

Qualis autem sit examinatio certa in eo libro qui ezic intitulatur quem ex arabico in latinum convertimus sermonem reperies.[18]

The date of Adelard's introduction appears as 1126 in the Bodleian MS.: [19]

Anno ab incarnatione domini .M°C°XX°VI°. die ianuarii .xxᵃvi. prima fuit dies Almuharran et feria tertia, annus autèm arabicus .Dᵐ'XX.

In the Corpus MS.[20] this is followed by a concordance for the eclipse of 2 August 1133:

Anno ab incarnatione Domini .M.C.XXVI. die ianuarii xxᵃviᵃ prima fuit dies Almuharram et feria tercia ali, annus arabicus .DX. planus .X. Anno igitur ab incarnatione Domini .M°.C°. XXXIII°. eclipsis solis ii° die augusti mensis feria .iiiiᵃ. ciclo .xix. x°iii°. luna vigesima viiᵃ., .ii. kal. novembris primus dies Elmuharram feria .iiiᵃ., annus arabicus adiunctus .DX. XVIII. planus. In anno sequenti .xii. kal. novembris .iᵃ. feria.

[18] McClean MS. 165, f. 83 v; Arundel MS. 377, f. 69. So f. 84 v, differing slightly from the Arundel text, which has, f. 70 v: 'Adhuc de umbris habeo que dicerem, sed quoniam in ezic [ed. Suter, pp. 21 f.] sufficienter diserta sunt labellum comprimam.' See also Arundel MS., ff. 71, 72 v.

[19] F. 159; Suter, pp. 5, 37, where the suggestion of 26 January as the date of composition is too precise, since this day (= 1 Almuharram) is given merely as a convenient starting point for reckoning. In the present form of the Bodleian MS., f. 159 follows the explicit on f. 158 v, but close examination shows that it was misplaced and in binding inserted at the end, whereas the text proves that it belongs after f. 99. The reference to the year 1126 is omitted in the Chartres and Mazarine MSS., which, however, announce in the second chapter a table 'per quam ab eo anno quo hic liber in nostrum sermonem translatus est usque in tempora infinita ex annis quotlibet romanis et mensibus cum diebus annorum et mensium et dierum arabicorum equalitas sumi queat.' The astronomical tables generally run to A. H. 570, but several of those in Corpus MS. (e. g., f. 121 = Suter, p. 128) stop in the original hand at A. H. 510 (= A.D. 1116), showing that they are not later than 1116–45.

[20] F. 141. Cf. the similar concordance for 1138 in the chronicle of John of Worcester (ed. Weaver, p. 53), who shows his acquaintance in this year with Adelard's version of the tables from the *Ezich* of 'Elkauresmus.' On the use of the Persian word *zig* for astronomical tables, see Nallino, *al-Battani*, p. xxxi; Suter, p. 32.

4 (?). *Liber ysagogarum Alchorismi in artem astronomicam a magistro A. compositus.* Bibliothèque Nationale, MS. Lat. 16208, ff. 67–71 (saec. xii); Milan, Ambrosian Library, MS. A. 3 sup., ff. 1–20 (saec. xii); Munich, Staatsbibliothek, Cod. Lat. 13021, ff. 27–68 v, Cod. Lat. 18927, ff. 31 ff.; Vienna, Nationalbibliothek, MS. 275, f. 27 (fragment).[21] This consists of an introduction, in five books, explaining the principles of arithmetic, geometry, music, and astronomy, hence in the Ambrosian MS. it is entitled *Liber ysagogarum Alchoarismi ad totum quadrivium.* The first three books of the introduction, which are interesting for the history of arithmetic, have been published by M. Curtze, in *Abhandlungen zur Geschichte der Mathematik*, viii. 1–27; the fifth shows plainly acquaintance with Hebrew chronology as well as with Arabic astronomy. As one of the Munich MSS. is of the middle of the twelfth century and the table of eras in book v is of the year 1115, this work belongs to Adelard's generation, and he is the only man bearing his initial who is known to have been at that time occupied with such translations. Moreover this same table of eras for 1115 recurs with a set of Adelard's Khorasmian tables in MS. 283 of Corpus Christi College, Oxford, f. 113, where at least part of the treatise claims Petrus Alphonsi as its author.[22] If Adelard is the author, Tannery[23] has suggested that the small knowledge of geometry shown in the introduction points to the period in his life when, although already familiar with al-Khwarizmi, he had not yet mastered the Arabic text of Euclid.[24]

5. The translation of Euclid's *Elements.* Numerous manuscripts,[25] showing considerable differences in text and arrangement. The preface (Bodleian, Digby MS. 174, f. 99), which contains an occasional Arabic word, treats chiefly of the place of geometry among the sciences and of its method and units of reckoning, and has little general interest. In

---

[21] This page, with its early forms of Arabic numerals, is published in facsimile by Nagl, *Zeitschrift für Mathematik und Physik*, xxxiv, sup., p. 129.

[22] Infra, Chapter VI, n. 11.

[23] *B. M.*, v. 416 (= *Mémoires scientifiques*, v. 344). The inclusion of the era of Spain (1153 = 1115) in the table points to the Spanish derivation of the treatise.

[24] There follows in the Munich MS. an astronomical treatise which Curtze connected with this introduction but which turns out to be the version of Zarkali by Gerard of Cremona (Steinschneider, in *Bullettino*, xx. 5 ff.) It begins and ends: 'Quoniam cuiusque actionis quantitatem temporis spacium metitur, celestium motuum doctrinam querentibus eius primum ratio occurrit investiganda. . . . Divide quoque arcum diei per 12 et quod fuerit erunt partes horarum eius, si Deus inveniri consenserit.'

[25] Several are indicated in *Bullettino*, xiv. 83.

working from the Arabic Adelard would seem to have made some use of an earlier version from the Greek, but his relations to this and to later versions require investigation, nor is it clear, pending a comparison of the manuscripts, whether in its original form his own work was an abridgment, a close translation, or a commentary.[26] It is, however, important to note what he himself says in the *Astrolabe*:[27]

Et omnium quidem supradictorum simpliciter expositorum siquis rationem postulaverit, intelligat eam apud Euclidem a quindecim libris artis geometrice quos ex arabico in latinum convertimus sermonem esse conniciendam.

Accordingly, whatever the manuscripts may show, Adelard translated the fifteen books in some form from the Arabic. Did he also write a commentary? The word is used loosely in mediaeval catalogues [28] and does not necessarily mean a commentary in our sense. Roger Bacon, however, cites on axioms a passage from the *Editio specialis super Elementa Euclidis* of 'Alardus Batoniensis,' [29] a work which Professor David Eugene Smith informs me he has not found mentioned elsewhere. The author can hardly be other than Adelard, although another writer of this name is indicated by the occurrence in a MS. of Corpus Christi College, Oxford, of a fragment of Jordanus *De ponderibus*, composed in the early thirteenth century, with marginal figures in the name of 'Alardus.' [30] Adelard of Bath was known as Alardus (e. g., Dresden, MS. Db. 87; Clare College, MS. 15, f. 185) or Aelardus (MS. lat. 18081, f. 196), as well as Adelardus or Athelardus, the last being doubtless the original form.[31]

[26] Weissenborn, in *Z. M. Ph.*, xxv, sup., pp. 141-166; Heiberg, *ibid.*, xxix, lit. sup., p. 21, xxxv, lit. sup., pp. 48-58, 81-86, and in the introduction to the Teubner edition of Euclid, v, pp. c-ci; Curtze, in *Philologische Rundschau*, i. 943-950, and in Bursian's *Jahresbericht*, xl. 19-21; Björnbo, in *B. M.*, vi. 239-248; Bubnov, *Gerberti opera mathematica*, p. 175, n.

[27] MS. Arundel 377, f. 71; MS. McClean 165, f. 84 v, with some differences.

[28] E. g., Delisle, *Cabinet des MSS.*, ii. 526.

[29] In the unpublished *De communibus mathematice*, cited from his forthcoming edition by David Eugene Smith in *Roger Bacon Essays* (Oxford, 1914), pp. 175 f. Cf. Thorndike, ii. 22, n.; Bridges, *The Opus Majus of Roger Bacon*, i. 6, n.; *B. M.*, xii. 98.

[30] MS. 251, ff. 10-13.

[31] We should not take too seriously the statement of a fragment on chiromancy in B.N., MS. lat. n. a. 693, f. 97: 'Sciendum est quod quedam ars reperta est naturalis a quodam philosopho Edmundo qui antea fuerat Saracenus et vocabatur Maneanus sed transtulit hanc artem magister Adulwardus de greco in latinum.'

6. *Questiones naturales.* A dialogue with his nephew in seventy-six chapters, purporting to explain what Adelard has learned from the Arabs. Twenty MSS.[32] and three early editions are known.[33] The Hebrew adaptation of the thirteenth century by Berachya under the title *Uncle and Nephew* has recently been edited by H. Gollancz, with a careless English version of the *Questiones* appended.[34] The printed text is poor;[35] a critical edition would be useful.

The treatise is dedicated to Richard, bishop of Bayeux, in an introductory letter [36] which speaks of Adelard's recent return to England in the reign of Henry, son of William. The nephew is reminded that the author left him and other pupils at Laon seven years before, in order to devote himself to the study of Arabic learning.[37] Since then Adelard has sojourned in the East, visiting specifically Tarsus and Antioch.[38] Now there were in this period two bishops of Bayeux named Richard, Richard Fitz-Samson, 1107–33, and his successor, Richard of Kent, 1135–42.[39] We should at first sight choose the former, as Adelard had begun to travel before 1116 and was at work on the Khorasmian tables

[32] B. N., MSS. lat. 2389, 6286, 6385, 6415, 6628, 6739, 14700, f. 273, 18081, ff. 196–210 v; Laurentian, MS. Gadd. Rel. 74, ff. 4–34; Escorial, MS. O. iii, 2, f. 72; Montpellier, École de Médecine, MS. 145; Rheims, MSS. 872, 877; Prague, MS. 1650, ff. 54–68 v, with cc. 72 and 73 added on f. 69; British Museum, MS. Cotton Galba E. iv, ff. 214–228; Bodleian, MS. 2596, ff. 108–127 (formerly also in MS. 3538); MS. Digby 11, ff. 97–102 v (incomplete); Oxford, Corpus Christi, MS. 86, f. 163; Oriel, MS. 7, f. 189 (extract); Eton, MS. 161, lacking about a page at the end. Contrary to the statement of an early librarian, there is no reason for thinking the Eton MS. to be Adelard's autograph; indeed its incorrect readings (e. g. 'constantiam' for 'inconstantiam' in the first sentence to the nephew) point to an opposite conclusion. Bale, *Index*, ed. Poole and Bateson, p. 9, cites an unknown text with introductory verses.

[33] Louvain, without date, but probably 1480, 1484, 1490 (Hain-Copinger, i, no. 85, ii, no. 26; Proctor, nos. 9219, 9260; Pellechet, no. 48).

[34] *Dodi Ve-Nechdi* (London, 1920); Steinschneider, *H. U.*, pp. 463 ff.

[35] Cf. Soury in *B. E. C.*, lix. 417; I have followed chiefly MS. lat. 6415 (saec. xii).

[36] Published by Martène and Durand, *Thesaurus anecdotorum*, i. 291.

[37] 'Meministi nepos quod septennio iam transacto cum te in gallicis studiis pene puerum iuxta Laudisdunum una cum cunctis auditoribus meis dimiserim, id inter nos convenisse ut Arabum studia ego pro posse meo scrutarer, tu vero gallicarum sententiarum inconstantiam non minus acquireres.'

[38] C. 32: 'Cum enim nuper a parte orientali venires qua causa studii diutissime steteras.' C. 16: 'Audivi enim quendam senem apud Tharsum Ciliciȩ.' C. 51: 'Cum semel in partibus Antiochenis pontem civitatis Manistrȩ transires, ipsum pontem simul ȩtiam totam ipsam regionem terrȩ motu contremuisse.'

[39] A copy of 'Adelermus Batensis' was in the library of the bishop in 1164: *Catalogue des MSS. des départements*, ii. 398, no. 112.

by 1126, if not by 1115, while he was certainly back in England in 1130. Richard of Kent, however, was a son of Robert, earl of Gloucester, and thus connected with the royal family, and Adelard's *Astrolabe* now shows him at work as late as ca. 1142.

The reference to Henry I is puzzling, since the king would naturally be taken for granted unless Adelard had left before his accession or returned after his death. In the former case the seven years' absence would place the treatise not later than 1107, while on account of the bishop's date it could not be earlier; in the latter case it would fall shortly after 1135, but, by reason of the seven-year period, at least as late as 1137. The first alternative would tend to place the *Questions* as early as the *De eodem*, whereas they show Arabic influences quite foreign to the *De eodem* and imply a longer period and wider range of travel.[40] On the other hand they show no Arabic words, such as are common in the *Liber ezic*, and no trace of Arabian mathematics or astronomy,[41] so that on internal grounds one would place them early, much earlier than a dedication to Richard of Kent would imply. The *Questions* quotes no earlier work, nor does Adelard refer to it, save in the undated treatise on falconry below.

7 (?). A treatise on the elements or on origins. The *Questiones naturales* concludes as follows:

In hac enim difficultate tractandi de Deo, de noy, de yle, de simplicibus formis, de puris elementis disserendum est, quę sicut propriam naturam compositorum excesserunt ita et de eis disputatio alias omnes dissertiones et intellectus subtilitate et sermonis difficultate precellit. Nos igitur quoniam quędam de compositis diximus, vespere iam somno suadente quiete naturali mentes reficiamus. Mane autem, si tibi idem sedet, conveniamus ut de inicio vel de iniciis disputemus. *Nepos.* Michi vero nichil magis sedet. De Deo etenim mentem instruere quoniam patrem omnium fatemur honestissimum de eodem etiam argute dicere, quoniam auctoritatem non recipio, difficillimum est. De his vero quę id ipsum comitantur discutere, quoniam multi multa inde turbaverunt, utillimum est. Quietis ergo refectionem libens accipio ut ad tractatum novum novi veniamus.

Such a sequel on primary and fundamental things would naturally follow a treatise devoted to compound substances and things; and the passage can hardly be put aside as a mere literary device to avoid these difficult problems.[42] At least one sequel to the *Questiones* has been found in the treatise on falconry, but no *De initiis* or similar work has

---

[40] Thorndike (ii. 44–49) discusses the order of the two works, tending to the same conclusion.

[41] Infra, p. 38. [42] As by Thorndike, ii. 28.

yet been identified. It is, of course, possible that the treatise was never written, but its obvious importance for Adelard's philosophical ideas justifies further search in the cosmological writings of the twelfth century, where it may lurk anonymous or without a title, even as did until recently the treatise on falconry.

8. On falcons. Anonymous in Vienna, MS. 2504, ff. 49–51; incomplete in Clare College, Cambridge, MS. 15, f. 186–186 v.[43] See below, Chapter XVII; and *E.H.R.*, xxxvii. 398–400. That this treatise follows soon after the discussion of *cause rerum* in the *Questiones* appears from the opening sentence:

> Quoniam in causis disserendis rerum animus noster admodum fatigatus sit, ad eiusdem relevationem id magis delectabile quam grave interponendum est.

This is the earliest Latin treatise on falconry so far known. It shows no trace of Arabic influence, but mentions English usage and English simples which suggest the Anglo-Saxon leechdoms. The citation of 'libri Haroldi regis' is further indication of Adelard's connection with the English royal court.

9. On the *Astrolabe*. Cambridge, Fitzwilliam Museum, McClean MS. 165, ff. 81–88;[44] incomplete at the beginning in British Museum, Arundel MS. 377, ff. 69–74. Apparently written at Bath, which is taken as the meridian for purposes of illustration.[45] The preface, found only in the McClean MS., reads as follows:

Incipit libellus magistri Alardi bathoniensis de opere astrolapsus

> Quod regalis generis nobilitas artium liberalium studio se applicat valde assentio, quod rerum gubernandarum occupatio ab eodem animum non distrait non minus ammiror. Intelligo iam te, Heynrice, cum sis regis nepos, a philosophia id plena percepisse nota. Ait enim beatas esse res pu[b]licas si a philosophis regende tradantur aut earum rectores philosophie adhibeantur. Huius rationis odore ut infantia tua semel [46] inbuta est in longum servat,[47]

---

[43] Perhaps the *De educatione accipitrum* ascribed by Tanner (p. 38) to Aluredus Anglicus.

[44] Saec. xii, formerly in the possession of Prince Boncompagni (see Narducci, *Catalogo*, no. 360). The portion corresponding to the Arundel MS. begins in the middle of f. 83; there are four finely drawn figures at the close, ff. 87–88 v.

[45] 'Verbi gratia ad natale solum: Quia enim Bathonia lii^{us} gradibus ab equinoctiali circulo et terra Ari distare cognoscitur, ideo et latitudo climatis eius totidem graduum esse perhibetur.' F. 82 v; cf. ff. 84 v, 85.

[46] MS. *senilis*.

[47] Cf. Horace, *Epist.*, i, 2, 69 f.:

> 'Quo semel est imbuta recens servabit odorem
> Testa diu.'

quantoque gravius exterioribus oneratur, tanto ab eisdem diligentius se sub-
trahit. Inde fit ut non solum ea que Latinorum scriptis continentur intelli-
gendo perlegas, sed et Arabum sententias super spera et circulis stellarumque
motibus intelligere velle presumas. Dicis enim ut in domo habitans quilibet,
si materiam eius et compositionem quantitatem et qualitatem sive distric-
tionem ignoret, tali hospicio dignus non est, ita si qui in aula mundi natus
atque educatus est tam mirande pulcritudinis rationem scire negligat, post discretionis annos indignus atque si fieri posset eiciendus est. His a te frequenter
ammonitus, licet meis non confidam viribus, tamen, ut nobilitati philosophiam uno nostre etatis exemplo coniungam, postulationi tue pro posse meo
dabo operam. De mundo igitur eiusque districtione quod arabice didici
latine subscribam, hoc prescripto nodo ut cum mundus nec quadratus nec
longilaterus nec alterius figure quam spericus sit, quicquid de spera dixero de
mundo dici intelligatur. Spera igitur globosum et rotundum corpus . . .

The treatise is accordingly dedicated to a young Henry, grandson
(or nephew) of a king. In the earlier part of the twelfth century this
can mean only Henry of Blois, bishop of Winchester, or Henry Fitz-
Empress. The allusions to secular government would have no point in
the case of Henry of Blois, who early became a Cluniac monk, and he is
also excluded by chronological considerations, for by 1126, the earliest
possible date for a treatise which cites the *Liber ezic*, he has become
abbot of Glastonbury and passed well beyond *infantia*.[48] To Henry
Plantagenet, on the other hand, early imbued with letters and receiv-
ing, perhaps, before the age of seventeen a collection of ethical maxims
compiled for his benefit by William of Conches,[49] the introduction is
entirely appropriate: he is a king's grandson, he is to become a ruler, he
divides his time between books and practical affairs. As he is still
infans and has not reached *discretionis annos*, this was doubtless
written before 1149, when he was knighted, and 1150, when he became
duke. If, as seems probable, the treatise was composed in England, it
would then fall between 1142 and 1146, while Henry, between the ages
of nine and thirteen, was living in his uncle's household at Bristol under
the tutorship of Master Matthew.[50] Adelard has not been elsewhere
found after 1130, but as he was then hardly more than fifty or there-
abouts, he may well have lived far into Stephen's reign. The *Astrolabe*
is one of Adelard's latest works. It cites the *De eodem*, the *Tables*, and
the Euclid, and thus serves to bind his work together.

[48] Adam of Domerham, pp. 304–315; John of Glastonbury, p. 165.

[49] Haskins, *Norman Institutions*, p. 131. Hauréau's argument to this effect I
now find less convincing.

[50] Gervase of Canterbury, i. 125. Cf. Miss Norgate, *Angevin Kings*, i. 334, 375;
Round, *Geoffrey de Mandeville*, pp. 405–408.

10. *Ysagoga minor Iapharis matematici in astronomiam per Adhelardum bathoniensem ex arabico sumpta.* Bodleian, Digby MS. 68, ff. 116–124; anonymous in British Museum, Sloane MS. 2030, ff. 83–86 v; formerly in Avranches MS. 235.[51] An astrological treatise,[52] evidently of abu Maʿashar Jaʿafar. Reference is made to the fuller treatment in the *Ysagoga maior*,[53] but it is not said that this has been translated.[54]

11. *Liber prestigiorum Thebidis (Elbidis) secundum Ptolomeum et Hermetem per Adhelardum bathoniensem translatus,* a treatise on astrological images and horoscopes by Thabit ben Korra. Lyons, MS. 328, ff. 70–74; formerly in MS. Avranches 235.[55]

12 (?). *Mappe clavicula,* dealing with the preparation of pigments and other chemical products. This work, which goes back to Greek sources and is of great interest for the history of technical processes, is printed in *Archaeologia,* xxxii. 183–244, from a manuscript of the twelfth century then in the possession of Sir Thomas Phillipps. The attribution to Adelard rests on the thirteenth-century table of contents (*Liber magistri Adelardi bathoniensis qui dicitur mappe clavicula*) in Royal MS. 15. C. iv of the British Museum; the treatise itself was missing from the manuscript as early as Tanner's time. Berthelot[56] has shown that Adelard cannot have been the author of the *Mappe clavicula* in its original form, for a version, free from Arabic elements, is found in a manuscript of Schlettstadt which goes back at least to the tenth century; but it is quite possible that Adelard is responsible for

---

[51] *Catalogue des MSS. des départements,* x. 114.

[52] 'Quicunque philosophie scientiam altiorem studio constanti inquirens. . . . Hec igitur sunt loca excessuum cum quibus finem institucionis faciemus.'

[53] 'Horum autem singula in ysagoga maiore dicta sunt, nunc autem compendiose introducendis propius dicetur.'

[54] On the translations of the *Ysagoga maior* ascribed to John of Seville and Hermann the Dalmatian, see Steinschneider, *H. U.,* pp. 568 f.; infra, Chapter III, no. 3.

[55] 'Quicunque geometria atque philosophia peritus astronomie expers fuerit ociosus est. Est enim astronomia omnium artium et re excellentissima et prestigiorum effectu commodissima. . . . Hec quidem omnia ceceraque circa principium enumerata in ysagogis exposita studiosa mente firmanda sunt, ut prestigiorum facultate artifex non decidat.' This translation is not mentioned in the list of Thabit's works given by Steinschneider (*Zeitschrift für Mathematik,* xviii. 331–338), nor identified in his discussion of the *Speculum* of Albertus Magnus (*ibid.,* xvi. 371), who cites it as a work of Hermes (*Catalogus codicum astrologorum Graecorum,* v. 100). Thorndike (i. 664) was the first to identify it.

[56] *La chimie au moyen âge,* i. 26–30; "Adalard de Bath et la Mappae clavicula," *Journal des savants,* 1906, pp. 61–66; and reprinted in his *Archéologie et science* (1908), pp. 172–177. Cf. Thorndike, i. 468, 765 ff., ii. 22 f.

the expanded form of the text, in certain chapters of which Arabic and English words occur.[57]

13 (?). Commentary on the *Spherica* of Theodosius. The *Bibliomnomia* of Richard de Fournival mentions 'Dicti Theodosii liber de speris, ex commentario Adelardi.'[58] No such treatise has yet been identified.

14 (?). Miscellaneous notes. In Warner and Gilson's *Catalogue of Western MSS. in the Old Royal and King's Collections* we read under MS. 7 D. xxv (saec. xii):

Chronological, philosophical, astronomical, medical, and other collections, in Latin: evidently the book, or more probably copies from the book, of a man of unusual learning. It seems worth suggesting that this scholar may be Adelard of Bath. . . . who studied at Laon, was something of a Platonist, travelled in the East, and in other respects coincides with the indications of the volume.

This interesting suggestion cannot be positively established from the contents of the manuscript, which, however, clearly represents Adelard's generation and circle of interests. The lunar cycle, as the catalogue points out, is that of 1136–54. A series of notes on ff. 53 and 54, giving various Platonic doctrines on the universe, cites Plato, Chalcidius, Macrobius, and Censorinus. One (f. 53 v) gives the three divisions of the brain as in c. 18 of the *Questiones*;[59] another (f. 54) reminds us of the Platonic theme of the *De eodem*:

Animam composuit Deus ex substantia et ex eodem et diverso, id est ex individuitate et vegetatione, ex mutabilitate et immutabilitate, anima ergo tercium genus nature est ex mutabilitate et immutabilitate mixtum.

The most curious passage is the following (f. 66), which occurs in the midst of a set of astronomical notes which have scattered Greek words:

Mons Amor reorum est locus medius mundi, ubi apposui mensuras et probavi per multa loca et posui lignum rea [*sic*] rotundum habens .xii. cubitos longitudinis et grossitudo illi cubitus unus et suspendi illum per funem et tantum commutavi eum de loco in locum in medio eius .vii. [60] kal. iulii donec

---

[57] Cc. 190, 191, 195–200. Cf. also the Saracen recipe in c. 289. The *Mappe clavicula* is also found, anonymous, in the Bodleian, MS. Digby, 162, ff. 11 v–21 v. A metrical version, made from the Arabic, is ascribed to Robertus Retinensis: Steinschneider, *E. U.*, no. 102 d; infra, Chapter VI, p. 122.

[58] Delisle, *Cabinet des MSS.*, ii. 526, no. 42; Birkenmajer, *Bibljoteka Ryszarda de Fournival* (Cracow, 1922), p. 53; infra, Chapter III, n. 42.

[59] Infra, n. 93; Chapter V, n. 59. The preceding passage suggests the *Questions*, c. 19, and there are other traces of the doctrines of Salerno.

[60] Read *xi*? In the last line we should read *exuperarer*.

suspendi illud in loco medii diei et residit suum cum splendor solis ex omnibus partibus et facta est umbra ipsius subtus cum rotunda sicut rotunditas ipsius ligni quod suspenderam; et de ipsa mensura cognovi quod medius mundus est in monte Amor reorum. Et tempore quo mensuravi hoc est annus .xxxviiii. et vinum non bibi, oculi mei somno satiati non fuerunt, ne exuperaveram in eo quod inquirebam.

In this corrupt Latin we have apparently the record of an observation about the time of the summer solstice undertaken to determine the place where the sun was directly overhead. The mount of Amor apparently means Mount Moriah; at least it was in Palestine, mediaeval tradition placing the *umbilicus terre* at Jerusalem.[61] Of course a vertical position of the sun could not really have been observed north of the tropics, but Palestine was the southernmost point in Christendom, and an observation in latitude 31° 45′ might approximate the desired result. In any case the painstaking character of the experiment is interesting, and it falls in with Adelard's habit of mind and his known travels in Syria. One cannot argue too closely from the cycle of 1136–54, which is in another hand and another portion of the manuscript; this would give 1115 as the latest date of the observation made thirty-nine years before. In any event, if Adelard is speaking, his visit to the East would fall in his youth.

It is not clear that the older bibliographers had other works of Adelard at their disposal. Tanner pointed out that the *De causis* and the *Problemata* are only other names for the *Questiones naturales*, and the incipit of the *De sic et non* indicates that it is probably a variant of the same treatise.[62] Similarly the *De septem artibus liberalibus* apparently has the incipit of the *De eodem et diverso*. The *Computus astro-*

[61] On the belief that Jerusalem was the navel and centre of the earth see W. H. Roscher, "Omphalos," in *Abhandlungen* of the Leipzig Academy, phil.-hist. Kl., xxix, no. 9, pp. 24–28 (1913); "Neue Omphalosstudien," *ibid.*, xxxi, no. 1, pp. 15–18, 73 f. (1915); A. J. Wensinck, "The ideas of the western Semites concerning the navel of the earth," in *Verhandelingen* of the Amsterdam Academy, xvii, no. 1 (1917). Different places were identified with the *umbilicus*, such as Bethel, Mount Moriah (infra, p. 339), and Garizim. Roscher, 1913, pp. 27 f., cites a passage of Gervase of Tilbury (ed. Leibnitz, p. 892; ed. Liebrecht, p. 1) to the effect that the well where Jesus conversed with the woman of Samaria was the centre of the earth since the sun at the solstice casts no shadow in it, a phenomenon which philosophers say occurs also at Syene (ca. lat. 24°). For Syene, see Macrobius, ed. Eyssenhardt, p. 600. In Adelard's Khorasmian tables (p. 1) the 'medius locus terre' is Arin.

[62] Or a continuation, as is suggested by the *incipit* given by Bale (1557, p. 184) and Pits: 'Meministi ex quo incepimus.' Without this *incipit* one would accept the suggestion of Poole and Bateson, in their edition of Bale's *Index* (1902), p. 8, that this is the well known work of Abaelard.

nomicus mentioned by Tanner is probably the Khorasmian tables; the *Compotus Adelardi*, formerly in the library of Christ Church, Canterbury,[63] may be either this work or, more probably, the *Liber abaci*. A treatise which follows the *Questiones* in a manuscript of the Laurentian library, which Bandini thought might have emanated from Adelard, belongs to the fourteenth century.[64] Jourdain conjectured that Adelard was the translator of the *Liber imbrium* of Ja'afar, but this is now known to be the work of Hugo Sanctallensis,[65] and the attribution to him of the translation of Euclid's *Optics* and *Catoptrics* is equally unfounded.[66] The cosmological treatise ascribed to Adelard in Cotton MS. Titus D. iv and analyzed by Thorndike (ii. 41–43) is the *De essentiis* of Hermann of Carinthia.[67] An interesting suggestion, made by Chasles and still awaiting confirmation, is that Adelard, as the translator of the Khorasmian tables,[68] is also the author of the translation of a treatise of al-Khwarizmi on Indian arithmetic, preserved in a unique manuscript at Cambridge,[69] which has an important bearing on the transmission of the Arabic system of reckoning to the West.

What can be gleaned from all this for Adelard's biography is disappointingly meagre. He was born in Bath, which he calls *natale solum*, and styles himself English;[70] but he early went to France, where he studied at Tours and taught at Laon. In this period of his life he found opportunity for travel, penetrating as far as Magna Graecia and, it would seem, Sicily before 1116 and probably before 1109. After leaving Laon he spent seven years in study and travel, and can be traced in Cilicia and Syria and pos-

[63] James, *Ancient Libraries of Canterbury and Dover*, p. 49. Contrary to Dr. James's conjecture (p. 508) this manuscript can hardly be Cotton MS. Caligula A. xv, part 2.

[64] MS. Gadd. rel., no. 74, f. 38 v: 'Anno gratie 1303 quo ego Petrus Paduanensis hunc librum construxi.'

[65] Infra, Chapter IV, n. 41.

[66] Infra, Chapter IX, n. 102. Dr. Dee also (James, *List*, no. 165) suggested Adelard as the author of the *De differentia spiritus et anime* of Costa ben Luca.

[67] Infra, Chapter III, n. 17.

[68] And, probably, of no. 4, above, p. 24.

[69] University Library, MS. Ii. vi. 5, f. 102, published by Boncompagni, *Trattati d'aritmetica*, 1 (Rome, 1857). See *Comptes rendus de l'Académie des Sciences*, xlviii. 1059 (1859); *Z. M. Ph.*, xxxiv, sup., p. 132; *Abhandlungen zur Geschichte der Mathematik*, x. 11; Cantor, i. 713, 906.

[70] *E. H. R.*, xxxvii. 398; supra, n. 45; infra, Chapter XVII. He also calls England his 'patria' in the dedication of the *Questiones*.

sibly, by 1115, in Palestine. By 1126 he is back in the West, occupied with making the astronomy and geometry of the Arabs
available to the Latin world.[71] Bath again becomes his residence,
and in 1130 as 'Adelardus de Bada' he receives 4s. 6d. from the
sheriff of Wiltshire.[72] His relations with the court, as well as his
account of his life as a student and his green cloak,[73] are quite
inconsistent with the common assertion that he was a monk; I can
find no contemporary authority for this statement, which doubtless owes its origin to a confusion with the monk Adelard of Blan-
dinium, who, a century earlier, wrote a life of St. Dunstan.[74] The
name 'Goth,' which is applied to Adelard in certain manuscripts
of the translation of Euclid,[75] I cannot pretend to explain; it may
be a mere corruption of Bath, or it may possibly refer to a sojourn
in northern Spain. It seems probable that Adelard visited Spain,
not only because this was the nearest abode of Saracen learning,
but because he used a Spanish edition of al-Khwarizmi, yet it is
always possible that he received this text indirectly. The date
of his death is unknown, though the discovery of his relations
with the future Henry II prolongs his activity at least as far
as 1142, later than has commonly been supposed. Here, as so
often, we have to lament the loss of the Pipe Rolls between 1130
and 1155.

Three bits of evidence connect Adelard with the Anglo-Norman court. First of all, the pardon of a murder fine of 4s. 6d. in
Wiltshire in the Pipe Roll of 1130 is not only made by royal writ,
but, as Poole has pointed out,[76] is the kind of favor customarily
granted to those in the employment of the court. Next, the dedication of the *Astrolabe* to the young Henry, his pupil; and, in the
third place, the mention of 'King Harold's books' in the treatise

---

[71] 'Nos vero latinorum studemus utilitati': MS. Chartres 214, f. 41; MS. Mazarine 3642, f. 83.

[72] Pipe Roll, 31 Henry I, p. 22.

[73] *Questiones*, c. 2.

[74] Stubbs, *Memorials of St. Dunstan*, p. xxx; cf. Tanner, p. 55.

[75] Bodleian MS., Selden Arch. B. 13; *Zeitschrift für Mathematik*, xxv, sup.,
p. 144; *Philologische Rundschau*, i. 946; *Centralblatt für Bibliothekswesen*, xvi. 262;
Hänel, *Catalogus Librorum MSS.*, col. 786.

[76] *The Exchequer in the Twelfth Century* (Oxford, 1912), pp. 56 f. Cf. also the
suggestion respecting the queen in n. 12, supra.

on falconry, itself a royal sport. Adelard may well, as Poole suggests, have been an officer of the Exchequer, where his arithmetical talent would have proved useful, but I see no reason for going on to associate him with the introduction of the abacus there, which seems to me of earlier origin.[77]

Of Adelard's other relations we know but little. One of his works is dedicated to the bishop of Syracuse, one to the bishop of Bayeux, another to a certain H. In three of them an unidentified nephew appears, though not necessarily the same person in each instance. The only reference to Adelard on the part of a contemporary is that of an enigmatical Ocreatus, possibly named John, who dedicates to Adelard the translation of an Arabic treatise on arithmetic which he has produced *iussus ab amico immo a domino et magistro*.[78] No other of Adelard's pupils is known, saving always Henry II. What we should most like to know is the extent and nature of Adelard's connections with the other translators and scholars of his age, but here we have little more than possibilities. His version of the Khorasmian tables seems in some way connected with Petrus Alphonsi, while it was in turn revised by Robert of Chester.[79] So his commentary on the *Spherica* of Theodosius recalls the citation of this work by Hermann of Carinthia.[80]

The range and variety of Adelard's interests can be judged from his writings, extending, as they do, from trigonometry to astrology and from Platonic philosophy to falconry, perhaps even to applied chemistry. He had a style of his own, easily recognizable by his readers, and a certain gift of apt illustration, while the treatise on falconry shows that he had none of the philosopher's disdain for the ordinary and the practical. Of the originality and

[77] Infra, Chapter XV.

[78] *Prologus N. Ocreati in Helceph ad Adelardum batensem magistrum suum*, edited by Henry in *Zeitschrift für Mathematik*, xxv, sup., pp. 129–139. Cf. Steinschneider, *E. U.*, no. 70; Cantor, i. 906, where the confusion with Bayeux rests upon an incorrect reading of the manuscript. Bernard, *Catalogi*, no. 8639, ascribes a version of Euclid to 'Ioannes Ocreatus,' but the first leaf of this MS. is now gone and the remainder bears no such indication. See Warner and Gilson's *Catalogue of the Royal MSS.* under 15 A. xxvii.

[79] Infra, Chapter VI, n. 31.

[80] Infra, Chapter III, n. 42.

profundity of his knowledge it is less easy to speak until his mathematical work has been more thoroughly sifted by specialists and its relations to his predecessors have been fixed. We now know him most fully as a philosopher, but his philosophical writings belong to his earlier years, and it is by no means clear that we here have him at his best.

In the *De eodem et diverso* Adelard speaks as a disciple of Plato, *princeps philosophorum*, from whose *Timaeus* he derived the theme of unity and diversity.[81] His Platonism is in general that of Chartres, and shows here no influence of Aristotle's science or of Arabic learning. In form the treatise reflects Martianus Capella and the *De consolatione* of Boethius. In the allegory which passes before Adelard's vision permanence is represented by Philosophy, surrounded by the seven liberal arts; change and decay by 'Philocosmia,' the love of this world, with appropriate companions. Philosophy, having won the debate, proceeds to explain briefly the nature of the seven arts in traditional fashion, though the more concrete temper of the author reveals itself at the end in an explanation of the geometrical determination of the height of a tower and in an account of a debate with a Greek philosopher of southern Italy on topics of natural philosophy. Adelard shows the influence of the atomic theory of Democritus. On the question of universals he seeks to reconcile Plato and Aristotle in the so-called theory of non-difference.

The *Questiones naturales* is written professedly to explain the new knowledge which Adelard has acquired from 'his Arabs,' under whose name it presents, as Thorndike has pointed out,[82] theories for which he does not care to assume personal responsibility. Although the *Questiones* is in no sense a systematic treatise, the seventy-six problems are taken up in a regular order. The first six chapters deal with plants: why they grow from earth where there are no seeds; how plants of opposite natures spring from the same soil; why the other three elements do not produce plants, and whether each of the four brings forth its appropriate

---

[81] See Willner's analysis, *Beiträge*, iv, no. 1; and cf. Ueberweg-Baumgartner[10], ii. 244, 311 f. Thorndike (ii. 48) is in error in seeking the source in Aristotle.

[82] ii. 25 f.

products; why fruit follows the graft rather than the trunk. The explanations are based upon the four elements and the four qualities of the Greeks as formulated by Galen, the so-called elements of our apprehension being in reality compounds, in which, however, the real element in each case preponderates. Then come chapters on animals (7–14), where the questions concern digestion, in ruminants and birds, and its products; the better sight of certain animals in the dark, explained by the humors of the eye; and the question whether animals have souls, a matter of current debate which Adelard decides in the affirmative, on the ground that they possess not only bodily sensation but the judgment which is a property of the soul. With chapter 15 we reach man, at first with the scarcely profitable question why mankind lacks horns or other bodily means of defence, and then with a brief note on the object of the network of muscles and veins. The following problems (cc. 17–32) are chiefly psychological: the relation of memory to mental ability, the parts of the brain allotted to memory imagination, and reason; hearing and sight and the other senses — with interspersed speculation as to the position of the nose above the mouth and the nature of baldness. Chapters 33 to 47 deal with the human body: breathing, the inequalities of the fingers, erectness in walking, food, the different temperaments of the sexes, and dead bodies. The remainder of the treatise (cc. 48–76) treats of meteorology and astronomy. How is the globe supported in the air? If the earth were perforated, how far would a body fall in the perforation, the author concluding correctly that it would stop at the centre. What is the cause of earthquakes and tides, of the saltness and constant volume of the sea, of the freshness of springs and rivers, of thunder and lightning and the course of the winds? Thunder is occasioned by the noise of hail and ice; the tides come, not from the moon, but from the flux and reflux occasioned by the meeting of waters from the several arms of the sea, a passage in which Adelard repudiates the influence of the moon and gives currency to the error introduced into the West by Macrobius.[83] At last (c. 69) we reach the upper world with the

---

[83] C. 52. Cf. Duhem, iii. 116 f. The text is not entirely clear. MS. lat. 6415 does not mention the moon but refers to the inundations of the Nile. The printed text

darkness and shadows of the moon, the course of the planets and the outer all-containing *aplanos*, and the life of the stars. The stars are alive, and so is the *aplanos*, though in one sense the *aplanos* may be called God. The nature of God, however, along with all questions of simple forms and pure elements, is, in conclusion, put off till another day.

In all this there is not much that comes from the Arabic, nor is any Arabic authority or phrase specifically quoted. Not only is the theory of innate ideas entirely Platonic,[84] but Plato is frequently cited, in one case in a long extract from the *Timaeus*.[85] We have references to the *Topica, Musica,* and *De consolatione* of Boethius.[86] Other Latins are Statius, Terence, Horace, and the *Saturnalia* of Macrobius.[87] So far we are within the same range of reading as in the *De eodem.* Now as to Aristotle: Adelard quotes *inter Aristotelicas sententias* [88] the principle that, when anything is added to anything, the whole becomes greater; he cites as Aristotle's a passage on motion which goes back ultimately to the *Physics*;[89] and he gives as authority for the localization of the three faculties in the brain *Aristotiles in Physicis et alii in tractatibus suis.*[90] Still more striking is the reminiscence of the *Physics* in a passage on motion where no authority is given.[91] In this sense he might be claimed as the first Latin writer of the Middle Ages to cite the Aristotelian physics,[92] but such scanty fragments

has 'Caribdis' in place of the Nile. Gollancz by an extraordinary slip renders this 'Caribbean'!

I cite chapters after the edition, folios in MS. lat. 6415. Thorndike gives an interesting summary (ii. 23–41).

[84] C. 28. See Hauréau, *Philosophie scolastique* (1872), i. 355.

[85] Cc. 23 (= *Timaeus*, cc. 45 f.), 24, 27, 28, 29.

[86] Cc. 20–23, 46.   [87] Cc. 35, 49, 53, 55.

[88] C. 34. Cf. in c. 10 the ascription to Aristotle of the theory of two entrances to the stomach.

[89] 'De actione itaque earum et notandum in quo non meam set Aristotilis accipe sententiam, immo quia ipsius ideo meam: quidquid enim movetur, ait, aut vi aut natura aut voluntate moveri convenit.' C. 74, f. 38 v; cf. *De physico auditu*, 8, 4, 1.

[90] C. 18.

[91] C. 60; cf. *De physico auditu*, 8, 5; and pp. 109 f. of the essay of Baumgartner cited below.

[92] Duhem, "Du temps où la scolastique latine a connu la Physique d'Aristote," in *Revue de philosophie*, xv. 163 (1909) (cf. *Système*, iii. 188–193), gives Thierry of Chartres as the first. by way of Macrobius.

hardly indicate a first-hand acquaintance. Indeed the only specific citation, that concerning the localization of the faculties, seems to come, not from Aristotle, but from Galen, from whom it and certain theories of the elements apparently reached Adelard and the later twelfth century via Constantine the African.[93] What, then, is most clearly of Arabic origin is the physiological part of the *Questiones*, and the sources for this were available to Adelard in southern Italy. There is no evidence that Adelard as yet knows Arabic or has assimilated the Arabic mathematics and astronomy for which he was later distinguished, and there are none of the Arabic words which appear freely in his astronomical works. From internal evidence, the *Questiones* belongs to Adelard's earlier rather than his later years, and there is nothing in it which he could not have found in Italy.

Adelard would probably have said that what he acquired from the Arabs on subjects of physics was not so much facts or theories as a rationalistic habit of mind and a secular philosophy. The recourse to observation and experiment, already evident in the *De eodem*, appears likewise in the *Questiones*, in spite of its reliance for the most part on *a priori* reasoning. The author knows that a distant blow is seen before it is heard;[94] he has stood on a bridge in Syria during an earthquake;[95] and he has watched the workings of a vessel in which water is held up by pressure of the air.[96] Indeed, in explaining the last phenomenon, he first enunciates the theory of the continuity of universal nature, as Thorndike has shown.[97] He also asserts the indestructibility of matter, but on the authority of an unnamed philosopher.[98]

---

[93] Werner, "Wilhelm von Conches," in Vienna *Sitzungsberichte*, lxxv. 387 (1873); Baumgartner, *Die Philosophie des Alanus de Insulis* (*Beiträge*, ii, no. 4), pp. 19, 94; Soury, in *B. E. C.*, lix. 417; infra, Chapter V, n. 60. On Constantine's influence on the medicine of the twelfth century, see Sudhoff, in *Archiv für Geschichte der Medizin*, ix. 348–356 (1916).

[94] C. 68.   [95] C. 50; supra, n. 38.   [96] C. 58.

[97] ii. 37–40; and in *Nature*, xciv. 616 f. (1915). Thorndike raises the question whether Adelard may have been acquainted with the *Pneumatica* of Hero. This was known in Sicily by 1156: infra, Chapter IX, n. 115.

[98] 'Unde phylosophus de mundo loquens ait, Nec quicquam ex eo recessit nec est addendi facultas, cunctis in se cohercitis, sed corruptela partium senescentium intra se vicem quandam obtinet cibatus. Et meo certe iuditio nichil sensibili mundo

"It is hard to discuss with you," Adelard tells his nephew, "for I have learned one thing from the Arabs under the guidance of reason; you follow another halter, caught by the appearance of authority, for what is authority but a halter?" Say what you please, for you will always find hearers who will demand no reason for an opinion but will accept anything on the weight of an ancient name. "If reason is not to be the universal judge, it is given to each to no purpose." Those who are considered authorities first reached that position by virtue of the exercise of their reason. Use reason first, then add authority, for authority alone cannot bring conviction to a philosopher.[99] Later he says: "I call myself a man of Bath, not a Stoic, wherefore I teach my own opinions, not the errors of the Stoics." [100] In like manner God is not to be used as a blanket explanation of things accessible to human understanding. At the outset Adelard reminds the interlocutor that, while plants spring from the earth by God's will, this does not act without a reason.[101] Human science must first be listened to, he says a little later, and "only when it fails utterly should there be

---

moritur nec minor est hodie quam cum creatus est, si qua enim pars ab una coniunctione solvitur non perit sed ad aliam societatem transit.' C. 4, f. 25.

[99] 'De animalibus difficilis est mea tecum dissertio. Ego enim aliud a magistris arabicis didici ratione duce, tu vero aliud auctoritatis pictura capistrum captus sequeris. Quid enim aliud auctoritas est dicenda quam capistrum? Ut bruta quippe animalia capistro quolibet ducuntur nec quo vel quare ducantur discernunt restemque quo tenentur solum secuntur, sic nec paucos vestrum bestiali credulitate captos ligatosque auctoritas scriptorum in periculum ducit. Unde et quidam nomen sibi auctoritatis usurpantes nimia scribendi licentia usi sunt, adeo ut pro veris falsa bestialibus viris insinuare non dubitaverint. Cur enim cartas non impleas? Cur et a tergo non scribas, cum tales fere huius temporis auditores habeas qui nullam iudicii rationem exigant, tituli nomine tantum vetusti confidant? Non enim intelligunt ideo rationem singulis esse datam ut intra verum et falsum ea prima iudice discernant. Nisi enim ratio universalis iudex esse deberet, frustra singulis data esset. Sufficeret enim preceptorum scriptori datam esse, uni dico vel pluribus, ceteri eorum institutis et auctoritatibus essent contenti. Amplius: ipsi qui auctores vocantur non aliunde primam fidem apud minores adepti sunt, nisi [MS. non] quia rationem secuti sunt quam quicunque nesciunt vel negligunt merito ceci habendi sunt.' C. 6, f. 25 v. The passage 'Quid . . . ducit' is quoted with approval by Roger Bacon (ed. Bridges, i. 5 f.), who had much in common with Adelard.

[100] C. 28, f. 30.

[101] 'Voluntas quidem Creatoris est ut a terra herbę nascantur, sed eadem sine ratione non est.' C. 1, f. 24.

recourse to God" as an explanation.[102] Proximate, not ultimate, causes are Adelard's theme, and his theories of God, mind, and matter are reserved for the *De initiis*.

The popularity of the *Questiones naturales* in the Middle Ages is attested by the twenty surviving copies, in which it frequently accompanies the *Naturales quaestiones* of Seneca.[103] It was probably used by Alexander Neckam and Thomas of Cantimpré.[104] It is quoted by Vincent of Beauvais and Roger Bacon in the thirteenth century,[105] and by Pico della Mirandola in the fifteenth.[106] Three editions appeared before 1500. It was the basis of the Hebrew dialogue *Uncle and Nephew (Dodi Ve-Nechdi)*.[107]

In significant contrast to the speculative and discursive character of the *Questiones* stands the *Astrolabe*, which was written in the later years of Adelard's life. Once more he explains 'the opinions of the Arabs' to an eager listener, this time concerning the sphere and the stars, though we must not take too seriously and literally the precocious interest of the young Henry or the author's references to philosophers as kings. Succinct, clear, and sharp, the treatise presents in systematic form the preliminary astronomical facts and the various applications of the astrolabe. Arabic terms are freely used, and for fuller discussion the reader is referred to Adelard's other works, the *De eodem*, Euclid, and especially the Khorasmian tables. Virgil, Horace, and Cicero are each quoted once, but without digression, and Ptolemy replaces Plato. We have once more the Adelard of the *Liber ezic*.

[102] 'Deo non detraho, quicquid enim est ab ipso et per ipsum est. Id ipsum tamen non confuse et absque discretione non [*sic*] est, quę quantum scientia humana procedat audienda est, in quo vero universaliter deficit ad Deum referenda est. Nos itaque quia nondum [non Deum?] in scientia pollemus ad rationem redeamus.' C. 4, f. 25.

[103] MSS. lat. 6286, 6385, 6628; and with other fragments of Seneca in MS. Reims 872 and MS. Prague 1650. See also MS. O. iii. 2 of the Escorial, whose contents should be compared with a volume given to Bec in 1164 (*Catalogue des MSS. des départements*, ii. 398, no. 112).

[104] Thorndike, ii. 196, 379.

[105] E. g., *Speculum naturale*, v, cc. 13, 31, vi, cc. 6, 7; *Opus maius*, ed. Bridges, i. 5 f.

[106] Duhem, iii. 116 f.

[107] Ed. Hermann Gollancz (London, 1920).

Adelard occupies a position of peculiar importance in the intellectual history of the Middle Ages. Standing at the point where the traditional knowledge of the cathedral schools meets the new learning of southern Italy and the Mohammedan East, his attitude was one of personal inquiry and not mere blind receptivity. The first, so far as we know, to assimilate Arabic science in the revival of the twelfth century, to him we owe the introduction of the new Euclid and the new astronomy into the West. Moreover he was a pioneer in more than a chronological sense. He went out to seek knowledge for himself by travel and exploration, penetrating as far as Sicily and Syria and, probably, Spain; and he showed a spirit of independent inquiry and experiment quite his own. Fragmentary as our information is, it reveals something of the originality and many-sidedness of the man; and if further research should lead to new discoveries concerning his life or writings, it will throw light on one of the most interesting and significant figures in mediaeval science.

CHAPTER III

HERMANN OF CARINTHIA

AMONG the scholars who in the twelfth century brought the science and philosophy of the Arabic world to western Europe, not the least important was Hermann of Carinthia, variously known as the Dalmatian, the Slav, or, to distinguish him from the earlier Hermannus Contractus, the Second. Somewhat younger than Adelard of Bath and less important than Gerard of Cremona, he must still be reckoned among the notable pioneers in the field of Saracen learning. *Acutissimi et literati ingenii scholasticus,*[1] he contributed to mathematics and philosophy as well as to astrology and astronomy, and in the case of one work of ancient science, the *Planisphere* of Ptolemy, his translation constitutes the sole intermediary through which this classical treatise has survived to later times. Moreover, while the origins of most of the other translators of this period remain unknown, Hermann's relations with the school of Chartres bring him into connection with the cathedral schools of the earlier twelfth century and link him with their Platonism as well as with the Aristotelianism of the Arabs. His real work, also, was long eclipsed by confusion with two others of the same name who wrote on similar themes, Hermannus Contractus, monk of Reichenau in the eleventh century, and Hermannus Alemannus, a translator of philosophical works from the Arabic in the thirteenth century;[2] and it is only in recent years that he has been disentangled, in part at least, from these and placed in his proper setting, while still more recently his authorship of the version of the *Planisphere* has been vindicated against his pupil Rudolf of Bruges. His work, however, has not

---

[1] Peter the Venerable, in Migne, *Patrologia*, clxxxix. 650.
[2] On Hermannus Contractus, see below. On Hermannus Alemannus, see the references in Chapter I, n. 51. Jourdain, *Recherches*, is still useful in distinguishing them.

heretofore been studied as a whole.[3] Let us begin with a list of his writings:

1. *Zaelis Fatidica,* or *Pronostica,* also known as *Liber sextus astronomie.* A translation of the *De revolutionibus* of the Jewish astrologer Saul ben Bischr (see Steinschneider in *Z. M. Ph.,* xvi. 388–390; *H. U.,* pp. 603–607; *E. U.,* no. 51): 'Secundus post conditorem orbis . . . minus fiunt efficaces.' Vatican, MS. Pal. lat. 1407, ff. 18–38; Metz, MS. 287, ff. 333–350 (saec. xv); University of Cambridge, MS. Kk. iv. 7, f. 102; Caius College, MS. 110, f. 295 (James, *Catalogue,* i. 115, ii. 542); Pembroke College, MS. 227, f. 133 (James, p. 205); Bodleian, MS. Digby 114, ff. 176–199. In all of these the translator is given as Hermann. For other possible MSS. see Thorndike, ii. 391. The date appears in the Metz and Digby MSS., the latter of which has, 'Explicit fedidica Zael Banbinxeir Caldei translatio Hermani 6ⁱ astronomie libri. Anno domini 1138. 3°. kal. octobris translatus est' (where by misreading '6ⁱ' as 'Gⁱ' Macray attributed the translation to Gerard of Cremona; and by misreading 'Hermani' as 'hec mam' Thorndike, ii. 84, makes matters worse). The phrase 'sixth book' apparently refers to some Arabic collection; it can hardly already be Hermann's sixth book. This is the earliest dated work of Hermann; the place is not indicated, and there is no accompanying preface.

2. (?) Translation of the Khorasmian tables. In Hermann's version of Albumasar we read:

in sectionibus	formis		tardis

Quorum plus fialcurdaget azerea secundum fialcurdaget albatia tractatur, que in translatione nostra zigerz Alchuarismi sufficienter exposuimus.'[4] So a note to his *Planisphere* speaks of 'Albatene et Alchoarismus quorum hunc quidem opera nostra Latium habet.'[5] As we already know of a version made by Adelard of Bath in 1126 and revised by Robert of Chester,[6] these statements do not simplify the problem, nor has any MS. been found with Hermann's name.

[3] The principal modern accounts are those of Wüstenfeld, pp. 48–50; Steinschneider, *H. U.,* pp. 534 f., 568 f., and *E. U.,* no. 51; Clerval, "Hermann le Dalmate," in proceedings of the *Congrès international des catholiques* of 1891 (also separately, Paris, 1891), and *Les écoles de Chartres* (Paris, 1895), pp. 188–191; Björnbo, in *B. M.,* iv. 130–133 (1903); Thorndike, ii. 84 f. Bosmans, in *Revue des questions scientifiques,* lvi. 669–672 (1904), I have not seen.

[4] Naples, MS. C. viii. 50, f. 43. Cf. Steinschneider, *H. U.,* p. 568.

[5] Ed. Heiberg, p. clxxxvii. Cf. Suter, *al-Khwarizmi,* p. xiii. Thorndike's proposal (ii. 85) to translate 'hunc' as 'the former' disregards Latin idiom without clarifying the situation.

[6] Supra, Chapter II, n. 16; infra, Chapter VI, n. 32.

3. Translation, in eight books, of the *Maius introductorium* to astrology of abu Ma'ashar Ja'afar al-Balki (Albumasar); a less slavish version than the contemporary one by John of Seville. See Steinschneider, in *Zeitschrift der deutschen morgenländischen Gesellschaft*, xviii. 170–172; *H. U.*, pp. 567 ff; *E. U.*, no. 51; and especially Dyroff, in Boll, *Sphära*, pp. 484 f. There is a copy of the twelfth century in the Biblioteca Nazionale at Naples, MS. C. VIII. 50, ff. 1–56 v[7]; also in Corpus Christi College, Oxford, MS. 95, f. 60; Bodleian, MS. Laud 594, ff. 144–153 (incomplete); Erfurt, MS. Ampl. Q. 363, f. 38; Florence, Conventi soppressi, J. II. 10, ff. 1–54 v (*B. M.*, xii. 195); Vatican, MS. Vat. lat. 4603; Parma, MS. 720, f. 1; Manchester, Rylands Library, MS. 67, f. 170; also formerly among the MSS. of Petau, Montfaucon, *Bibliotheca Manuscriptorum*, p. 87 b. Apparently the "Albumasar minor Hermanni" of the Sorbonne catalogue (Delisle, *Cabinet des MSS.*, iii. 68). Printed at Venice, in 1489, 1495, 1506; see Steinschneider, *H. U.*, p. 568. Cited under Hermann's name by Roger Bacon, *Opus tertium*, ed. Brewer, p. 49. This translation is probably of 1140[8] and in any case anterior to 1143, being cited in the introduction to the *De essentiis*[9] and probably in the *Planisphere*.[10] The introduction,[11] addressed to Robertus Ketenensis, reads as follows:

Liber introductorius in astrologiam Albumazar Albalachi[12]

Apud Latinos artium principiis quedam ars extrinseca prescribi solet. Librum autem iniciis non scripto ullo autentico quidem ego in ea lingua invenerim, set doctorum sua cuiusque sententia aditus paratur. Apud Arabes contra. Duorum siquidem primum nec advertisse videntur umquam, ta-

---

[7] The subscription which Björnbo declared illegible (*B. M.*, iv. 133) reads: FINIT. ΘΡΑΚΘΟC. ΑΒΟΜΑΤΖΑΡ. ΑΛΒΑΛΑΚΓΤ. FHPMΑΝΝΤ. CΗΚΟΝΔΤ. ΘΡΑΝCΛΑΘΤΩ. ΦΗΛΤΚΤΘΗΡ. Θ is of course Τ. The small λ is confused by the scribe with a. I have rendered by F the peculiar form for Roman H, an F without the upper stroke.

[8] So the printed text as cited in Duhem, iii. 175 f. The Naples MS. (f. 32) omits the current year.

[9] 'Quas Abumaixar in annalibus suis usque ad .iii. milia numerat, quem numerum nec nos in eiusdem libri translatione pretermisimus.' MS. Naples, C. viii. 50, f. 70; MS. Corpus Christi 243, f. 105; MS. Titus D. iv, f. 112 v. Albumasar is frequently cited in this work: MS. Naples, ff. 61, 63, 65, 67, 70, 74, 74 v, 75 v; MS. Corpus, ff. 94 v, 97, 99, 101, 104 v, 108 v, 109, 111 v.

[10] 'Ad imitationem alterius translationem nostre.' Heiberg, p. clxxxiii, line 8. Cf. the mention of Albumasar as amplifying the *Quadripartitum* on p. clxxxv.

[11] Steinschneider, *H. U.*, p. 568, cites various remarks of Hermann inserted in the text, which Dyroff calls a 'Bearbeitung' rather than a mere translation.

[12] Corpus Christi College, Oxford, MS. 95, f. 60, with some variants from Naples, MS. C. viii. 50, f. 1.

metsi particulatim nonumquam ac sparsim assumant, nostro tamen iudicio non parum necessarium. Secundum vero commenticium quidem illis nec scripto dignum visum est tanquam egregium aliquod inventum scripture commendatum. Ab hoc igitur secundo genere huius operis auctor incipiens, .vii., inquit, sunt omnis tractatus inicia: auctoris intentio, operis utilitas, nomen auctoris, nomen libri, locus in ordine discipline, species inter theoricam et practicam, partitiones libri. Quod apud nos quinquipertito sufficeret, operis videlicet titulo, auctoris intencione, finali causa,[13] modo tractandi, et ordine, que omnis fere tam [14] tractatus quam orationis exordio et necessaria et sufficere videntur, suam tamen singulis reddit causam. Que cum ego, prolixitatis exosus et quasi minus attinencia cum et hunc morem Latinis cognoscerem preterire volens, ab ipso potius tractatu exordiri pararem, tu mihi studiorum omnium [15] specialis atque inseparabilis comes rerumque et actuum per omnia consors unice, mi Rodberte, si memor es, obviasti dicens: "Quamquam equidem nec tibi pro more tuo, mi Hermanne, nec ulli consulto aliene lingue interpreti in rerum translationibus a Boecii sentencia quadam ullatenus divertendum sit, ita tamen alienum intersequendum videtur nec precuratur presertim ne [16] qui librum hunc in arabica lingua legerit si in latina non ab exordio suo qua[m] primum legentis intuitus inciderit inceptum videat, non industriam set ignoranciam putans et operis forsan integritatem detrimenti et nos devie digressionis arguat." Parui quidem, cum ipsum etiam laborem tuo potissimum instinctu aggressus sim, ut siquid ex hoc nostro studio latine copie adiciatur, non mihi maius quam tibi meritum rependatur, cum tu quidem et laboris causa et operis iudex et utriusque testis certissimus existas. Expertus quippe tu nichilominus quam grave sit ex tam fluxo loquendi genere quod apud Arabes est latine orationi congruum aliquod commutari atque in hiis maxime que tam artam rerum imitationem postulant. Hiis habitis, ne longius differatur, ab ipsius verbi tractatus inicium sumamus. Intentionis, inquit, exposicio rei summam breviter et absolute proponens discentis animum attentum parat et docilem utilitatis promissio laborem allevians internum animi quendam affectum adaptat. Auctoris nomen duabus de causis necessarium est, tum ut opus autenticum reddat tum ne alii dum vagum et incerti sit nominis immerito ascriptum iniustam parat gloriam. Libri nomen intentionis testimonio accedit, locus in ordine discendi animum discentis, quo lectio quid legendum sit instruens ad disciplinarum intellectum non inconsulte dirigit. Scientie genus partitionumque numerus et exposicio attentum item reddunt et docilem. Quoniam igitur inter omnes huius artis scriptores nullus hactenus inventus est qui contradicentibus responderet vel approbantibus argumentum daret, ad hec nec ullus qui plenarie totam scriberet artem, nostra quidem in hoc opere intencio et illis resistere et hiis firmamentum dare et integram divino auxilio artem tradere, unde non minimam hanc utilitatem consequi manifestum sit, ne qui deinceps operam huic artificio dederint, quia diversa ex diversis operibus

---

[13] *finali causa* inserted from Naples MS.   [14] MS. *tum.*

[15] Not *olim*, as in the printed text.

[16] So Naples MS. The Corpus MS. has *precurratur presertim nec.* For the method of Boethius see Chapter XI, n. 37.

adminicula necessaria sint, vel desistant vel deficiant. Tantum igitur opus certis et auctoris et libri nominibus confirmari necessarium duximus, quem titulum prescribentes dicimus introductorium in astrologiam Albumasar Albalachi, qua de causa etiam post astronomiam in astrologiam primo loco legendus sit, in theoricam scilicet huius artis partem principaliter atque generaliter editus, .viii. partitionum numero discretus, queque suis differentiis subdivisa. Partitionis prime capitula .v.: primum de invencione astrologie, secundum de siderei motus effectu, tertium de effectus qualitate, quartum de confirmatione astrologie, quintum de utilitate astrologie.

4. Two polemical treatises against Mohammedanism: 'De generatione Mahumet et nutritura eius quam transtulit Hermannus Sclavus scolasticus subtilis ingeniosus apud Legionem Hispanie civitatem'; 'Doctrina Mahumet que apud Saracenos magne auctoritatis est ab eodem Hermanno translata cum esset peritissimus utriusque lingue latine scilicet et arabice.' Bodleian, MS. Selden supra 31, ff. 16–32; Cambridge, Corpus Christi College, MS. 335, f. 57. Printed in Bibliander's edition of the Latin Koran (Basel, 1543), i. 189–212. Cf. Steinschneider, *Polemische und apologetische Literatur in arabischer Sprache* (Leipzig, 1877), pp. 227–234; id., *E. U.*, no. 51, where the *Chronica mendosa Saracenorum* should, on the authority of the MSS. just cited and others, be transferred to Robertus Ketenensis. These versions were doubtless prepared in conjunction with the Latin translation of the Koran for which Peter the Venerable, abbot of Cluny, engaged the services of Robertus Ketenensis and Hermann in 1141 and which was completed, with an accompanying letter in Robert's name, in 1143. See Migne, *Patrologia*, clxxxix. 650–674, 1073–76; Steinschneider, *E. U.*, no. 102.

5. Translation of Ptolemy's *Planisphere*, completed at Toulouse 1 June 1143 and dedicated to Thierry of Chartres. This version, of which nine MSS. and three early editions are known, is based on the Arabic text of Maslama and is the only medium through which Ptolemy's treatise has come down to us. Critical edition by Heiberg, *Ptolomaei opera astronomica minora* (Leipzig, 1907), pp. xii f., clxxx– clxxxix, 225–259; the preface is on pp. clxxxiii–vi. Formerly attributed to Hermann's pupil, Rudolf of Bruges, this has been restored to Hermann by Clerval, *Les écoles de Chartres*, p. 190; Steinschneider, *H. U.*, pp. 534, 569; and especially Björnbo, in *B. M.*, iv. 130–132 (1903). See below, n. 68. The identification of the Tolosa of the MSS. with Toulouse, rather than with the unimportant Tolosa proposed by Steinschneider and Björnbo, is strengthened by the fact that the *De essentiis* was written in the same year at Béziers.

6. *De essentiis*, a philosophical treatise discussed below. Three MSS. are known: one of the twelfth century (N) in the Biblioteca Nazionale at Naples, MS. C. viii. 50, ff. 58–80; one of the fifteenth century (C) in Corpus Christi College, Oxford, MS. 243, ff. 91–115 v; and a third (L), incomplete at beginning and end, in Cotton MS. Titus D. iv, ff. 75–138 v, of the British Museum (saec. xiv), attributed by a modern hand to Adelard of Bath.[17] Planned, but evidently not completed, when the preface to the *Planisphere* was written, 1 June 1143 (Heiberg, p. clxxxv, l. 10); finished at Béziers later in the same year:

> De essenciis Hermanni Secundi liber explicit anno domini millesimo centesimo quadragesimo tercio Byterri (MS. N Biternis) perfectus.

It was dedicated to Robert de Ketene, as appears from the following preface: [18]

> Athlantidum his diebus me crebro murmure concitum gravis et insupeɪ agit admiracio. Quisnam casus [19] queve novitas inmotum te hactenus nunc demum tibi ipsi subduxerit, ut relicto videlicet altero te a communi munere omnis vite nostre nova qualibet occasione [20] secesseris? An quemadmodum Hercule substituto Athlas terrena rigavit, tu quoque [21] similiter tocius muneris onere [22] mihi relicto quasi respirandum tibi interim censueris,[23] fortasse quia securus secretorum ociis vaces [19] dum ego publicis gimnasiis expositus insidiosos colluctantium impetus sustineam? Queruntur dee pariter iniuriarum agentes meque tanquam pignore obligato collegam inpacienter requirunt. Excuso, responsa differo, cupiens potius redditum defendere quam reddendum excusare. Nunc quoniam ipsa divina manus te voto meo reddidit, presentiam dearum ne dubites. Nec enim valde metuo quidquid cause fuerit dum te ipsum habeant, nisi forte malis me tanquam advocatum premittere quam ipse excusandus prodire. Consulte agendum censes, optime Rodberte, eamque mihi semper apud te gratiam sentio, seu quod prudenti animo cuncta circumspicis et provides seu quod individua nobis vita mens eadem atque omnino una anima. Ego itaque si recte memini causam ordine exponam. Meministi, opinor, dum nos ex aditis nostris in publicam Minerve pompam prodeuntes circumflua multitudo inhianter miraretur, non tanti personas pensans quantum cultus et ornatus spectans quos ex intimis Arabum thesauris diutine nobis vigilie laborque gravissimus acquisierat, subiit me gravis admodum pietas super his qui hec forinseca tanti habebant, quanti pensarent si interulas ipsas contueri liceret. Que cum nobis nocte iam cubili receptis me minime sineret valde ex adverso obstante Numenii metu criminis, ecce cuncta somno tenente desuper adveniens [24] altissima dea ver-

---

[17] Under whom it is discussed, with some hesitation, by Thorndike, ii. 41–43.
[18] Robert is also mentioned in the body of the work: see below, pp. 60 f.

[19] Om. C.	[22] N, *honore.*
[20] C, *actione.*	[23] N, *consueris.*
[21] N, *tuque o.*	[24] N, *advenientis.*

ticem meum dextra tetigit, cuius visione tanquam subito irradiante sole cum primum vehementer attonitus deinde paulatim assuefierem, Expergisce, inquid, et respice. Quam cum [25] cognovissem provolutus pedibus dive, Innue, inquam, o numinum omnium regina, quicquid alumpno possibile videas. Surge, inquid, et sequere me. Cui cum nisi previo te nichil mihi licere pretenderem, illa quidem, Unum, inquit, hominem mihi ex utroque vestri factum ab initio putavi, nec putes vel sine illo [26] munus hoc institutum cum nichil inter vos divisum vel sine suprema illius manu tibi possibile quem ipsa rerum omnium et actuum auctorem tibi prestitui. Quippe in quo mihi [27] complacuit quem ipsa mihi inter universas delitias meas archane conscientie delectum singulari cura summoque studio educavi demumque nec diffiteor certe nequaquam [28] repugnantem livida furia tocius viris et consilii mei privilegio dato et tibi universe familie mee certissimum ducem presignavi. Et verum est, inquam. Impera, obtempero. Evolat igitur in summum maiestatis sue solium, quo [29] cum in angustissimo receptaculo consedisset, preposita [30] in medio universa substantie sue materia pariter et huiusmodi instrumentis appositis,[31] primo loco calculis et radio deinde equilibri dipondio postremo lucifera quadam lampade cuncta penetrante, Hec, inquid, suscipe, hoc muneris iniungo nec particulatim, ut hi qui miseros auditorum animos vario diripientes tante tibi pietatis causa sunt, datumque larga manu distribue nichil dubitans; opes enim nostre largitate crescunt nec [32] indigno animo ullo modo possibiles. Suscepi tandem et ecce munus ipsum [33] offero rude quidem ac tuo ipsius antequam in publicum prodeat examine castigandum, quod ubi perspexeris non me dearum ministerio defuisse cognosces.

At hoc unum opinor, mi anime, quod non solum excusationi verum maxime [34] approbationi sufficiat quod tam necessaria de causa tamque honesta occasione institutum est.[35] Magnum quippe nec a primo seculo de quoquam mortalium auditum.[36] Fac ergo ne differas atque ab ea potissimum materia exorsus sacre institutionis legem prosequere, ego, ut equi cognitoris est, orationis seriem attente et cum summa benivolentia amplectar.

7. *Liber ymbrium quem edidit Hermannus.* Clare College, Cambridge, MS. 15, f. 1-2 (cf. James, *Catalogue*, p. 29); Dijon, MS. 1045, f. 187; Vienna, MS. 2436, f. 134 v; anonymous, in Corpus Christi College, Oxford, MS. 233, f. 122; St. Mark's, Cl. xi, 107, f. 53 (Valentinelli, iv. 285); MS. Boncompagni 4, f. 63 (Narducci, *Catalogo*, p. 5). Inc. 'Cum multa et varia de imbrium . . . .' The various treatises on meteorological predictions current under this and similar titles have

---

[25] N, *quantum.*
[26] N, *ullo.*
[27] C, *pristinum. Quippe in quo nichil.*
[28] N, *nec quicquam.*   [29] N, *qua.*
[30] C, *preposui.*
[31] C, *prepositis.*
[32] N, *nunc.*
[33] C om.
[34] C, *etiam.*
[35] C, *necessaria opinione. Magnum.*

[36] N, *auditur.* C then has *Fac igitur.* In this paragraph the speaker is obviously Robert.

not yet been clearly separated; see Steinschneider, *E. U.*, nos. 36, 51, 54, 68 *p* 4; infra, Chapter IV, n. 41.

8. *Commentary on Euclid.* In the *Biblionomia* of Richard of Fournival [37] we find at the head of the mathematical section, 'Euclidis geometria arithmetica et stereometria ex commentario Hermanni secundi.' Birkenmajer [38] has shown reasons for identifying this MS., in part, with MS. LVI. 48 of the old catalogue of the Sorbonne, now MS. Lat. 16646 of the Bibliothèque Nationale.[39] This consists of the first twelve books of Euclid's *Geometry*, in a Latin version different both from that of Adelard of Bath and from that ascribed by Björnbo to Gerard of Cremona.[40] Its abbreviated character indicates closer affinities with that of Adelard. It begins:

> Septem sunt omnis discipline fundamenta in quibus omnium rerum ad mathematice studia pertinentium firma essentie conceptio certusque veritatis intellectus in quadam quasi materia et causa fundata existunt. Sunt autem hec: Preceptum, exemplum, alteratio, collatio, divisio, argumentum, et finis. . . .

It would be interesting to have the whole of this version, and still more interesting if Hermann's preface could be recovered. The appearance of *essentia* here and in the text suggests the preface to the *Planisphere* and still more the *De essentiis*.

9. Arithmetical works. Other mathematical treatises appear in Richard's *Biblionomia*: 'Item liber de invenienda radice, et alius Hermanni secundi de opere numeri et operis materia.' [41] The MS. has not been identified, nor have other copies been found.

10. *Liber de circulis.* In a passage in his *Planisphere* Hermann says (Heiberg, p. clxxxvii): 'Nos discutiendi veri in libro nostro de circulis

---

[37] Delisle, *Cabinet des MSS.*, ii, 526, no. 37.

[38] *Bibljoteka Ryszarda de Fournival*, in *Rozprawy* of the Cracow Academy, lx, no. 4, pp. 49–52 (1922); cf. *Isis*, v. 215.

[39] Delisle, *Cabinet*, iii, 68, no. 48; 108 folios, 13th century.

[40] *B. M.*, vi. 242–248 (1905). In the parallel passages here cited MS. 16646 agrees in 2, 1 more nearly with Gerard, in 5, 1, and 10, 1, more nearly with Adelard, but in no instance exactly. It has a few Arabic words, e. g., 'alalem' = vexillum (2, 1, f. 13 v); 'mut kefia, id est mutue . . . anint ale chelkatu wa tahtit, id est in elevatione et lineatione' (6, 19, f. 39 v). At the end of Book ix we read (f. 64):
'Perfectus siquidem numerus cunctis partibus suis equalis individua natus origine eadem proportione compactus nichil extraneum assumens nichil sui relinquens gemina proprie essentie plenitudine integer ad omnem rerum perfectionem aptissimus est. W.'a delitah^ine aradene enne beienne W.' hed horatu.'

[41] Delisle, *Cabinet*, ii. 526, no. 45; identified by Birkenmajer with Sorbonne LVI. 32 (*ibid.*, iii. 68).

rationem damus.' This treatise, which has not been identified, would seem distinct from his versions of Euclid and Theodosius.

11 (?). The *Spherica* of Theodosius. Two Latin versions seem to have been current in the Middle Ages, and are ascribed respectively to Plato of Tivoli and Gerard of Cremona.[42] It appears, however, that Hermann and Robert had something to do with this treatise, for Hermann cites it in his *De essentiis*,[43] while Robert speaks of the *Cosmometria* of Theodosius as one of the treatises on which he hopes to work.[44] If either of them produced a Latin version, it has not yet been identified.

12 (?). MS. Dijon 1045, ff. 148–172 v, contains "Hermannus de ocultis," beginning, 'Astronomie judiciorum omnium bipertita est via' . . . .[45]

The *Astronomia* and *Astrologia* cited in the *De essentiis*[46] may refer merely to the translations of al-Khwarizmi and Albumasar.

Two other works have been ascribed to Hermann which require some consideration:

a. On the astrolabe. The name of Hermann is associated with three treatises on this subject preserved in numerous MSS.[47] and printed by Pez, *Thesaurus*, iii, 2, pp. 94–139, whence they are reprinted by Migne, cxliii. 379–412. The second of these (Migne, coll. 389–404) has been separated from the others by Bubnov and ascribed conjecturally to Gerbert. The third (Migne, coll. 405–412) is probably by Hermann,

---

[42] Boncompagni, *Platone*, pp. 251 ff.; Steinschneider, *E. U.*, nos. 46 (39), 98 *a*; Björnbo, in *B. M.*, iii. 67, xii. 210; supra, Chapter II, n. 58.

[43] 'Sic enim et Theodosius in Sperica: Super hunc, inquit, movetur totum, ipse vero immotus. Quo facto educit ex eodem centro in utramque partem lineam rectam usque in intrinsecam planiciem spere acutis hinc inde angulis ut secundum Eratostenem Ptolomeus describit ad quadrantem ferme recti anguli': MS. C, f. 97 v; MS. N, f. 63 v; MS. L, f. 88.

[44] Infra, p. 121.

[45] In MS. Avignon 1022, f. 209, the 'Centiloquium Ptolomei cum expositione Her[emani]' is evidently an emendation for 'Her[metis].'

[46] MS. C, f. 100: 'Tum fere circa centrum *a*, ut in astronomia firmavimus, describetur epiciclus Veneris circulus,' where *firmavimus* may mean merely that he has verified the statement. F. 108 v: 'Quod quale sit de sole in aeris temperie de luna in aquarum motu in astrologia plane exposuimus.' F. 114: 'Quippe cum generales quidem diversitates vulgares scribant girographi, speciales vero nos ipsi in astrologicis satis exposuimus.'

[47] For the MSS. of the several treatises, see Bubnov, *Opera Gerberti*, pp. 109–112.

though it bears no name.[48] The first (Migne, coll. 381–390) is addressed: 'Hermannus Christi pauperum peripsima et philosophie tironum asello, immo limace, tardior assecla, B. suo jugem in Domino salutem.' No date or other indication of authorship is given in the text, so that the treatise has been claimed both for Hermann Contractus, the lame monk of Reichenau (1013–1054),[49] and for Hermann of Carinthia. Both were interested in astronomy, and no copy has been found clearly anterior to the time of Hermann of Carinthia.[50] In his favor [51] have been argued, not only the silence of the biographer of the Reichenau monk, but also the numerous Arabic terms which appear in the treatise, words which would be familiar to him and quite unfamiliar to a German monk of the eleventh century, cut off from travel by his infirmity. If we read in the preface 'Turonum' with one MS. (Mazarine 3642, f. 55) or 'Tyronum' with certain others (Vatican, Ott. lat. 309, f. 152; B. N., Lat. 16208, f. 84; Avranches, 235; British Museum, Royal 15 B. ix, f. 51; Caius College, 413, f. 9), the B. or Ber.[52] of the dedication becomes Bernard of Tours, with which school Hermann is ranged by his preface to the *Planisphere*, addressed to Thierry of Tours and Chartres.

Tempting as is the identification, the temptation must, I believe, be resisted. The style of the preface is quite foreign to Hermann of Carinthia, whereas its extreme monastic humility reappears in a tract on lunar months ('H. pauperum Christi abortivum vile') in which the references to Bede and Notker of St. Gall plainly indicate Hermannus Contractus.[53] We now know from Bubnov that Arabic words in conjunction with the astrolabe were current by the eleventh century,[54] so

[48] The main reason for the identification is (Cantor, i. 886 f.) the coincidence of ch. 3 with a letter addressed to Hermann by Meinzo, *scolasticus* of Constance: *Neues Archiv*, v. 202–206.

[49] On Hermannus Contractus, see particularly Bubnov, pp. 109–114, 124–126; Wattenbach, *Deutschlands Geschichtsquellen im Mittelalter*,[6] ii. 41–47; Cantor, i. 885–889; and now Manitius, *Lateinische Litteratur*, ii. 756–777.

[50] To any one familiar with the difficulty in distinguishing MSS. of this period it is not surprising that MS. Royal 15 B. ix, dated 'saec. xi' by Bubnov, should be placed at the end of saec. xii by Warner and Gilson's *Catalogue*.

[51] This ascription is favored by Clerval, *Les écoles de Chartres*, pp. 169, 190, 239; Langlois, in *B. E. C.*, liv. 248–250; and B. Lefebvre, *Notes d'histoire des mathématiques* (Louvain, 1920),p. 146.

[52] MS. Ott. lat. 309, f. 152, has 'Ber.' MS. Selden supra 25 of the Bodleian has 'Be.' MS. Arundel 377, f. 35 v, has ' B'.'

[53] Bubnov, p. lxx; ed. in G. Meier, *Die sieben freien Künste* (Einsiedeln, 1887), ii. 34–36 (Manitius, ii. 767).

[54] Supra, Chapter I, nn. 20, 21; Thorndike, ch. 30.

that a Latin work containing them might then have reached Reichenau. Moreover a gloss in the Bodleian (MS. Digby 174, f. 210 v; Macray, p. 186; Bubnov, p. 113) states that Hermann wrote the tract to supplement Gerbert at the request of a certain Berengarius. In that case it would fall in line with the Gerbertian tradition, which Hermann of Reichenau, in the generation succeeding Gerbert, upholds with his *Abacus* and *Rithmomachia*, as well as with his *Compotus* and *Prognostica*.[55] *Eximius doctor*, his *Astrolabe* is found in more than thirty MSS.,[56] and he is even portrayed, astrolabe in hand, in a position of equal honor with Euclid.[57]

b. Translation of Ptolemy's *Almagest*. The Louvain MS. of the *Astrolabe* has at the head of the treatise, in a hand of the thirteenth century, the following note:[58] 'Hermannus iste astrologus fuit natus de Karinthia, non Contractus de Suevia, et transtulit Almag.' This is confirmed by one of the four MSS. of the version of the *Almagest* made from the Greek in Sicily,[59] MS. Vat. Pal. 1371, where we read in a hand of the fifteenth century: 'Translatus in urbe Panormi tempore regis Roggerii per Hermannum de greco in latinum.' Like the author of the *Astrolabe* and Adelard of Bath, the author of the preface to this work calls himself *philosophie tardus assecla* and implies that he has written other things.[60]

It does not, however, seem possible to reconcile this version of the *Almagest* with the known facts of Hermann's career. The Sicilian translator tells us that he was pursuing the study of medicine at Salerno when he heard that the copy of the *Almagest* had been brought from Constantinople to Sicily by Aristippus, an envoy of the Sicilian king, whereupon he sought out Aristippus, and after long study of the advanced works of Euclid, his mind, *scientie siderum expers*, was brought to the point of turning Ptolemy's work into Latin. Now obviously Hermann, who in 1143 translated the *Planisphere* and wrote

---

[55] Bubnov, pp. cix f. The *Compotus* is also in Arundel MS. 356, f. 28.

[56] To the twenty-six cited by Bubnov, *Opera Gerberti*, pp. 109–112, should be added MS. Bodley 625 (Bernard, 2180); MS. 413 (630) of Caius College; MS. Vat. Ott. lat. 309, f. 152; MS. Chigi, F. iv. 48; and the MSS. in Manitius, ii. 765.

[57] Ashmolean MS. 304, f. 2 v. The *Experimentarius* which follows does not, however, indicate his connection with Bernard of Chartres and Tours (Langlois, in *B. E. C.*, liv. 248–250), but is subsequent to 1164. See Chapter VII, infra.

[58] See the facsimile of this page in Reusens, *Eléments de paléographie* (Louvain, 1899), p. 236. The MS., no. 217, formerly no. 51, is attributed by Bubnov (p. xxxix) to the twelfth century.

[59] See below, Chapter IX.

[60] See the preface in full, infra, Chapter IX, end.

the *De essentiis*, could not then speak of himself as ignorant of astronomy, and there was no such royal embassy to Constantinople before the negotiations of 1143–44. Moreover, even if we could assume that the MS. arrived earlier, we cannot place such a translation before 1143. In the preface to the *Planisphere* Hermann says: 'Quorum almagesti quidem Albeteni commodissime restringit,' so that he evidently then knew the *Almagest* only in the compend of al-Battani, and it is in the light of this statement that we must regard the citations of the *Almagest* in the *De essentiis*.[61] These all refer to a single portion of the *Almagest* (5, 16–18) in connection with the relative size of earth and sun and the parallaxes of the moon at Ptolemy's four terms, and there is nothing in them which involves a direct use of that treatise, whose contents were then known through various Arabic intermediaries. Moreover, neither here nor elsewhere does Hermann show a knowledge of Greek, and the style of the Sicilian preface is not his. Its author apparently wrote after 1158.

Very likely the attribution of the version of the *Almagest* to Hermann of Carinthia arose simply out of a confusion with the *Planisphere*. It is at the same time entirely possible that the author of the Sicilian translation should have been named Hermann.

For Hermann's biography, the evidence accordingly consists of these titles and prefaces to his works, the preface of Robert of Chester to al-Kindi, to be printed later,[62] and the letter of Peter the Venerable.

A native of Carinthia[63] and, if we may trust the names generally

[61] MS. C., f. 100: 'Quemadmodum in Almagesti probamus, in primo quidem termino lxiiii, in secundo lix, in tercio xliiii, in quarto xxxix, quarum singule equales semidiametro globi terreni.' F. 100 v: 'Quemadmodum in Almagesti geometrica demonstratio constituit solem terra centies et septuagies fere maiorem.' F. 101: 'Primum quidem in Almagesti ex diversitate videndi lunam quaterna eius distancia per quatuor terminos reperitur.' Cf. f. 101 v. 'Sortiatur secundum diligentissimam Ptolomei observationem puncta tantum xxxiii de diametro circuli per polos circuli lunaris ipsiusque globi centrum transeuntis, diametro (LN, diametros) vero umbre nisi (LN, ubi) minima partem unam puncta xxiiii de cxx partibus eiusdem diametri.'

[62] Chapter VI.

[63] In the version of Albumasar (MS. Naples C. viii. 50, f. 38 v), he says: 'Istrie .iii., maritima et montana, in medio patria nostra Kaunthia.' So the Louvain MS. cited above under *b* has 'natus de Karinthia; cf. MS. Dijon 1045, ff. 187, 191: 'de Kanto'? He is called 'Sclavus' in the heading of one of the anti-Mohammedan tracts. The name Dalmatian is twice applied to him by Peter the Venerable (Migne, clxxxix, coll. 650, 671).

applied to him, of Slavic descent, Hermann early came under the teaching of Thierry of Chartres, whether at Chartres or Paris we cannot say; and it may well have been the influence of this powerful personality, fundamentally Platonist but quick to assimilate the new Aristotle and whatever of new knowledge came its way, that turned Hermann toward the Arabic sources of philosophical and scientific learning.[64] How early Hermann reached Spain is not known, but by 29 September 1138 he was already sufficiently familiar with Arabic to produce his translation of Zael, and in 1141 he was still engaged in astrological studies when Peter the Venerable found him and his companion Robert in the region of the Ebro, "both skilled in the two languages." To these years should doubtless be assigned the translations of al-Khwarizmi and Albumasar (1140), while the *Planisphere* and the *De essentiis* were completed by 1143. In or about 1142 he was in Leon, as we learn from the tract against the Saracens. By 1 June 1143 he is at Toulouse, and later in the same year at Béziers. Doubtless he also visited Toledo, which he uses for geographical illustration,[65] but we know nothing of his relations with the school of Arabic studies which flourished there, nor can we follow him or his writings subsequently to the *De essentiis*.

Of the literary partnership and close friendship with Robertus Ketenensis there is, however, abundant evidence. Peter the Venerable found them together in 1141 and engaged them in a joint labor of translation. Hermann receives the dedication of Robert's translation of the *Iudicia* of al-Kindi;[66] to Robert, *unicus atque illustris socius studiorum omnium, specialis atque inseparabilis comes rerumque et actuum per omnia consors unice*, Hermann dedicates the version of Albumasar and the *De essentiis*. It appears from the preface of the last-named work that their studies in the inner treasures of Arabic learning were at first carried on in secret

---

[64] Hermann addressed Thierry in the preface to the *Planisphere* as a second Plato and 'Latini studii patrem.' On Thierry see Hauréau, in *Mémoires de l'Académie des Inscriptions*, xxxi. 2, pp. 77–104; Clerval, *L'enseignement des arts libéraux d'après l'Heptateuchon de Thierry de Chartres* (Paris, 1889); id., *Les écoles de Chartres*, pp. 169–172, 188 ff.; Hofmeister, "Studien zu Otto von Freising," in *Neues Archiv*, xxxvii. 135 (1912); Poole, in *E. H. R.*, xxxv. 338 f. (1920).

[65] Infra, n. 203.   [66] On Robert see Chapter VI, infra.

and only brought before the world after long vigils and severe labor. It also appears that Robert had recently withdrawn for a time from the common task to a life of quiet leisure, perhaps on the occasion of his appointment as archdeacon of Pamplona, while Hermann kept up the struggle *in publicis gimnasis* and by his teaching doubtless earned the title of *scolasticus* which is given him by Peter the Venerable.[67] Any list of Hermann's writings must take account of Robert's collaboration, and *vice versa*.

No disciple of Hermann's is known save Rudolf of Bruges, whom we know only from the description of an astronomical instrument of Maslama which as *Hermanni secundi discipulus* he dedicates to John of Seville: [68]

> Cum celestium sperarum diversam positionem stellarum diversos ortus diversosque occasus mundo inferiori ministrare manifestum sit huiusque varietatis descriptio ut in plano representetur sit possibile, prout Ptholomeo eiusque sequaci Mezlem qui dictus est Aloukakechita [69] visum est, pro posse suo huius instrumenti formulam dilectissimo suo Iohanni David Rodulfus Brugensis Hermanni secundi discipulus describit.
>
> Primum igitur huius instrumenti est postica . . . formulam tenaci memorie commendet.   Explicit.[70]

As the *De essentiis* is the only independent work of Hermann which has so far been identified, we must depend mainly upon it for light on his philosophic and scientific ideas. It belongs, as we have seen, to 1143, when he has already translated Zael and Albumasar and has just completed his version of the *Planisphere*, as well as a *primus liber* on astronomical topics which may be al-Khwarizmi. Its subject is the five essences — cause, motion, place, time, *habitudo* — which have a permanent, unchanged existence. There is no connection apparent with the much briefer *De quinque essentiis* of al-Kindi, later translated into Latin by

[67] Migne, clxxxix. 650. The word in the title of the *De generacione Mahumet* (supra, no. 4) may be copied from Peter.

[68] Bubnov, *Opera Gerberti*, pp. 114 f.; Steinschneider, *E. U.*, no. 104, where MS. Naples C. viii. 50 should be added to the MSS. By confusion with this treatise Hermann's translation of the *Planisphere* was formerly attributed to Rudolf; Steinschneider's conjecture of Hermann was confirmed by Björnbo, *B. M.*, iv. 130–133, and by Heiberg, preface to *Planisphere*, p. clxxxvii. Cf. Jourdain, pp. 100, 104; Bosmanns, in *Biographie Nationale de Belgique*.

[69] Maslama's name was ben Ahmed el-Magriti Abul-Quasim: Suter, no. 176.

[70] Naples, MS. C. viii. 50, f. 80.

Gerard of Cremona, whose essences do not coincide and are more clearly Aristotelian, namely, *hyle, forma, motus, locus, tempus.*[71] Its approach may be seen from the opening pages, where we find a curious mixture of the Platonism of Chartres, the Aristotelian physics, and the Neo-Platonism of Hermes Trismegistus:

Esse quidem ea dicimus que simplici substantia eademque materia [72] immota nichil alienum nichil alterum unquam paciuntur. Diversum quippe in motu illa que [73] in eodem semper sunt nature sue statu prorsus ignorant. Ea vero sunt que in subiectis sibi rebus mobilibus consistencia subiecti quidem inconstancia quodammodo agitantur, nullam tamen proprie et naturalis constancie sue iniuriam paciencia. Nec enim est simpliciter quod est et non est, proprie vero ea que [74] semper sunt. Hec igitur cum huiusmodi sint proprio nomine essencie nuncupantur, que cum per [75] species quidem innumera [76] sint, quinque principaliter [77] generibus comprehendi posse [78] videntur. Sunt autem hec: causa, motus, locus, tempus,[79] habitudo. Hec [80] etenim huiusmodi sunt plane ut proprie nimirum essencie dicantur nec extra hec aliquid quod eo nomine recte designari queat. Quippe que in substantia sua perfecta naturaque absoluta [81] genituram quidem omnem [82] ad esse conducunt, ut nec sine horum aliquo ulla constet[79] geniture integritas nec preter hoc extraneum aliquod necessarium sit adminiculum, unde necesse sit ipsa in se eiusdem esse nature perfecteque integritatis sine ulla alteritatis contagione, cum omnis diversitatis inter inequalitatis et dissimilitudinis species radix sint et origo nec ex imperfectis prorsus ulla sit perfectionis absolutio. Tria sunt enim,[83] ut philosophis placet, omnis geniture principia. Primum [84] est causa efficiens; secundum est id ex quo aliquid fit; tercium in quo totidemque adminicula ad omnem rerum effectum usu [85] quodam [86] communi quadam ratione cuncta continente. Atque id quod in quo vel de quo fit, quoniam tanquam matris pacientis vice supervenienti [87] virtuti ad omnes motus patet, recte rerum materia nominatur, forma vero id ex quo, quoniam informem illam [88] necessitatem agentis virtutis motibus in varios effingit eventus. Sic enim apud Hermetem Persam, Forma quidem ornatus est materie, materia

---

[71] *Die philosophischen Abhandlungen des Ja'qub ben Ishaq al-Kindi,* ed. Nagy (*Beiträge*, ii, no. 5, 1897).

[72] N, *natura.* The extracts printed from the *De essentiis* reproduce the text of C except where one of the other MSS. seems to have preserved the best reading, but I must leave the emendation of obscure passages to philosophical experts, with such aid as they may get from the variants.

[73] C, *immotum.* N, *quam illa.*

[74] Om. N.

[75] Om. N.

[76] N, *innumera quidem.*

[77] C, *specialiter.*

[78] Om. N.

[79] Om. C.

[80] N, *Nec.*

[81] C, *quia dissoluta.*

[82] C, *omne.*

[83] Om. N.

[84] N, *prima.*

[85] N, *uno.*

[86] Om. N.

[87] C, *superveniente.*

[88] N, *illam informem.*

vero forme necessitas. In omni siquidem rerum constructione sustinens in primis est necessarium, postremus est operis eventus et perfectio. Dat quidem materia massam ipsam informem et inordinatam que nisi[89] presto sit nec habet ubi assit forma, que cum supervenit propositum ordinata quadam explanacione absolvit. Horum igitur principalis motus rerum omnium est generacio. Est enim is motus moderata[90] quedam forme cum materia coeuntis habitudo, ita quidem ut in ipsa movendi ratione vis et causa movens recte demum cognoscatur,[91] in qua, quoniam omnis motus ratio constat ut ea que proposita sunt ex integro constituantur, tractatus ordo abhinc instituendus videtur. Sic enim, opinor, decet ut quid de essenciis instituitur ab ea si qua est que[92] cunctis aliis origo procedat et in ea tamquam rursus in girum expleto cursu tandem terminetur.

Constat plane nichil genitum sine causa genitrice naturaque vetitum ne quid sibi ipsi geniture sit origo seque ipsum efficiat. Sic igitur in omni generacione auctorem generantem causamque moventem intelligi necesse est, prout omne posterius id infert quod prius est. Sic contentum continens species[93] genus individuum genus et speciem, unum autem plane omnium principium intelligi necesse est. Duobus namque prius est unum, nisi enim precedat unum nichil est quod duo constituat.[94] Atque ubi duo[95] unum est necessario, non vero convertitur ut si unum est et duo fore necesse sit. Duo itaque principia qui vel existimari possint dum utrumque prius esse laborans neutri principalem sedem relinqueret. Nisi enim alterutrum altero prius esset, nequaquam primum omnium existeret, dum vel unum complendo omnium numero deesset. Quemadmodum igitur omnis geniture effectus principaliter bipertitus[96] est, prout loco suo[97] explicabitur, sic causa gignens et efficiens primo loco bimembri differentia dividitur, in primam scilicet et secundariam. Prima quidem una et simplex que ipsa quippe immota cunctis aliis causa movendi[98] est et ratio stabilisque manens dat cuncta moveri. Ita siquidem habet ratio ut omne motum id quod immotum est antiquitate precedat omni quoque genito causa genitrix antiquior. Sic igitur quod cuncta alia movet id[99] primam omnium et efficientem esse causam necesse est, quam, ut Thimaeus ait,[100] tam invenire difficile est quam inventam digne profari impossibile.

The author has something to say about the Incarnation and the Trinity, against the Mohammedans,[101] quoting certain Arabic words and citing Hermes and Astalius as well as his own translation of Albumasar, but on all these questions he refers to the

[89] C, *non.*

[90] N, *his;* C, *motus motus moderata.*

[91] N, *cognoscant.*

[92] Om. C.

[93] N, *speciem.*

[94] C, *constituunt.*

[95] N, *duo et.*

[96] Om. C.

[97] N, *suo loco.*

[98] N, *movendi causa.*

[99] N, *id ad.*

[100] *Timaeus,* c. 28.

[101] Cf. his polemic activity, supra, no. 4. He later cites the Koran in one passage: 'Quod et ipse auctor eorum in lege sua fatetur dicens se missum in gladio ad fidem suam ferri virtute et argumento persuadendum.' C, f. 105; N, f. 70 v; L, f. 113.

Fathers for fuller treatment. He then returns to the main theme as follows: [102]

Deinde videndum utrum ne idem ipse auctor ille universitatis est facta sunt utpote que moventur. Omnis vero motus undecumque inceperit aliquando necesse est, incepisse vero temporis est nec tempus eternum cum et ipsum in motu. Est autem omnis motus aut localis aut alteritatis aut translacionis. Locus quidem extra sensibile non est, consistit enim et [103] ipse in subiectis quo res insensibiles Boetius prohibet. Omne vero sensibile compositum, nam quod tangitur ex materia est, quod videtur ex forma, per se quidem nichil huiusmodi prestantibus cum nisi in subiectis non consistunt,[104] sed licet in communem habitudinem unitis in proprie tamen nature partem familiarius accedentibus.[105] Alteritas autem in augmento est aut detrimento aut permutacione, quorum priores duos motus a certa semper quantitate incipere necesse est, permutacio vero in [106] alterutrum semper consequitur. Cum enim ex calido fit frigidum ex alterius detrimento alterius augmentum [107] procedit, que nullatenus accidunt nisi in compositis. Quidquid enim compositum est sine parcium proporcione stare non posset, dum nullum videlicet interesset medium societatis vinculum, proporcio vero nisi inter maius et minus nulla est. Infert itaque diversitas sedicionem quam quandoque alterius partis incrementum alterius detrimentum consequi necesse est.

Translacionis autem motus nec existimari quidem potest nisi circa ea que fiunt. Amplius: nam hec [108] ipsa quidem alterius [109] sumi possunt, ab ipsa videlicet conditoris differentia et eorum que condita dicimus. Conditor etenim siquidem eternus ideoque a seipso est, quidquid in se habet idem ipse est, sic sapientiam bonitatem beatitudinem ut idem ipsa sapientia bonitas beatitudo. In hiis longe aliter. Inest enim mundo pulcritudo rotunditas motus, que cum illi per accidens insunt nec aliquid eorum ipse mundus est. Hec igitur et huiusmodi cum in eis que condita dicimus ex diverso composita videamus, omnis vero composicio actio quam auctorem suum habere necesse est. Si quidem huius modi ab eterno fuisse credantur, fingat qui potest quis [110] hec tam diversa coniunxerit.

Quoniam ergo facta sunt auctoritas [111] facti ei necessario relinquitur qui solus preerat, omnis autem operis modus et finis in arbitrio auctoris. Licet igitur ex omnibus concludere quod unus ipse primus et [112] novissimus unus omnipotens unus tocius universitatis auctor, omnis quidem in essencie sue integritate motus extraneus, omnis namque motus eius in opere eius, quemadmodum virtus quidem in auctore semper eadem et componens et resolvens. In subiecto tamen alia composicio alia resolucio nec simul eiusdem. Amplius: semper quidem creator non vero [113] semper creata, in illo quidem eadem potencia semper eadem semper voluntas creatrix. Circa hec autem

[102] C, f. 92 v; N, f. 59 v; L, f. 75.
[103] N, *ut.*
[104] N, *consistant.*
[105] N, *accidentibus.*
[106] Om. N.
[107] N, *vel alterius augmento.*
[108] N, *nec.*
[109] C, *alcius.*
[110] N, *qui.*
[111] N, *auctoris.*
[112] Om. C.
[113] Om. N.

opposita, nunc scilicet [114] creari nunc [115] minime, legem quippe imponit opifex operi non opus opifici.

Duo sunt igitur cause primordialis omnium motuum genera, creacio et generacio, cetera [116] secundarie [117] ministre obsequentis arbitrio prime. Creacio quidem a primordio principiorum ex nichilo, generacio autem rerum ex antedatis principiis usque nunc, neque enim preerat materia de qua fierent [118] cum solus omnium sit principium nec de seipso quorum tanta ab ipso differentia sed a seipso. Quod enim ex ipso vel de ipso [119] est idem Deus est ideoque non factum a Deo sed genitum vel procedens. Omne vero opus gemina auctoritate constituitur, artificis videlicet et instrumenti, at [120] creacioni quidem idem extitit artifex et instrumentum. In generacione vero, quoniam secunde dignitatis est, aliud sibi aptavit [121] artifex [122] instrumentum. Quod ipsum et secundariam causam si quis in eodem pariter intelligat, eum recte existimare opinor. Ita quidem ut per [123] se ipsum prima effecerit, secunda vero sicque per ordinem tertia et quarta ministre sue cause secundarie moderacione et instituto suo exequenda commiserit. Hec est igitur bipertita illa divisio cause in primam et secundariam. Prima namque et efficiens causa universitatis est ipse prudentissimus artifex et auctor omnium Deus, secundaria vero instrumentum eius de ipsis eiusdem operibus sed prime sedis prelateque [124] auctoritatis. Hec sunt que eius quod de essenciis instituitur integritatem absolvant, si quis recta via nemo quidem ad plenum sed quantum homini fas est assequatur.[125] Quippe que in se quidem absoluta rerum omnium effectum constituunt, videtur autem omnino necessarium ut inter inicia ipsius tanquam thematis [126] fiat ordinata particio,[127] quo facile amplectamur animo quid quo loco expectandum sit neque id passim atque lege incerta [128] verum ipsa naturali consequencie serie. Cum enim de prima et movente causa quantum locus exigebat expeditum sit, a motu qui proximus ceterisque prior et generalior est consequenter inchoandum videtur ac potissimum ab eo qui primus eorum que ceteris principia sunt, id est forme et materie, postquam de ceterorum habitudine locique [129] receptaculo temporisque spacio, ut undique propositis ex quo et in quo ubi et quando qua demum lege quidque fiat, postremo ipsa instrumenti ratio subiuncta in ipso prout institutum est universitatis opifice facto demum reditu consistat.

Hec que dicta sunt hercle sine Deo dici possent nec de eis que restant despero quin quemadmodum ex ordinacione [130] tractatus intelligi datur mirandum altissimi numinis munus debita opera exequaris. Illud vero consulte nec sine summa industria factum.[131] Videtur quidem a vera [132] divinitatis fide primordium operis sumis quippe que omnium bonorum inicium ne quemad-

[114] N, *nichil.*
[115] N, *non.*
[116] N, *ceteri.*
[117] N, *secundario.*
[118] N, *faceret.*
[119] *Vel de ipso* om. N.
[120] C, *a;* N, *ac.*
[121] N, *adaptavit.*
[122] N, *opifex.*   [123] Om. C.

[124] N, *planteque.*
[125] C, *assequitur.*
[126] Here L begins.
[127] L, *paratio.*
[128] C, *certa.*
[129] N, *loci quoque.*
[130] N, *ordine.*
[131] L, *facere.*
[132] L, *qui antea.*

modum temerariis hominibus visum est sed plane intelligatur extra veram divinitatis fidem locum [133] sapientie nullum esse. Fac ergo quem arreptum tenes ne moreris.[134]

After this paragraph, evidently a dialogue between Hermann and Robert, there begins what in the Naples manuscript is entitled the second book, although no trace of division into books appears elsewhere:

Optimus auctor omnium [135] Deus summeque beatus nequaquam invidit quin aliquid [136] sibi gracie sue [137] tanteque glorie consors efficeret. Scimus enim nichil invitum fecisse cum nulla necessitas cogeret, consors autem eius qui posset esse quicquam mortale quod numquam desiturum esset. Sane mortale quidem omne id quod non ex integris perfectisque principiis firmissimo demum [138] nexu vinculisque perpetuis atque indissolubili nodo compactum constaret. Firmus vero societatis nexus neque intus [139] penitus eadem neque intus [139] prorsus diversa, primum igitur [140] necessaria fuit huiusmodi fabrice eiusdem diversique proposicio, diversum porro nichil primum. Iecit itaque semina commiscendi potencia virtutisque generative que [141] per se quidem eiusdem nature ac substantie individue collata vero adinvicem diverse nec umquam commixtione [142] sui quietem eiusdem essencie admittencia. . . .

The necessity of *actio* and *passio* in generation then comes in. The four elements are mentioned as a subject of dispute among philosophers, then the four modes "to which Aristotle added a fifth," the whole bringing us now into line with the *De generatione et corruptione*. Before long, however, we are back with Albumasar and "the most weighty authority of Hermes." [143] The disagreements of Plato and Aristotle are emphasized later: [144]

Multa quidem veteris prudencie studia, mi Rodberte, in hiis que agimus consumpta nec ulli ad integritatis evidenciam consecuta videmus. Sic Plato proposita generacione primaria tandem ad extremum [145] enisus [146] partem dedit pro toto, Aristotiles vero totum [147] item amplexus extremitates demum sine mediorum contextu terminavit. Michi autem nulla ratione universitatis constructio absoluta videatur si minus sit quod solum in omni composicione compaginis retinaculum est. . . . Recte quidem quale Plato [148] diffinit

---

[133] N, *nullum locum sapientie.*

[134] N, *ne moveris.* C adds *etc.*

[135] L, N, *omnium auctor.*

[136] L, *aliud.*

[137] Om. C.

[138] Om. N.

[139] L, N, *inter.*

[140] L, N, *ergo.*

[141] C, *etiam.*

[142] L, *in commixtionem.*

[143] C, f. 97; N, f. 63; L, f. 87.

[144] C, ff. 102 v, 106; N, ff. 68 v, 71 v; L, ff. 105 v, 117 v.

[145] *tandem ad extremum* om. C.

[146] *est* inserted by N.

[147] C, *totus.*

[148] Om. C.

Aristotiles describit. Plato quidem in Cadone:[149] Anima est, inquit, incorporea substantia corpus movens. Aristotiles vero in libro de anima sic: Anima est, ait, perfectio corporis naturalis instrumentali potencia agentis.[150] Et alibi: Anima est perfectio corporis agentis et viventis potencia.[151]

There is a fair amount of astronomical and geometrical illustration, with four astronomical figures,[152] and references to his own treatise on such matters.[153] The most noteworthy geographical passage is the following, where we find brought together the classical names of geography, Arin, and the North:[154]

Triplex est universa dimensio,[155] in longum, latum, et [156] altum. Quoniam igitur omnis corporis sedes in fundamento suo terra vero tocius mundi fundamentum, multo pocius mundane prolis ex substantia [157] collecte sedem terram esse [158] necesse est. Eius pars quedam a terra in altum crescit, alia vero super terram in altum elevatur tocius fomentum hic spiritus terreni vaporis pinguedine crassus,[159] sine quo nulla huius geniture [160] vita per aliquot horarum spacium possibilis.[161] Hic autem vapor, ut per altitudinem Olimpi concipit Aristotiles,[162] a terre superficie non plus quam .16. stadiis exaltatur.[163] Hic ergo terminus videtur in altum omnis nostre habitabilis. Videtur fortasse huius altitudinis mensura sumi [164] posse vel per [165] arcum yris que secundum Ipparci [166] descriptionem ab ipsis [167] nubibus usque in superficiem[168] terre perveniat. Sed quoniam nec ipsa descriptio constans nec ipsius arcus ad semicirculum habitudo, propterea nos id cuilibet probandum relinquimus.

Latitudo vero terrarum est ab equinoctiali [169] circulo in alterutrum polum distantia ac nostra [170] quidem in borealem qui, cum ab eo circulo per .90.[171]

---

[149] *De senectute*, c. 21. Chalcidius (c. 226) corresponds more closely to this definition. L. has 'substantia incorporea.'

[150] N, L, *viventis.*

[151] *De anima*, 2, 1, 6, 7 (p. 412). So just before this passage he promises to summarize 'quod Aristotiles vix tribus integris libris explicavit.'

[152] N, ff. 66, 67 v, 68 v, 78; L, ff. 98, 101 v, 104.

[153] Supra, n. 46.

[154] C, f. 112; N, f. 76 v; L, f. 136.

[155] N, *diversio.*   [156] Om. L.

[157] L, *substantie.*   [158] Om. L.

[159] C, *grassus*; N, *cssus*; L, *cursus.*

[160] L, *geniture huius.*   [161] L, *spacia possibile.*

[162] L, N, *Aristotiles per altitudinem.*

[163] On the contrary Aristotle omits Olympus from his list of the highest mountains (*Meteorology*, 1, 13). The usual figures for the highest mountains vary in Greek writers from ten to fifteen stades. See W. Capelle, *Berges- und Wolkenhöhe bei griechischen Physikern*, in Boll's Στοιχεῖα, v (Berlin, 1916), especially pp. 13, 34.

[164] L, *summi.*   [168] C, *superficie.*

[165] Om. L.   [169] C, *equali*, L, *equabili.*

[166] N, *Parci.*   [170] L, *nostram.*

[167] N, *his.*   [171] C, 20.

gradus distet, in principio Arietis illic oriri solem in principio Libre occumbere necesse est, secundum quod in primo libro diximus orizontem illic esse ipsum circulum equinoctialem, sicque ab eo polo in austrum perpetuo gelu [172] inhabitabiles fere .30. gradus relinquuntur usque prope montes Ripheos [173] silvasque Rubeas [174] atque paludes Meotidas.[175] Nec enim longe plus .12. gradibus ultra terminos septimi climatis, unde et Scitie fines ei [176] termino [177] contiguos Scitica lingua Ysland [178] nominat, quod [179] latine sonat terra glacialis. At vero circa equabilem circulum non parum item intollerabili estu intractabile pariter et abinde cum [180] arenis siccitate sterilibus ut Libice et inter quas Nilus occultatur.[181] Insulas tamen habitatas sub ipso eodem circulo Tamprobanem,[182] Arin, et .vi.[183] Fortunatas girographi tradunt satis possibiliter. Duplici namque ratione probat Ptolomeus eas terrarum partes aptissimas habitacioni: nec enim, ait, vel estum eis [184] exasperari [185] patitur velox illic solis in latum transitus nec validum admittit frigus haut longinqua ab eo circulo solis remotio. Unde si prosequatur dubitacio cur ergo non pateat transitus vel usque in alteram [186] temperatam, dicimus quia Sagittarius impedit. Unde totius habitabilis nostre latitudo fere .60. graduum relinquitur.

Longitudo vero quanta a principio Indie [187] usque in finem Libie inventa est graduum fere .clxxx. Illinc per occeani insulas sub ipso equinoctiali .15. fere gradibus usque [188] ultra Meroen [189] insulam Niliacam [190] sub Tauro et Leone sitam haut procul a superiori Egypto. Hinc vero per Amphitritis sinus ab Athlante Libico Strixisque influxu [191] per littora Gaditana per confinia Thiles prope [192] Temiscirios campos e vicino portibus Caspiis [193] usque ad Caucason [194] et Ethiopici Gangis [195] effluxus. Sic enim astronomia [196] demonstrat circa meridiem [197] Arin solem simul primis Indie partibus occidere atque ultimis Libie finibus oriri, que ratio utriusque termini populos antipodas adinvicem constituit utpote integra fere terreni orbis diametro interposita.

[172] C, *gere.*

[173] L, *proprie montes Rumpheos.*

[174] N, L, *Rebeas.* On the *silvae Rubeae* cf. Pliny, *N. H.*, 4, 13, 27.

[175] L. *Meoridas.*

[183] N, *vi. insulas.*

[176] L, *ex.*

[184] Om. C; L, *eius.*

[177] N, *terminos.*

[185] N, *esperari.*

[178] N, *Islanil*; L, *Island.*

[186] C, *aliam.*

[179] Om. C.

[187] Om. N.

[180] N, *in.*

[188] Om. C N.

[181] N, L, *occulatur.*

[189] L, *Merorem.*

[182] C, *Tampropebanen*;
N, *Tamprofanem*;
L, *Tamprobamen.*

[190] N, *oceani magni.*

[191] L, *inflexu.*

[192] L, *proprie.* On the 'Amphitritis sinus' as the ocean encircling the globe, cf. Grosseteste, *De sphaera*, ed. Baur (*Beiträge*, ix), pp. 24 f. 'Strixis' is puzzling; can it be a deformation of Septa, which appears in many corrupt forms for the Straits of Gibraltar (Nallino, *al-Battani*, p. 18, n. 7)? Themiscyra is unusual in this connection.

[193] L, *campiis.*

[195] N, *Gangisque.*

[194] L, *Ciucason.*

[196] N, *in astronomia.*

[197] Om. C.

Cum ergo dimidium per .vi. partem [198] multiplicatum [199] tocius duodecimam conficiat, tota demum terreni globi porcio, ut Albeteni visum est, universe nostre habitacioni relicta est.

Here the first paragraph harks back ultimately to Greek meteorology, through channels which require investigation. The two following paragraphs correspond in general to the doctrine of the Arab astronomers, particularly al-Battāni, whom Robert of Chester translated.[200] The doctrine of the habitability of the equatorial regions seems to have come into Europe in the twelfth century from Arabic sources.[201] The land of ice, "called in the Scythian speech Ysland," is evidently as little known to Hermann as Arin or the Blessed Isles to the west. In another instance, however, he seems to make use of his more direct knowledge of the Iberian peninsula to elucidate a point in geography, though he ends with 'Amphitrite' and the terrestrial paradise. The essence of the argument is clear, even without the accompanying diagram. If the distance from Toledo west to the ocean at Lisbon is 4° or eight days' journey, then the remaining 22° [202] or forty-four days' journey represents the distance thence to the western prime meridian or one-half the width of the surrounding ocean stream of Amphitrite: [203]

Cuius demonstrationi describimus exempli gratia Toleti circulum paralellum ysemerino meridianum in supraposita figura [204] secantem ad punctum y gradibus fere 40 a puncto e versus c in punctis quidem a sinistris q a dextris z transeuntem per primo descriptum orizontem. In quo designamus punctum o loco Toleti metropolis Hyspanie gradibus a puncto y occidentem versus 62. Tum [205] ubi circulus qyz secat circulum nrk signamus nota [206] i loco civitatis Ulixispone que sita est qua Tagus a Toleto descendens occidentali oceano influit eadem distantia ab ysemerino, a puncto vero y gradibus 66. Cum igitur o distet ab i gradibus quatuor i vero a puncto z gradibus [207] 22, sitque oi linea recto tramite itineris dierum fere viii, procedit spatium inter i et z

---

[198] L, *sex partes*.   [199] N, *multiplicant*.

[200] See Nallino, *al-Battani*, pp. 15 ff., and the editor's notes.

[201] For information on these matters I am indebted to my former pupil, Dr. J. K. Wright of the American Geographical Society, who discusses them in his book, *Geographical Lore of the Time of the Crusades.*

[202] Why not 24°, since Toledo is 62° west of Arin? On mediaeval reckonings of latitude and longitude, see J. K. Wright, in *Isis*, v. 75–98 (1923).

[203] C, f. 113 v; N, f. 78 v; L breaks off just before this.

[204] The figure appears only in N, f. 78; C has a space for its insertion on f. 113

[205] N, *Tu*.   [206] N, *notam*.   [207] Om. N.

dierum 44 que secundum quod ratio tribuit est dimidia latitudo Amphitritis, tota [208] videlicet itineris terrestris equabilis dierum fere 88. Tantum ergo spatii vel etiam aliquanto plus que ratio hucusque transnatari prohibuit nondum audivimus nisi forte illa quam [209] exposuimus. In ea tamen parte non modica est opinio eam esse regionem quam paradisum vocant, cuius indicio [210] sunt signa tam ab oriente quam ab occidente.

The *De essentiis* concludes thus: [211]

Recepto siquidem semente statim ad retinendum accedit virtus Saturnia, retentum digestione salubri Iupiter nutrit, demum Mars consolidat, post hunc Sol informat, informato Venus reliquias temperate expellit, expulsioni Mercurius moderate obvians necessaria retinet,[212] postremo Lucina succedens gemina virtute partum maturum absolvit. Que ipsa continuo tenerum fetum suscipiens eo usque tuetur quoad cessante materie nutrimentique lunaris illuvie utpote in corporis augmentum diffusa ac digesta excitato paulatim sensu animeque semitis adaptis Mercurius in racionabilem institucionem succedens usque in Veneream adolescentiam provehat; hinc temperata iam voluptuosa levitate in Phebeam iuventutis plenitudinem conscendit qui usque in Martie virtutis statum provehit, hic virili animo roborato Iovialis auctoritas succedit. Postremo est etas Saturnia nature orbe expleto finem origini continuans, cui quod ab ipso cepit reddito quod ultra ipsum est minime legibus hiis subditum ideoque cognacionis sue racionem superstes per alterutrum Pitagorici bivii tramitem, aut tanquam devium et aberrans usque in posterum nichil descendere, aut naturali circuitu servato ad summi [213] triumphi coronam originalem videlicet patrieque sedis arcem demum conscendere necesse est, qua beati evo sempiterno fruuntur in gloriam regis altissimi cui virtus honor et potestas in infinita secula.

De essenciis Hermanni Secundi liber explicit anno Domini millesimo centesimo quadragesimo tercio Byterri perfectus.[214]

Hermann's sources in the *De essentiis* give an idea of the range of his education. Of the Latins he cites Cicero and Boethius, Pliny, in one instance Seneca, Macrobius, Apuleius, Aratus, Vitruvius, verses of Hesiod in a Latin version;[215] Augustine and the annals of St. Jerome each appear once. He shows the Platonism of the school of Chartres in his citations of the *Timaeus* and 'Plato

[208] N, *totam.*
[209] N, *que.*
[210] N, *indicia.*

[211] C, ff. 115–115 v; N, ff. 79 v–80.
[212] Om. C.
[213] N, *sumam.*

[214] N has: 'DE ESSENTIIS liber HERMANNI Secundi explicit. ANNO DOMINI. M⁰. C⁰. xl⁰. iii⁰. Biternis PERFECTUS.'

[215] C, f. 108 v; N, f. 73 v; L, f. 124 v: 'Sic enim et Esiodo revelatum ferunt et ab ipso conscriptum hiis versibus in libro de etatibus animalium,

Ter binos deciesque novem superextit in annos '

in Catone,' but he is largely Aristotelian. He seems to know the *Logic* only through Boethius, and the only Aristotelian work cited directly is the *De anima*,[216] but he is familiar with the subject matter of the *De generatione et corruptione* and the *Meteorology*, if not that of the *Physics*. Euclid he quotes, and the *Spherica* of Theodosius; Ptolemy is cited frequently but not necessarily at first hand; the references to Eratosthenes, Archimedes, and Hipparchus are of course borrowed. Al-Battani, thrice cited, Hermann would know, as he was translated by Robertus Ketenensis. Albumasar, whom Hermann himself translated, is cited most frequently of all. The other astrologers mentioned are Hermes Trismegistus, Apollonius, Messehalla, Druvius, 'Iorma Babilonicus,' and 'Tuz Ionicus.' [217] The bare references to Galen and Hippocrates are unimportant.

The total impression is rather confusing, a conglomerate rather than a fused whole, but we must remember that Hermann stands at the meeting-point of diverse currents of thought and tradition, where only a distinctly superior mind could achieve consistency. He clearly lacks the originality and experimental habit of Adelard of Bath, but he has mastered a considerable portion of the new mathematics and astronomy, and has a respectable place among the transmitters of the twelfth century. The list of his works is impressive, and it is to be hoped that others may yet be recovered.

[216] See note 151.
[217] On Hermann's astrology, see Thorndike, ii. 41–43.

CHAPTER IV

THE TRANSLATIONS OF HUGO SANCTALLENSIS [1]

In the history of culture in the Romance countries of mediaeval Europe an important place must be given to the movement which it is becoming common to call the renaissance of the twelfth century. This revival of learning had many aspects, according as we consider it from the point of view of classical literature, of law, of natural science, or of philosophy and theology; but on its philosophical and scientific sides it owed its significance to the influx of a great body of new knowledge, coming in some measure from direct contact with Greek writers in the Norman kingdom of Sicily and elsewhere,[2] but derived for the most part through the intermediary of Arabic and Jewish sources as these were made accessible in central and northern Spain. Here the chief centre was Toledo, where a large amount of Arabic literature survived the Christian conquest of 1085 and whence in the course of the twelfth and thirteenth centuries an active school of translators spread over western Europe the Latin versions of Aristotle, Ptolemy, Euclid, Galen, Hippocrates, and their Arabic expositors and commentators which constituted the basis of study and teaching in the mediaeval universities. The impulse to this movement may have come in the first instance from Raymond, archbishop of Toledo from 1126 to 1151;[3] but it would be a mistake to regard it as confined to Toledo. The Toletan translators were in relations, how close we do not know, with a group of scholars from other lands, including Plato of Tivoli, Robert of Chester, Hermann of Carinthia and his pupil Rudolf of Bruges, who worked, mainly on astronomical subjects, in various cities of northern Spain and, probably, southern France. Plato, who is found in Spain as early as 1134, is con-

[1] Revised from *The Romanic Review*, ii. 1–15 (1911).
[2] Infra, Chapters VIII–X.
[3] On the Toletan and other Spanish translators see Chapter I and the works there cited.

nected particularly with Barcelona; Hermann and Robert first appear in 1141 as students of astrology on the banks of the Ebro, and one or both of them can be traced at Segovia, Leon, Toulouse, Béziers, and Pamplona, where Robert became archdeacon. It is the purpose of this chapter to call attention to an active and hitherto unknown centre of such studies at Tarazona, in Aragon, and to examine the work of a contemporary translator, Hugo Sanctallensis, of whom exceedingly little has hitherto been known.

In the course of this movement more than one version of the same work might be made, whether from the Arabic or from the Greek, and it was not always the earliest or the most accurate which secured the widest circulation.[4] Thus in the case of Ptolemy, his *Planisphere* was translated from the Arabic by Hermann of Carinthia in 1143;[5] the Latin version of the *Optics*, which has survived the loss of both the Greek and the Arabic texts, was made from the Arabic in Sicily about the middle of the century; while his great work, the *Almagest*, became known at first only through the translated compend of al-Fargani[6] and passed into general use, not in the first and more faithful version made from the Greek in Sicily about 1160, but in the translation from the Arabic which Gerard of Cremona completed at Toledo in 1175.[7] On the other hand, Ptolemy's astrological treatise, the *Quadripartitum*, was the first of his works to be translated into Latin, in the version produced by Plato of Tivoli in 1138,[8] and the abridgment of this, the *Fructus* or *Centiloquium*, which was ascribed to Ptolemy throughout the Middle Ages, was translated somewhat earlier. The Latin rendering of the *Centiloquium* bears in most of the manuscripts the date of 1136, and while it was formerly ascribed to Plato of Tivoli, it is now, on the authority of

---

[4] Björnbo, "Die mittelalterlichen lateinischen Uebersetzungen aus dem Griechischen auf dem Gebiete der mathematischen Wissenschaften," in *Archiv für die Geschichte der Naturwissenschaften*, i. 387 (= *Festschrift Moritz Cantor anlässlich seines achtzigen Geburtstages*, Leipzig, 1909, p. 95), suggests that the first translation made after the revival of the eleventh and twelfth centuries was the one which held the field; but the opposite was true in the case of the *Almagest*, as appears below.

[5] Supra, Chapter III, no. 5.

[6] On which see Steinschneider, *H. U.*, p. 554; *E. U.*, no. 68 *h*; infra, p. 369.

[7] Infra, Chapters V and IX.

[8] Steinschneider, *E. U.*, no. 98.

an Erfurt manuscript, generally assigned to John of Seville.[9] Whether this attribution is correct and how many versions of the *Centiloquium* were made, only a comparison of the numerous copies can determine, but in any event there is extant in the Biblioteca Nazionale at Naples[10] and at Madrid[11] a translation prepared by Hugo Sanctallensis for the bishop of Tarazona, as appears from the following preface:

Incipiunt fructus Ptolomei, liber scilicet quem Grecorum quidam centum verba appellant, Hugonis Sanctelliensis translatus. Prologus eiusdem ad Michaelem Tirassonem [*sic*] antistitem.

De hiis que ad iudiciorum veritatem actinent, cum in illis totus astronomie consistat effectus secundum arabice secte verissima[m] inquisicionem et tam Grecorum quam Arabum qui huius artis habiti sunt profexores famosissimi auctoritatem, volumina decem in hiis de tam multimoda auctorum copia eligendis diucius obversatus, ne tante expectacionis fructus minor tantique laboris merces in aliquo deficere videretur, de arabico in latinum translatavi sermonem. His enim quot sufficiunt ut decent preiacentibus, tota huius artis structura atque series dignissimo gaudebit effectu. Ut enim Aristotiles in libro de signis superioribus asseruit, Siquis prudentissimus faber sive architectus in construenda cuiuslibet hedificii machina congruis et quot sufficiant careat instrumentis, totam fabricam vacillare aut aliquit minus perfectum inveniri necesse est. Quod si nec desit huiusmodi sufficiencia cum opificis industria, non aliud postulat examen, unde et quasi sese comitancia sunt et aliud alio indigere videtur. Nec ab huius ordinis serie declinat quod in prologo Rethorice dicitur sapiencia sine eloquencia parum prodesse civitatibus, eloquencia sine sapiencia prodesse nunquam, obesse plerumque. Quia ergo Ptholomeus inter ceteros astronomie professores precipuus habetur interpres et auctor post Almagesti et Quadripartitum hunc solum de iudiciis astrorum reliquid tractatum, ut tue, mi domine Tirassoniensis antistes, satisfiat iubsioni, eius translacionis fructum ego Sanctelliensis adporto, hac videlicet occasione compulsus ne dum in portu iudiciorum navigas in cimba locatus vasa saxosa formides et ne de tanti preceptoris operibus quippiam abesse queratis. Hic enim, si quelibet hucusque circa huiusmodi negocium fuerat ambiguitas, poterit aboleri, si quelibet disgressionis circuicio, poterit breviari,

---

[9] Leclerc, *Histoire de la médecine arabe* (Paris, 1876), ii. 374; Steinschneider, *H. U.*, pp. 527–529; id., *E. U.*, no. 68 *a*; Nallino, *Albatenii opus astronomicum* (Milan, 1903), i, p. lvii; Pelzer, in *Archivum Franciscanum historicum*, xii. 59 f. (1919).

[10] MS. D. viii. 4, copied at Naples in the fifteenth century. The text proper begins: 'Verbum primum. Astrorum sciencia de te et de illis. Hoc in sermone de te et de illis videtur velle Ptholomeus duplicem esse astrorum scienciam. . . .' Still another version of the *Centiloquium* was used by Albertus Magnus: *Catalogus codicum astrologorum Grecorum*, v. 97; Steinschneider, in *Z. M. Ph.*, xvi. 383.

[11] Biblioteca Nacional, MS. 10009, ff. 85–105 v, which lacks the heading but offers a better text.

quidquid tandem hians vel minus perfectum hiis centum verbis poterit re-
parari. Unde ex ipsius auctoris edicto tuam non incongruum video exortari
diligentiam ne tante sapiencie archana cuilibet indigno tractanda commic-
tas et ne quemlibet participem adhibeas qui pocius gaudet librorum numero
quam eorum delectetur artificio.

The dedication to Bishop Michael establishes an approximate
date. Of unknown origin, this prelate was placed over the see of
Tarazona in 1119, immediately after the recovery of that region
from the Moors by Alfonso VII and seven years before Raymond
became archbishop of Toledo, and he continued in office until
1151. His labors for the establishment of his authority and the
restoration of the ecclesiastical organization throughout his dio-
cese are attested by a number of contemporary documents,[12] but
he has not hitherto been known as a patron of learning. From the
preface just quoted we see that the translation of the *Centiloquium*
was made by his command, to serve as a guide to the voluminous
body of astrological literature which had already been placed at
his disposal; and, while we must make due allowance for the high-
sounding praise of his learning and wisdom in the prefaces printed
below, the mere list of the translations made at his orders shows
that the *insaciabilis filosophandi aviditas* ascribed to him [13] is no
empty phrase. If he likes compendious treatises, he wishes them
to be correct,[14] nor does he desire mere rule-of-thumb manuals
which do not explain their reasons.[15] He cannot have been very
familiar with Arabic, else there would have been no need of Latin
versions for his use, yet he searches for Arabic manuscripts on his
own account, one of the texts translated having been found by him
in Rotensi armario et inter secretiora bibliotece penetralia.[16] *Rotensis*

---

[12] Lafuente, in *España sagrada*, xlix. 125–142, 330–368; Moret, *Annales de
Navarra* (Pamplona, 1766), ii. 285–446; Arigita y Lasa, *Colección de documentos
inéditos para la historia de Navarra* (Pamplona, 1900), i. 75, 259, 264; Villanueva,
Viage literario, xv. 369–378; Bruel, *Chartes de Cluni*, v. 397, 454. The necrology of
Monte Aragon (*Neues Archiv*, vi. 280) fixes his death 16 February era 1188, which
must be interpreted as 1151 since he attests a charter of 23 August 1150 (Férotin,
Chartes de l'abbaye de Silos, no. 51); cf. *España sagrada*, xlix. 368. He was not a
monk of S. Juan de la Peña: *ibid.*, p. 125.

[13] Infra, p. 73.

[14] See the preface to the *Liber imbrium*, infra, p. 77.

[15] P. 73.   [16] P. 73.

at first sight suggests Roda, in Aragon, then seat of the bishop of Lérida,[17] but, as these were Arabic manuscripts, there is something to be said for the Moorish stronghold of Rota, now Rueda Jalón, between Tarazona and Saragossa, to which the Moors retreated for a time after the fall of Saragossa in 1118.

The author of this preface, Hugo Sanctallensis, though not previously connected with the *Centiloquium* by bibliographers, has been known as the translator of certain other astrological works, but his time and place have not before been determined. The principal authorities on the occidental translations from the Arabic, Wüstenfeld [18] and Steinschneider,[19] make Michael a French bishop and are inclined to place Hugo in the latter part of the Middle Ages, and while the late Paul Tannery would seem to have reached correct conclusions on these matters, he died before presenting any evidence in support of them.[20] As at least two manuscripts of Hugo's translations are of the twelfth century,[21] he cannot be put later, and the mention of Bishop Michael in the prefaces fixes him definitely in the second quarter of this century and in Aragon. His surname appears in various forms — Sanctelliensis, Sanctellensis, Sanctallensis, Sanctaliensis, Sandaliensis, Satiliensis, Strellensis, and, in Provençal, de Satalia [22] — without any indication of the country. None of these forms suggests France or Italy, while they all point to Santalla, a place-name

[17] On Roda see *España sagrada*, xlvi; Villanueva, *Viage literario*, xv. 131 ff.; Beer, *Handschriftenschätze Spaniens*, no. 392.

[18] Pp. 22, 120.

[19] *H. U.*, pp. 566–567, 574; *E. U.*, no. 54. Steinschneider's list of Hugo's writings, which is so far the most complete, enumerates al-Fargani, the Pseudo-Aristotle, the *Liber imbrium*, the *Geomantia*, and the *De spatula*.

[20] The materials for this chapter were collected in 1910 and the conclusions drawn before I discovered that Tannery, shortly before his death, had placed Hugo between 1120 and 1150 (*B. M.*, ii. 41). An earlier note of the same author, while assigning him to Aragon, gave as his date the first half of the eleventh century, an obvious impossibility (*Comptes-rendus de l'Académie des inscriptions*, xxv. 529, 1897). His posthumous memoir, primarily concerned with geomancy, will be found in his *Mémoires scientifiques*, iv. 295–411 (1920). For Hugo's date see pp. 334 f.; no new works are indicated.

[21] MS. Selden Arch. B. 34, in the Bodleian, containing the translation of al-Fargani; B. N., MS. Lat. 13951, containing Apollonius.

[22] For the Provençal form see Paul Meyer, in *Romania*, xxvi. 247.

common in the northwest of Spain, especially in Galicia.[23] A reference to the Gauls in one of his prefaces — *Gallorum posteritas tua benignitas largiatur* [24] — suggests that Bishop Michael, and perhaps Hugo, had some connection with France. Michael may well have been of French origin, one of the French ecclesiastics brought into Spain in the course of the *reconquista*;[25] and in any case it is very likely that copies of these translations were sent beyond the Pyrenees in the same way as those of the Toledo school. Nothing is known of Hugo's relations with the other translators of his age, nor have we any external evidence for his biography; the most that we can do is to examine the treatises upon which he worked, and in these, it is plain, he was closely under the orders of his patron bishop.

So far as the preface to the *Centiloquium* throws light on Hugo's literary labors, it shows him as a student of astrology and divination. From books dealing with these subjects, which he regards as the real justification for the study of astronomy, he has selected and turned into Latin ten volumes which exhibit the principles and applications of the art in all its aspects. The titles of these treatises are not given, but an examination of the numerous translations preserved under his name enables us to identify eight extant versions of astrological and similar works, besides the *Centiloquium*, while in these reference is made to at least five others. From an astronomical point of view, the most important of these is a treatise with the following introduction: [26]

[23] According to Madoz é Ibañez, *Diccionario geográfico-estadístico-histórico de España* (Madrid, 1846–50) there are twenty places of this name in the province of Lugo, one in the province of Coruña, and one, the largest, in the province of Leon. There is also a Santalle in the province of Oviedo and a Santalha in Traz os Montes.

[24] Infra, p. 77. This is the passage that misled Wüstenfeld and Steinschneider into thinking Michael a Gallic bishop.

[25] Note that French crusaders were established in Tudela, over which Bishop Michael claimed jurisdiction, and that he confirmed the neighboring church of Santa Cruz to the abbey of S.-Martin of Séez: Ordericus Vitalis, v. 1–18; *Gallia Christiana*, xi. 720; *España sagrada*, l. 399 (Jaffé-Löwenfeld, *Regesta*, no. 8803); B. N., MS. Fr. 18953, pp. 38, 220, 259.

[26] Bodleian Library, MS. Selden Arch. B. 34, ff. 11–62 v, of the twelfth century. Also in MS. Savile 15, f. 205, saec. xv; and in Caius College, Cambridge, MS. 456, saec. xiii (James, *Catalogue*, p. 531).

Incipit tractatus Alfragani de motibus planetarum commentatus ab Hugoni Sanctaliensis [*sic*].

Quia nonnullos nec inmerito te conturbat quod priscorum astrologorum intentio multas et varias in suis voluminibus, in his precipue que de stellarum collocatione et situ descripta Arabes azig appellant, videtur protulisse sententias, nullam tamen quare potius sic aut sic agere eorum suaderet tradicio protulere rationem, unde huiusmodi minus plena perfectaque volumina pro auctoris defectu lectoris sensum et intelligentiam corrumpunt. Que cum ita se habeant, nichil obstare videtur artis istius emulos, hos de quibus loquimur, gemino urgere incommodo, ut videlicet ex ignorantia aut ex invidia hoc factum fuisse coniectent. Nam inter multiplices antiquorum tractatus, de quorum videlicet prudentia ac discretione nulla est hesitatio, nonnulla legimus ea ratione fuisse descripta que tamen ut preceptori sic et lectori inutilia totius posteritatis clamat assertio. In libro autem Alhoarizmi quoniam huiusmodi diversitates te repperire confiteris, eum ex invidia ut supradiximus aut ex ignorantia suspectum esse palam est, sed etiam quendam Alfargani librum de rationibus azig Alhoarizmi imperfectum nec sufficientem te asseris repperiri, ubi videlicet que facilia sunt expediens que intricata et difficilia ad intelligendum fuerant pretermisit. Quia ergo, mi domine Tyrassonensis antistes, ego Sanctelliensis tue peticioni ex me ipso satisfacere non possum, huius commenti translationem, quod super eiusdem auctoris opus edictum in Rotensi armario et inter secretiora bibliotece penetralia tua insaciabilis filosophandi aviditas meruit repperiri, tue dignitati offerre presumo. Habet enim ex tantis astronomie secretis ut placeat et ut ad omnium ex eadem materia voluminum expositionem ex sui integritate sufficiat. Quamvis tamen Alfargani edicione[m] minus plenam perfectamque cognoscam, cum ex aliis suis operibus perfectus et sapiens comprobetur, hec quam subscribam mihi videtur fuisse occasio. Potuit enim fieri ut morte preventus talem relinqueret, aut si perfectum atque emendatum eadem intercessit occasio ne id divulgaret, unde aliquid inde corrumpi aut ab invidorum manibus ut eius auctoritati quicquam derogarent abici satis liquido constat argumento, vel forsitan hic idem Alfargani, quod prudencioris cautele est, tante subtilitatis archana aggredi formidans difficillima pretermittens cetera reseravit. Nemo enim ad huius exposicionis intelligentiam accedere potest nisi geometrie institutis et universo mensurandi genere quasi ad manum plenissime instruatur. Ne itaque antiquorum vestigiis penitus insistens a modernis prorsus videar dissentire, non per dialogum, ut apud Arabes habetur, verum more solito atque usitato hoc opus subiciam, ac deinceps non solum Quadripertiti atque Almaiezti ab Alkindio datam expositionem sed etiam quoddam Aristotilis super totam artem sufficiens et generale commentum, si vita superstes fuerit et facultas detur, te iubente aggrediar.

Ad ingressum cuiuslibet arabici mensis, ut ait Alhoarizmi . . .

As here given from the Selden manuscript, the title of this work is misleading and should be corrected from the other copies to *Hamis Benhamie Machumeti frater de geometria mobilis quantitatis et azig, hoc est canonis stellarum rationibus.* What we have is not

al-Fargani's explanation — this indeed the bishop has found insufficient — of the astronomical tables of al-Khwarizmi, which
go back apparently to the Indian astronomers, but a commentary
on al-Fargani written, with the aid of the tables and geometrical
methods of Ptolemy, by a later astronomer who has recently been
identified with Mohammed ben Ahmed el-Biruni.[27] A Hebrew
translation of this commentary, preserving the questions and
answers of the original, was made by Abraham ibn Ezra at Narbonne about 1160,[28] with an introduction which shows certain
parallelisms with that of Hugo, but no Latin version has hitherto
been identified.[29] The discovery of such a version, by facilitating
a comparison with the translation of the Khorasmian tables made
by Adelard of Bath in 1126,[30] may be expected to throw some
light on the relations between Greek, Indian, and Arabian astronomy. It would be interesting to know in what form the
bishop, whose knowledge of Arabic must have been inadequate
for the free use of the works which he had Hugo translate, used
the Khorasmian tables and the explanation of al-Fargani.

Of the two other works which Hugo has here promised to translate, the commentary of al-Kindi seems to have been lost,[31] but
the *generale commentum* of Aristotle is doubtless contained in two
manuscripts of the Bodleian [32] under the high-sounding title:
*Liber Aristotilis de .255. Indorum voluminibus universalium ques-
tionum tam generalium quam circularium summam continens.* The
attribution to Aristotle will deceive no one,[33] but the account of

[27] Suter, "Der Verfasser des Buches 'Gründe der Tafeln des Chowarezmi,'" in
B. M., iv. 127–129, where the utility of a comparative study is suggested.

[28] Steinschneider, in *Zeitschrift der deutschen morgenländischen Gesellschaft,* xxiv.
339–359, xxv. 421; *H. U.,* pp. 572–574.

[29] Steinschneider, *H. U., l. c.;* Suter, in *Abhandlungen zur Geschichte der mathematischen Wissenschaften,* xiv. 158.

[30] Supra, Chapter II.

[31] A commentary on the Almagest appears in the Arabic catalogue of his works
(Flügel, in *Abhandlungen für Kunde des Morgenlandes,* i, 2, p. 27, no. 123) but has
not been identified among those extant (Suter, in *Abh. Gesch. Math.,* x, 25).

[32] MS. Digby 159; MS. Savile 15, f. 185.

[33] With Thorndike (ii. 256 f.), I find no other mention of this compilation. For
other pseudo-Aristotelian works on astrology, magic, and divination, see *Catalogus
codicum astrologorum Graecorum,* i. 82 f., v. 92, 96, 102; Steinschneider, *E. U.,* no.
141; *Centralblatt für Bibliothekswesen,* Beiheft xii. 87–91; and especially Thorndike,
in *Journal of English and Germanic Philology,* xxi. 229–258 (1922).

the books upon which the compilation is based may contain something of interest for students of ancient astrology. The prologue, being chiefly devoted to an account of the two hundred and fifty volumes from which the work is compiled, yields no new information for the translator's biography. The opening and closing portions are:

Ex multiplici questionum genere et ex intimis philosophie secretis, quibus frequenter mee parvitatis aures pulsare non desinis, subtilissime tue inquisitionis archanum et celebris memorie intrinsecam vim et purissime discretionis intelligentiam, ad quam videlicet nostri temporis quispiam aspirare frustra nititur, manifestius licet attendere. Quare quod ex libris antiquorum percepi aut experimento didici aut existimatione sola credidi aut exercitio comparavi, et assidua scribere cogit exortatio et imperitie veretur formido. Ad graviora transcendere subtiliora penetrare novis etiam affluere tanta preceptoris daret auctoritas, si congrua ociandi daretur facultas. Nam humani generis error, ut qui inscientie crapula sui oblitus edormit stulticie nubibus soporata iudicio philosophantium sectam estimans lacivienti verborum petulantia, sicut huius temporis sapere negligit, sapientes et honestos inconstantie ascribit, veritatis concives imperitos diiudicat, verecundos atque patientes stolidos reputat. Ego tamen, quoniam auctoritate Tullii ad amicum libera est iactancia,[34] amore discipline cui semper pro ingenii viribus vigilanter institi Arabes ingressus, si voto potiri minime contigisset, Indos autem Egiptum pariter adire, si facultas unde libet [35] subveniat insaciata philosophandi aviditas omni metu abiecto nullatenus formidaret, ut saltem, dum ipsius philosophie vernulas arroganti supercilio negligunt, scientie tamen quantulamcumque portionem vix tandem adeptam minime depravari contingat sed potius ab eius amicis et secretariis venerari. Nunc autem, mi domine antistes Michael, sub te tanto scientiarum principe me militari posse triumpho, quem tocius honestatis fama et amor discipline insaciatus ultra modernos vel coequevos sic extollunt ut nemo huius temporis recte sapiens philosophi nomen et tante dignitatis vocabulum te meruisse invideat. Unde fit ut hoc duplici munere beatus, dum hinc amor hinc honestas tercium quod est amor honestus constituant, non modicum probitatis habes solacium. Ego itaque Sanctellensis Hugo tue sublimitatis servus [36] ac indignus minister, ut animo sic et corpore labori et ocio expositus dum et mentis corporis torporem excitando pulsas oblivionis delens incommodum, quoniam id assidua vult exortatio quod a nullo modernorum plenissime valet explicari, ne plus videar sapere quam oportet sapere, quodque a meipso haberi scientie negat viduitas ab aliis mutuari priscorum multiplex suadet auctoritas, hunc librum ex arabice lingue opulentia in latinum transformavi sermonem. Sed quoniam, ut ait quidam sapiens, tam secretis misticisque rebus vivaciter pertrac-

---

[34] Doubtful; *iactantia* is not Ciceronian.

[35] The Savile MS. has 'unde libri.'

[36] Dr. Craster of the Bodleian and Professor Thorndike (ii. 85 f.) have corrected my earlier reading of 'serus' in the first printing of this text.

tandis multimoda sunt auctoritatum perquirenda suffragia, istius auctor
operis ex .cc.l. philosophorum voluminibus qui de astronomia conscripserunt
hoc excultum esse asseruit, a quorum nominibus serio conterendis proprie
narrationis duxit exordium. . . .

Hunc ergo, mi domine, ex tot ac tantis philosophorum voluminibus et
quasi ex intimis astronomie visceribus ab eodem, ut iam dictum est, excepi,
tamen et si mea de arabico in latinum mutuavit devocio supprema, tamen
tue tam honeste ammonicionis optatos portus dabit correptio. Explicit prologus. Incipit Aristotilis comentum in astrologiam. Primo quidem omnium
id recte atque convenienter preponi videtur . . .

Among more special works on astrology, we learn that Hugo
translated four treatises on nativities, one of these, from the Arabic
of Mashallah, beginning as follows: [37]

Liber Messehale de nativitatibus .14. distinctus capitulis Hugonis Sanctalliensis translacio. Prologus eiusdem ad Michaelem Tirassone antistitem.

Libellum hunc Messehale de nativitatibus, etsi apud nos Albumazar et
Alheacib Alcufi ex eodem negocio et nostre translacionis studio plenissime
habeantur, ob hoc placuit transferri ut quemadmodum ex eius secretis et
iudiciorum via et ceteris astronomie institutis tua, mi domine antistes Michael, pollet sciencia tuumque pre ceteris studium nec inmerito gloriatur, sic
et in genezia, nativitatum dico, speculatione tanti preceptoris certa imitando
vestigia copiosius triumphet. Hoc igitur ego Sanctelliensis, non tam meo
labore faciente quam auctoris testimonio confisus, ut placeam mitto compendium, quendam alium librum de eadem materia a quodam Messehale
discipulo Abualy Alhuat nomine editum deinceps tractaturus, ut et supra
nominatis voluminibus hoc attestante maior insit auctoritas et tanquam
variis diversarum opum ferculis tua in hoc negocio sacietur aviditas. . . .
Ut alio sicut idem asserit Messehala nullatenus videatur indigere. Explicit
prologus. Incipit textus. Quamvis librum istum ex ordine a libro secretorum
assumpto per .14. capitula dividendum proposuerim . . .

Of the authors of the two versions which are here mentioned
as already completed, Albumazar is, of course, abu Ma'ashar
Ja'afar, author of a number of works on astronomy and astrology,
including one on nativities which has not yet been specially
studied; [38] Alheacib Alcufi I have not identified, unless the name
be a corruption of el-Chasîb.[39] Various manuscripts of abu Ali's
work on the same subject exist, all of them anonymous except

[37] Bodleian, MS. Savile 15, f. 177 v. This translation is unknown to the bibliographers.

[38] On his writings see Steinschneider, H. U., pp. 566 ff., and E. U., no. 165; Suter,
no. 53; Houtsma, Encyclopaedia of Islam, i. 100.

[39] Suter, no. 62. Professor Suter suggested this identification to me in a letter
of 16 May 1911.

one in the Bodleian which ascribes the translation to John of Seville.[40]

Hugo's translation of another work of Albumazar dealing especially with meteorological predictions is found in a score of manuscripts [41] and two early editions. The preface reads:

Incipit liber ymbrium ab antiquo Indorum astrologo nomine Iafar editus deinde vero a Cillenio Mercurio abbreviatus. Superioris discipline inconcussam veritatem . . . Quia ergo, mi domine antistes Michael, non solum compendiosa sed etiam certa et ad unguem correcta te semper optare cognovi, hunc de pluviis libellum ab antiquo Indorum astrologo Iafar nomine editum, deinceps a Cillenio Mercurio sub brevitatis ordine correctum, tue offero dignitati, ut quod potissimum sibi deesse moderni deflent astrologi Gallorum posteritati tua benignitas largiatur. Incipit series libri. Universa astronomie iudicia [42] . . .

Hugo is not mentioned in the text but is found in the margin of one of the manuscripts.[43] Two similar treatises, ascribed to Mashallah and al-Kindi, appear as having been translated by a Master Drogo or Azogo, which has been conjectured to be a corruption of Hugo; [44] but as these are not accompanied by prefaces, the question must for the present remain open.

Those who look for signs in the heavens are likely also to look for them on the earth, and we are not surprised to find that Hugo was the author of an elaborate treatise on geomancy, based upon the work of an unknown Tripolitan (Alatrabulucus) and sufficient to give him a certain reputation among vernacular writers

[40] MS. Laud 594. See Steinschneider, *E. U.*, no. 68 *m*; and in *B. M.*, 1890, pp. 69 f.

[41] Besides those mentioned by Steinschneider, *H. U.*, p. 566, see MS. Bodl. 463, f. 20 (= Bernard, No. 2456); MS. Savile 15; Corpus Christi College, Oxford, MS. 233; Clare College, Cambridge, MS. 15; Bibliothèque Nationale, MS. Lat. 7329, f. 66 v, MS. Lat. 7316, f. 167 (extract only); Leyden, Scaliger MS. 46, f. 36; Madrid, MS. 10063, f. 43; Vatican, MS. Reg. 1452, f. 29; MS. Borghese 312, f. 43; Venice, Cl. xi. 107, f. 53 (Valentinelli, iv. 285). Printed at Venice in 1507 with al-Kindi, *De pluviis*; also at Paris, 1540, from which edition is copied MS. 529 of the University of Coimbra. On Jafar cf. Tannery, pp. 337 f.

[42] Bodleian, MS. Savile 15, f. 175 v.

[43] Steinschneider, *l. c.*

[44] Leclerc, *Histoire de la médecine arabe*, ii. 476 (where MS. Lat. 7439 should be 7440, and 10251 is incorrect); Steinschneider, *H. U.*, pp. 564, 600; *E. U.*, nos. 36, 54 *d*; Suter, no. 8.

as an authority on this art,[45] which he seems to have introduced into Latin Europe.[46] The copy in the Bibliothèque Nationale begins: [47]

Incipit prologus super artem geomantie secundum magistrum Ugonem Sanctelliensem interpretem qui eam de arabico in latinum transtulit.

Rerum opifex Deus qui sine exemplo nova condidit universa, ante ipsam generationem de illorum futuro statu mente diiudicans, hec quidem etiam que de sue universitatis thesauro rationali creature largiri dignatur singulis prout ipse vult distribuit. Unde universa creatura tam rationalis quam irrationalis vel inanimata eidem exibet obedientiam ac, licet in vita ad secularium ordinem dilapsa, eum saltem ex sola unitate veneratur. Imaginarie priusquam fierent cuncta habens eorundem noticiam archano cordium quasi suspectam et intellectualem infudit. Habite tandem creature hic modus consistit ut summitates atque venerandos scriptorum institutores atque huiusmodi computationis industria quasi quadam compagine sociaret, ut ablata tocius alterationis rixa rationale alias positiva iusticia nexu equabili federaret adinvicem. Cum igitur universos stolidos videlicet tanquam sapientes ad philosophandum pronos fore contigisset, eruditior prudentium secta ad computandi artem et astronomie secreta rimanda mentis oculum revocans, astrorum loca cursus directos retrogradationes ortus occasus sublimationes depressiones et que sunt in his alterationes atque admiranda prodigia attendens, astrologorum minus prudentium multiplicem cognovit errorem. Hac igitur ratione cogente compendium hoc certissimum ex his omnibus prudens adinvenit antiquitas. Denique aput universos philosophie professores ratum arbitror et constans quicquid in hoc mundo conditum subsistendi vice sortitum est haut dissimile exemplar in superiori circulo possidere, quicquid etiam hic inferius motu quolibet agitatur superioris regionis motus sibi congruos imitari. Sicque manifestum est quia huiusmodi figure quas hic prosequi volumus signorum pariter et lunarium mansionum formas omnino sequuntur . . . Quia huiusmodi artificium antiquissimum fore et apud sapientum quamplurimos dignos et indignos in usu fuisse philosophorum antiquitas refert, ego Sanccelliensis [48] geomantie inscriptionem aggredior et tibi, mi domine Tirasonensis antistes, ex priscorum opulentia huiusmodi munusculum adporto, aeremantia et piromantia quas audivi sed

---

[45] Paul Meyer, "Traités en vers provençaux sur l'astrologie et la géomancie," in *Romania*, xxvi. 247–250, 275. Cf. Steinschneider, *E. U.*, no. 54 c. On geomancy in general, see Thorndike, ii. 110 ff.

[46] Tannery, iv. 329.

[47] MS. Lat. 7354, written in the thirteenth century, apparently in Spain or southern France. Also in Vatican, MS. Pal. Lat. 1457; Bodleian, MS. Digby 50, f. 1; MS. Bodley 625, f. 54; Cambridge, Magdalene College, MS. 27; Vienna, MSS. 5327, 5508 (the last three I owe to Thorndike, ii. 86). The treatise of Hugo on geomancy preserved in the Laurentian and studied by Meyer has a different *incipit* and may be another work. See Tannery, pp. 324–328, 339–344, 373–411.

[48] Vat. Pal. Lat. 1457 has 'Hugo Sanctalliensis.'

minime contingit reperiri postpositis, deinceps idromantiam tractaturus
. . . Que quidem disciplina sub quadam existimatione potissimum manat
ab antiquorum peritissimis, ut iam dictum est, qua ipsi noverint ratione
certis experimentis usitata. Explicit prologus.

Arenam limpidissimam a nemine conculcatam et de profundo ante solis
ortum assumptam . . .

Whether Hugo ever wrote on hydromancy or succeeded in in-
forming himself on aeromancy or pyromancy, we cannot say; but
while searching the heavens above and the earth beneath and the
waters under the earth, he did not disdain the humbler form of
divination which draws its inferences from the shoulder-blades of
animals, and we have under his name two short treatises on spatulamancy. The first, which claims to go back ultimately to
Greek sources, begins: [49]

Refert Ablaudius babilonicus inter antiquissima Grecorum volumina
cartam vetustissimam in qua de spatule agnitione nonnulla continebantur
precepta apud Athena[s] se invenisse. . . . Hunc igitur librum, cuius auctor
apud Caldeos Anunbarhis (?) apud Grecos Hermes fuisse legitur, et tante
antiquitatis arkana et latinum aggrediar sermonem. . . . Quia igitur, mi
domine antistes Michael, tuo munere tuaque munificentia ut me ipsum
habeo, sic et philosophantium vestigii desidia et ignorantia gravatus insisto,
ne ceteris compensatis istius expers inveniaris discipline, hoc tibi de spatula
mitto preludium. . . .

In medio itaque cartilaginis foramen ultra eminens repertum pecoris
domino pacem nunciat . . .

This is followed by a similar *Liber Abdalabeni Zolemani de
spatula Hugonis translatio.*[50]

Another translation of Hugo Sanctallensis, not mentioned in
his prefaces or by modern writers, appears in the Bibliothèque
Nationale in a manuscript of the close of the twelfth century formerly at St.-Germain-des-Prés, where it received the title in a
modern hand, *Hermetis Trimegesti Liber de secretis naturae et*

---

[49] Bodleian, Ashmolean MS. 342, f. 38, headed "Tractatus de spatula" and referred to in the margin as "Hugonis translatio"; B. N., MS. lat. 1461, f. 68. The
tract in MS. Canon. Misc. 396, ff. 106–110, mentioned by Steinschneider (*E. U.*, no.
54 *e*) is different, beginning, 'Incipiam adiutorio Dei.' Steinschneider curiously
fails to understand the meaning of *spatula*.

[50] Ashmolean MS. 342, f. 40 v; MS. lat. 1461, f. 71 v. Cf. Tannery, *Mémoires*,
iv. 340. The references to MS. lat. 1461 I owe to the kindness of Dr. Birkenmajer.

occultis rerum causis ab Apollonio translatus. It begins and ends: [51]

> Incipit liber Apollonii de principalibus rerum causis et primo de celestibus corporibus et stellis et planetis et etiam de mineriis et animantibus, tandem de homine.
>
> In huius voluminis serie eam principaliter tractatus sum disciplinam ex qua philosophorum antiquissimi suscepte narrationis protulerunt exordium, ut meę intentionis agnita prudentia et ad vestram aspirare valeat intelligentiam et intimam pulsare discretionis naturam. Cuiuscumque ergo naturalis intentio huius sermonis capax extiterit eam accidentalis vel quasi extranee sollicitudinis incursu liberam velud a sompno excitari palam est. . . . Quod videlicet Hermes philosophus triplicem sapientiam vel triplicem scientiam appellat. Explicit liber Apollonii de secretis nature et occultis rerum causis, Hugonis Sanctelliensis translatio .vi. particionibus discretus.

As a result of this investigation we now have, as against the five previously known, nine extant translations by Hugo, not counting those ascribed to Drogo and Azogo, besides two others which have been lost or are still to be identified [52] and three which he promises but may not have completed.[53] None of these are dated, but the *Centiloquium* is one of his later efforts, since ten have been produced before it, while the Khorasmian commentary is evidently early, being anterior to the Pseudo-Aristotle. It would seem that both translator and patron gave chief attention first to astronomy and later to astrology, but to draw a sharp line between these subjects would be contrary to the spirit of mediaeval, if not of Greek, learning, to which they were simply the pure and the applied aspects of the same subject. There is no evidence on Hugo's part of initiative or power of adaptation, indeed he expressly disclaims the ability to elucidate these problems from his own knowledge; he was a translator, rather than a compiler or popularizer. There is, at the same time, no indication of any connection with the other translators of his age,

[51] MS. lat. 13951, ff. 1–31. This translation is analyzed by F. Nau, in *Revue de l'Orient chrétien*, xii. 99–106 (1907). On Apollonius see also Thorndike, i. 267, ii. 282 f.

[52] The *De nativitatibus* of Albumazar and of Alheacib Alcufi. Tannery has shown that there is no reason for assigning to our Hugo the *Practica Hugonis*, a geometrical treatise of the twelfth century: *B. M.*, ii. 41; *Mémoires scientifiques*, iv. 331–333.

[53] Abu Ali, *De nativitatibus*; al-Kindi, *Expositio Quadripertiti atque Almaiesti: Idromantia*.

and the fact that certain of the treatises at which he labored were also translated by John of Seville indicates that they worked independently. That Hugo's versions nevertheless obtained a certain currency is shown by the number and wide distribution of the existing manuscripts, and the range and quantity of his work entitle him to a respectable place among the Spanish translators of the twelfth century.

CHAPTER V

SOME TWELFTH–CENTURY WRITERS ON ASTRONOMY

THE growth of astronomical knowledge in western Europe in the twelfth century constitutes an interesting chapter of intellectual history. The century opens with the traditional learning of the older encyclopedists and the standard manuals of *computus*. Then comes a definite revival of the Platonic cosmology, chiefly in conjunction with the school of Chartres, so that Platonic influences are clearly marked in the first exponents of Arabic astronomy in the second quarter of the century, as illustrated by the *Questiones* of Adelard of Bath and the *De essentiis* of Hermann of Carinthia. These, however, are accompanied and followed by translations of the tables of al-Khwarizmi and al-Zarkali, and the treatises of al-Fargani and al-Battani, as well as by a mass of astrological literature. The translation of Ptolemy's *Almagest* from the Greek ca. 1160 and from the Arabic in 1175 made possible the full reception of ancient astronomy. Meanwhile the Aristotelian physics had begun to filter in through Arabic writers, and the conflict of this with Plato and Ptolemy sorely puzzled an age which desired at all costs to reconcile its standard authorities. The new knowledge, the new controversies, and the more exact observations long occupied some of the best minds of the age, whose activity is reflected in a fairly abundant body of literature, both anonymous and ascribed to known authors; and the sharp contrast between the astronomical writings of the beginning and end of the century helps us to measure the intellectual progress of the intervening years.

The history of this phase of European thought has still to be written. The late Pierre Duhem made an admirable beginning as a part of his comprehensive survey of cosmological theories in antiquity and the Middle Ages;[1] but, valuable as is his analysis on the scientific side, it rests, for the twelfth century, on a quite

[1] *Le système du monde de Platon à Copernic* (Paris, 1913–17).

inadequate examination of the material. For some unexplained reason he never saw the *Questiones naturales* of Adelard, though it is available in three editions as well as a score of MSS.; he explored but little the large number of unpublished treatises; and he left untouched many problems of date and authorship. Thorndike's new material [2] is chiefly concerned with magic and astrology; the relevant volumes in Baeumker's *Beiträge* are primarily philosophical in content. As a contribution to the general history of astronomy, it may be worth while to describe certain unpublished treatises which I have come upon, illustrating as they do the various ways in which the new learning made itself felt.

COMPUTISTS

One of the clearest indications of intellectual revival in the early twelfth century is the large number of manuscripts of that period or shortly before which deal with the elements of arithmetical and astronomical reckoning. On the side of arithmetic these take the form chiefly of treatises on the abacus, carrying on the tradition of Gerbert and the Lotharingian abacists of the eleventh century and elaborating this in its practical aspects.[3] On the side of astronomy activity is seen partly in copies and excerpts from the older manuals of Bede and Heiric of Auxerre,[4] occasionally with extracts from Isidore and Hyginus and, in the more ambitious works, with illustrations; [5] partly in new compilations. A good example of the learning of this period is contained in a manuscript of St. John's College, Oxford, made up of material copied in the

---

[2] *History of Magic and Experimental Science* (New York, 1923).

[3] See Chapter XV, below; Bubnov, *Opera Gerberti*, app. vi.; Cantor, i. ch. 40.

[4] See the list of MSS. in the admirable study of Traube, "Compotus Helperici," *Neues Archiv*, xviii. 73–105, 724 f. Duhem (iii. 71 f.) unfortunately overlooked Traube's work. Helperic's treatise will be found in Migne, cxxxvii. 15; Bede's, *ibid.*, xc. 293. For other examples of collections of excerpts, cf. Arsenal, MS. 371, ff. 75 v–87; Evreux, MS. 60 (from Lire); Dijon, MS. 448. Cf. the extracts from Bede (*De natura rerum*, c. 45) and the *Geometry* of Gerbert at Tortosa, MS. 80 (*Revue des bibliothèques*, vi. 16).

[5] Cf. Saxl, in Heidelberg *Sitzungsberichte*, 1915, no. 6–7. A good example is the Ripoll MS. of 1056: Saxl, pp. 45–59; supra, Chapter I, n. 18. There are some good figures of constellations in early English MSS., e. g., Cotton MS. Tiberius, C. i, ff. 21–32 v; Harleian MS. 647, ff. 2 v–13.

later eleventh and early twelfth centuries.[6] Besides Bede and
Heiric of Auxerre, with a preface by Brithferth, monk of Ramsey,
it contains astronomical tables and excerpts, extracts from the
arithmetic of Boethius, treatises on the abacus as late as that of
Gerland,[7] and some scattered medical and grammatical notes.
'The present time' is given as 1110 on f. 3 v; the lunar tables on
f. 29 begin with 1083 and contain marginal entries from 1085 to
1111 which show that in these years this part of the manuscript
was in possession of Thorney Abbey. Other examples of this age,
which we shall examine in other connections,[8] are the writings of
Thurkil the computist and the *Comput* of Philip de Thaon (1119).

The ecclesiastical preoccupations of the close of the eleventh
century are illustrated in the discussions of the basis of the Christian era. Marianus Scotus the chronicler, who died at Mainz in
1082, made a determined effort to supplant the current Dionysian era as twenty-two years too late, and the argument was developed in England by a 'learned Lorrainer,' Robert, bishop of
Hereford from 1079 to 1095; but the new system found few ad-
herents.[9] A similar theory appears in an anonymous *Liber decen-
nalis in modum dialogi compositus* preserved in the Biblioteca
Angelica at Rome,[10] where the author, arguing from the astronomical cycles, finds a discrepancy of twenty-one years in the Dionysian era, so that the current year of 1092 is corrected to 1113.[11]

[6] MS. 17. See the detailed description in Coxe's *Catalogus*, and Bubnov, pp.
lii f.; and cf. Singer, "Byrhtferd's Diagram," in the *Bodleian Quarterly Record*, ii,
no. 14 (1917); and in *Proceedings of the Royal Society of Medicine*, x. 118-160 (1917);
R. L. Poole, *The Exchequer in the Twelfth Century*, p. 47, note.

[7] Ff. 50-52, not identified by Bubnov, printed in *Bullettino*, x. 597-607.

[8] Infra, Chapters XV, XVI.

[9] See Robert's treatise in the Bodleian, MSS. Auct. F. 3. 14 and Auct. F. 5. 19
(2148), f. 1; and cf. W. H. Stevenson in *E. H. R.*, xxii. 72 ff.

[10] MS. 1413, ff. 1-24 (saec. xii): 'Cum temporum scriptores diversi quamvis
diverse . . .'

[11] 'Presens autem annus secundum veraciorem evangelio congruentem numerum
est ab incarnatione Domini annus millesimus centesimus tercius decimus habens
concurrentes iiii^{or} cum bissexto, epactas .viiii., terminum paschalem versum 'Sene
kalende titulant ternos,' diem dominicum .v^o. kal. aprilis, indictionem xv. Secundum Dionysium autem est annus ab Adam quinque millesimus quadragesimus tercius, ab incarnatione vero Domini millesimus nonagesimus secundus, distans annis
xxi^o ab ea consequentia paschalis compoti quam superius posui et que presenti anno
competit' (f. 21).

Another critic of Dionysius in this period was Gerland in his *Computus*, who works out an era seven years earlier than the Dionysian. Author likewise of treatises on the abacus [12] and on ecclesiastical matters, Gerland has usually been identified with a canon of Besançon who appears in documents of 1132–48, in which latter year he and Thierry of Chartres, *duos fama et gloria doctores*, accompanied the archbishop of Trier down the Rhine.[13] It would seem, however, that this is a different person from the computist, who specifically gives the year of his treatise as 1081,[14] whose 'floruit' is given as 1084 at Besançon by Albericus,[15] and who is cited as early as 1102.[16] His *Computus*, in twenty-seven chapters, while criticizing Dionysius and Helperic, purports to follow closely Bede,[17] who is cited by chapter. The author adds

---

[12] Published in *Bullettino*, x. 595–607. "Gerlandus ex libro magistri Franconis Legiensis" in MS. 107 of the University of Edinburgh, ff. 62 v–68, turns out to be neither mathematical nor astronomical, but is evidently the same as the "De ligno crucis" at Trinity College, Dublin, MS. 517.

[13] See Boncompagni, *ibid.*, pp. 648–656; Cantor, i. 898; *Histoire littéraire*, xii. 275–279; T. Wright, *Biographia literaria*, ii. 16; and in *Transactions of Royal Society of Literature*, ii. 72–75 (1847); U. Robert, in *Analecta juris pontificii*, xii. 596–614 (1873).

[14] 'Ab incarnatione domini modo sunt .i. lxxxi ᵘˢ. annus': B. N., MS. lat. 11260, f. 15 v. This occurs in what may be supplementary to the treatise proper, which differs considerably in the different MSS., and in some (e. g., MS. lat. 15118, f. 39) has what purports to be a second book. The tables of the earlier part, however, clearly belong to the close of the eleventh century. Thus in MS. lat. 15118 they begin at 1082 (f. 37), mention the eclipse of 23 September 1093 (f. 50), and have notes on eclipses added after 1102 (ff. 31 v, 32); cf. the reference to 1094 on f. 33. In MS. Rawlinson C. 749 of the Bodleian, f. 11 v, we have eclipses, 'nostris temporibus,' of 1085–95, i. e., 1093–1103 by the ordinary reckoning, including the eclipse of 23 September 1093, the difference of era being here reckoned as eight years.

[15] *Scriptores*, xxiii. 800.

[16] Infra, Chapter XV, n. 37.

[17] 'Sepe volumina domini Bede de scientia computandi replicans et in eis quedam aliter quam tradicio doctorum ostenderet presentium repperiens, Dei fretus auxilio Deum invocans preesse meo operi que visa fuerunt mihi utilissima inde pro captu ingenioli mei defloravi et deflorata cum quibusdam aliunde conquisitis in unum congessi. Queso itaque, si unquam hec compositionis fimbria, hec stili ariditas, huius scientię gutta ad alicuius intuitum pervenerit, ne statim in morsum livoris dentes exacuat nec antequam perlegat preiudicet, ne si quid in toto notandum invenerit pro parte totum, ut nonnulli solent, vituperet, quandoquidem, ut ait non insipientium quidam, nichil ex omni parte beatum. Non equidem me latet quosdam qui Helpericum legerunt et tabulam Dionisii viderunt aliter in quibusdam sentire

quotations from the Fathers, Pliny, Virgil, and 'Cingius,'[18] and works out various lunar calculations by the awkward methods of Roman fractions and the subdivision of the hour into points, moments, and atoms. To Gerland the *computus* is based partly on nature and partly on authority,[19] but in the long run authority proved too strong for him. Various copies of his treatise survive, and he is often cited with respect,[20] but most writers of the twelfth century look askance at him as contravening the settled usage of the church.[21]

In the later twelfth century writings on the *computus* conserve much the same character. Examples are the treatises of 1159 and 1161 just cited in connection with Gerland;[22] the *Summa magistri Wilelmi de compoto* of 1163;[23] a treatise of 1169;[24] and the *Compotus Petri* of 1171.[25] The treatise of Michael, monk of Dover, now in Glasgow, is undated,[26] as is also an anonymous set of

quam ego. Sed si quis Bedam perlegerit et naturalem compotum tenere voluerit, hic ut arbitror partim auctoritati partim artis naturę acquiescens non indigne feret hic quedam esse posita que obviare videntur Dionisio, quedam que Helperico. Nec tamen eos censeo redarguendos per omnia si in aliquam partem somnus obrepserit, quia [ubi] spiritus vult spirat, aliquando autem ut ardentius queratur subterfugit.' B. N., MS. lat. 11260, f. 1 v. Cf. f. 11 v: 'Venerabilis Beda cuius fere verba per totum hoc opusculum dispersimus.'

[18] 'Cingius' or 'Zingius' appears in Philip de Thaon, *Li cumpoz*, l. 744; and the *computus* of 1102 in MS. Vat. lat. 3123, f. 47 v; and B. N., MS. lat. 11260, f. 25 v. The reference is to the *Fasti* of L. Cincius as quoted by Macrobius, *Saturnalia*, I, 12, 12.

[19] MS. lat. 11260, f. 7 v.

[20] E. g., 'Gerlandus vero Lotherencus in extremis omnes alios correxit et scripta vilissima cum tabula abiecit': anonymous treatise in Cotton MS. Titus D. vii, f. 14 ('Quid in compoto doceatur . . .') Cf. Philip de Thaon, infra, Chapter XV.

[21] 'Liber Gerlanni non legitur quia longo usui et doctissimorum auctoritati obviavit': B. N., MS. lat. 2020, f. 198, a treatise of 1171. See further the passages cited by T. Wright, in *Transactions of the Royal Society of Literature*, ii. 74 f.

[22] MS. Cotton Vitellius A. xii, ff. 101–103 v, 105–106. See note 30.

[23] B. N., MS. lat. 10358, ff. 273 v–283 v; described in *B. E. C.*, xvii. 403. In MS. Digby 56, ff. 202–219 v, the treatise is dated 1164 (f. 219). Inc. 'Annorum duo sunt genera . . .'

[24] Laon, MS. 71, which I know only from the catalogue.

[25] Bibliothèque Mazarine, MS. 3642, ff. 13–49 v, incomplete at the beginning (the date occurs on f. 44). Apparently the same as a computus in MS. lat. 2020, f. 198, also dated 1171, and beginning 'Sunt in aliis artibus . . .'

[26] Hunterian Museum, MS. 467, where three treatises are attributed to him, though the first looks like a copy of Helperic. The printed catalogue gives this

eighty-five *Regule de compoto* in Brussels.[27]   Three copies of a *Computus constabularii* were formerly at Canterbury.[28]

In such conservative circles it was natural that Arabic astronomy should penetrate slowly, and we are not surprised that Roger of Hereford should inveigh against the ignorance of the computists as late as 1176.[29]   An early example of the introduction of Arabic influence into such works is seen in an anonymous treatise of thirty-nine chapters composed in 1175, apparently in England.[30]   The author is an admirer of Gerland, whom he imitates in the opening sentence and whom he proposes to follow except where ecclesiastical usage would be contravened:

Sepe auctorum volumina qui de compoto vel principaliter vel incidenter egerunt studiose revolvi, inter quos invenio quosdam iuniores in arte calculatoria non mediocriter eruditos longo usui ecclesie rationibus vehementer, ut videtur, acutis obviare.   His quidam nostrorum modernorum applaudentes nuper ausi sunt cartulis pascalibus suas novitates inscribere et sanctorum patrum vestigia preterire.   Sunt enim quidam novitatis venatores et antiquitatis improbi calumpniatores qui etiam in doctrina Christiana locum ab auctoritate tanquam inartificiosum superciliose repudiant et de suo confidentes ingenio aliter quam tota ecclesia soli sentire volunt ut soli scire videantur.   Sed, quod deterius est, vidi equidem doluique videre scripto quoque commendatum quedam aliter se habere secundum ecclesiam, aliter secundum veritatem.   Te quoque, dilectissime, timor Domini et reverentia fidei catholice vehementer abhorrere fecerunt veritatem et ecclesiam in aliquo posse reperiri contrarias.   Quoniam igitur rationes illorum nobis vise sunt posse non irrationabiliter infirmari, quod proprio consilio non audebam, tuo propulsus instinctu illis respondere aggressus sum. . . .   Ceterum propter instructionem aliorum et precipue G. mei quem in omni scientia et virtute proficere cupio, universum apposui percurrere compotum quatenus singula que mihi dubitabilia visa sunt explanarem.   Noveris etiam preter ceteros auctores Geralandum quoque imitatum et etiam imitandum in omnibus exceptis hiis in quibus obviat usui ecclesie, nam ubi bene dicit nemo melius. . . .

portion of the MS. as saec. xii, but the algorism (no. 3) there ascribed to Michael is not earlier than the thirteenth century, to judge by its contents. I have a photograph of no. 3 only.   Cf. the musical writer, Tenred of Dover: *E. H. R.*, xxx. 658–660.

[27] Bibliothèque Royale, 2194, ff. 8 v–48 v (saec. xii): 'Si invenire volueris per quam feriam . . .'

[28] M. R. James, *Ancient Libraries of Canterbury and Dover*, p. 49.

[29] Digby MS. 40, ff. 21–21 v.

[30] Cotton MS. Vitellius A. xii, ff. 87–97 v, with tables appended: 'Sepe auctorum volumina . . . .'   The date appears from ff. 90 v, 93, 94.   The reference to England is on f. 96: 'Quando est luna distans a sole paulo minus quam xxix gradibus in Anglia non apparebit maxime circa equinoctium autumpnale.'

The author has a broader education than most computists, as he cites, with references, passages from Hippocrates, Solinus, Pliny, and the *Digest*. The astronomers, cited in the later chapters chiefly respecting the date of the vernal equinox, are Ptolemy, Hipparchus, Thabit, al-Battani, al-Zarkali, and al-Fargani.

The School of Chartres

The persistent influence of Plato is one of the curious facts in the intellectual history of the Middle Ages. If we accept Schlegel's dictum that every one is either Platonist or Aristotelian, then the Middle Ages were clearly Aristotelian, but with lapses into Platonism and resultant efforts to reconcile the two systems. Until the translation of the *Meno* and *Phaedo*, ca. 1156,[31] the only work of Plato directly known to the western Europe of the Middle Ages was the *Timaeus*, or rather the first fifty-three chapters as translated and commented upon by Chalcidius in the fourth century.[32] This in itself is a curious fact, for "of all the writings of Plato," says Jowett,[33] "the *Timaeus* is the most obscure and repulsive to the modern reader, and has nevertheless had the greatest influence over the ancient and mediaeval world." Accordingly, mediaeval Platonism was largely concerned with the vague and mystic cosmogony of this dialogue. The other principal source of Platonism was the fifth-century commentary of Macrobius on the *Somnium Scipionis* of Cicero.[34] Revived in the tenth century, this contained a considerable amount of ancient astronomy and geography; and it served as the vehicle for transmitting an important fragment of non-Platonic astronomy, the hypothesis respecting the movement of Venus and Mercury about the sun which is commonly ascribed to Heraclides of Pontus. Neo-Platonism concerns us less at this point, as its influence becomes

---

[31] Infra, Chapter IX.

[32] Ed. Wrobel, Leipzig, 1876; the commentary is examined by Switalski, *Beiträge*, iii, no. 6 (1908).

[33] *Dialogues of Plato*, ii. 455.

[34] Duhem, iii. 47 ff.; M. Schedler, *Die Philosophie des Macrobius und ihr Einfluss (Beiträge*, xiii, no. 1, 1916); and, for the hypothesis of Heraclides, J. L. E. Dreyer, *History of the Planetary Systems* (Cambridge, 1906), ch. 6.

important only with the thirteenth century.[35] There are also bits of Platonism in the astronomical part of Martianus Capella, from which an extract beginning 'Mundus igitur ex quatuor elementis . . .' is sometimes found in manuscripts of the period.[36]

How Martianus was copied and conflated in this period is illustrated by a treatise which masquerades under the title of *Liber Iparci*,[37] but has no direct relation to Hipparchus, whose influence, under the form Abrachis, must be sought rather among translators from the Arabic. Beginning with a rearrangement of extracts from Bede's *De naturis rerum*, the author soon [38] picks up the eighth book of Martianus, which he follows through the climates, inserting a bit on climates from the sixth book of Pliny and the discussion of tides in Bede's *De temporum ratione* (c. 29). Closer search might reveal scattered passages from other sources.[39]

In the twelfth century there was a definite revival of Platonism in the school of Chartres.[40] Its chief exponents were William of Conches, Bernard Silvester, and Thierry of Chartres, with whom may be grouped such writers as Adelard of Bath and Hermann of Carinthia, the latter a pupil of Thierry. Thus much has been made clear by Hauréau and others,[41] while the general course of

[35] Note, however, the Hermetic citations in Hermann of Carinthia (supra, Chapter III, pp. 57–66); and the question of the first traces of the *Liber de causis*: Duhem, iii. 168; Bardenhewer, *Die pseudo-aristotelische Schrift über Das reine Gute.*

[36] E. g., Montpellier MS. 145, ff. 94–102 (= pp. 302–331 of Eyssenhardt's edition), following the *Questiones* of Adelard of Bath.

[37] Bodleian, Rawlinson MS. G. 40, ff. 1–30, of the late twelfth century: 'Terra fundata est super stabilitatem suam . . . aut in latitudine declinare aut retrogradiari facit. Explicit.' Dr. Craster, to whom I am indebted for suggestions respecting the contents of this MS., calls my attention to a fragment of the treatise in Bodleian MS. Auct. F. 1. 9 (another MS. of English origin, on which see below, Chapter VI, n. 6), ff. 160–162, entitled 'Liber Yparci de cursu siderum' and beginning at f. 22 v of the Rawlinson MS. Another copy is at Cambridge, McClean MS. 165, ff. 1–16 v. Curiously, the two mentions of Hipparchus in this portion of Martianus (ed. Eyssenhardt, pp. 304, 322) are omitted in the conflated text (Rawlinson MS., ff. 6, 18 v).

[38] Rawlinson MS., f. 5: 'Mundus igitur ex quatuor . . .' Cf. note 36 above.

[39] On f. 24, the lettering of an omitted figure shows traces of Greek influence: a, b, Γ, d.

[40] A. Clerval, *Les écoles de Chartres* (Paris, 1895), book iii; R. L. Poole, "The Masters of the Schools of Paris and Chartres in John of Salisbury's Time," in *E. H. R.*, xxxv. 321–342 (1920).

[41] Hauréau, *Histoire de la philosophie scolastique* (1872), i, chs. 16–18; idem, in

the movement has been sketched by Baeumker.[42] Nevertheless we still lack a detailed study of the range and depth of Platonic influences in this period, as measured in the lesser writers and as manifested in the various anonymous treatises which have not yet been collected or explored;[43] nor do we know, apart from the general fact of the school's efforts to harmonize Plato and Aristotle, what reactions the newer knowledge produced upon the older habits of thought.[44]

The decline of this Platonic cosmogony came with the reception of the Ptolemaic astronomy and the Aristotelian physics, as transmitted by the Arabs of Spain. For this it is not easy to give precise dates. Thus at Chartres a manuscript of the cathedral preserves a treatise on astrology containing Arabic words which dates from 1135, with notes added from 1137 to 1141;[45] and another manuscript of the twelfth century contains Adelard's version of the Khorasmian tables.[46] Hermann of Carinthia's version of the *Planisphere* was, as we have seen, dedicated to Thierry of Chartres in 1143.[47] Yet Thierry's *De sex dierum operibus* is a daring piece of Platonism,[48] and the trace of Aristotelian physics found therein carries us no farther than Macrobius.[49] Some time before his death ca. 1155[50] Thierry drew up in his *Eptatheuchon* a summary of the seven liberal arts composed of extracts from

Notices et extraits des MSS., xxxii, 2, pp. 169–186; R. L. Poole, *Illustrations of the History of Mediaeval Thought* (London, 1920), ch. iv; Duhem, iii. 87 ff., 184 ff.; M. Grabmann, *Geschichte der scholastischen Methode*, ii. 407–476; M. De Wulf, *Philosophie médiévale* (1912), pp. 210–217; Ueberweg-Baumgartner[10], ii. 306–327 (1915); A. Schneider, in *Beiträge*, xvii, no. 4, pp. 3–10.

[42] *Der Platonismus im Mittelalter* (Munich, 1916), and its numerous references.

[43] For one example, the Pseudo-Bede, see Duhem, iii. 76 ff. So a treatise of this period on semitones, perhaps by Ralph of Laon, begins 'Quoniam et Macrobii et Platonis auctoritate' (B. N., MS. lat. 15120, f. 41).

[44] Some one with easy access to the manuscripts ought to attack this problem.

[45] Chartres MS. 213, ff. 63–141. F. 116 has: 'In hoc anno quando erant anni a nativitate Christi M.C.XXXV. in kal. iulii fuit Venus incensa in Cancro.'

[46] Chapter II, no. 3.

[47] Chapter III, no. 5.

[48] Hauréau, in *Notices et extraits des MSS.*, xxxii, 2, pp. 167 ff.; Duhem, iii. 184 ff.

[49] Duhem, iii. 188–193; and in *Revue de philosophie*, 1909, pp. 163–178.

[50] Clerval is much too positive in placing it ca. 1141 on the ground that Thierry ceased to teach about that year.

forty-five authorities, the original, in two large volumes, being still preserved at Chartres.[51] Yet the mathematics and astronomy of the *Eptatheuchon* show no certain trace of the new learning.[52] The geometry is that of the *agrimensores*, Gerbert, and the Pseudo-Boethius; arithmetic is represented by Boethius and the abacus of Gerland and others, astronomy by the fables of Hyginus and the canons of Ptolemy, followed, it is true, by a set of tables which require closer examination.[53] The tone throughout is that of the earlier Middle Ages; even the *Posterior Analytics* is as yet unknown.[54] The main peculiarity of the school of Chartres lay in its "reverent dependence on the ancients"; [55] it stressed the *trivium* rather than the *quadrivium*, and with the decline of humanism in the second half of the twelfth century its fall was rapid, so that Chartres never became a centre of the new science.

A survival of the school of Chartres may be seen in the *Microcosmographia* which a certain William dedicates to William, archbishop of Rheims from 1176 to 1202, and previously (1164-1168) bishop of Chartres.[56] Preceded in the manuscript by an astro-

---

[51] MS. Chartres 497–498, which I examined in 1919. See the detailed account by Chasles, *Catalogue des MSS. de la ville de Chartres* (Chartres, 1840), pp. 30–36; the *Catalogue général*, xi. 211–214; Bubnov, *Opera Gerberti*, p. xxvi. Cf. Clerval, *Ecoles*, pp. 221 ff.; and his detailed analysis in *L'enseignement des arts libéraux à Chartres et à Paris d'après l'Heptateuchon de Thierry de Chartres*, read before the Congrès scientifique des Catholiques in 1888, and separately, Paris, 1889. What these writers say of the introduction of Arabic numerals needs to be read in the light of more recent discussion; cf. D. E. Smith and L. C. Karpinski, *The Hindu-Arabic System of Numerals* (Boston, 1911).

[52] F. 141–141 v, which was once considered a fragment of Adelard's version of Hypsikles, is identified by Bubnov (pp. xxvi f.) as a part of the geometry of the Pseudo-Boethius.

[53] There is no basis for Clerval's assumption (*L'enseignement*, pp. 21 f.) that the *Canons* were translated from the Arabic by Hermann of Carinthia; indeed the numerous Greek words in the Chartres text (ff. 174–184) would point to a quite different conclusion. The *Canons* are also in MS. Chartres 214, f. 1, likewise translated from the Greek (Björnbo, in *Archiv für die Geschichte der Naturwissenschaften*, i. 393); the date and author of this version have yet to be determined.

[54] Infra, Chapter XI, nn. 11, 34.

[55] Poole, *Illustrations*, p. 102.

[56] Preface and contents in Martène, *Veterum scriptorum amplissima collectio*, i. 946, from a MS. which is now no. 1041 (1267), ff. 3–43, of the Stadtbibliothek at Trier. I have collated the preface by means of photographs, but have not been able

logical table for 1178, the treatise may well have been written in
1177. It is not, as we might expect it to be, a work of astronomy
or cosmology, but a comparison of human and animal nature,
discussing intelligence, free will, and the senses, and based upon a
collection of the *opiniones antiquorum*, among whom Plato duly
figures.

TREATISES ON THE ELEMENTS

Respecting the arrival of the Aristotelian physics the chronological evidence is less definite than in the case of astronomy. The
De physico auditu makes its appearance as a whole, in versions
from both Greek and Arabic, toward the year 1200.[57] Yet Duhem
has shown one of its doctrines, derived through Macrobius, in the
De operibus sex dierum of Thierry of Chartres,[58] and we have seen
other traces of its teaching still earlier in Adelard of Bath, who
seems to have got them through Galen and Constantine the
African.[59] Certainly this was the source for William of Conches,
who seeks to reconcile with Plato Constantine's definitions of the
elements,[60] and who cites from Constantine the same passage on
the place of the faculties in the brain [61] which Adelard cites from
Aristotle. His further references to Johannitius and Theophilus [62]
confirm the conclusion that the school of Chartres was acquainted
with the early translations of medical writings from the Arabic.
To what extent and through what channels the ideas of the *Physics* affected the writers of the latter half of the twelfth century is
a problem which awaits investigation. One body of writings may
be indicated as a field of inquiry, namely the various treatises on

to secure a rotograph of the whole treatise, which would evidently repay examination. The author cannot be William of S. Thierry (Clerval, *Ecoles de Chartres*, p.
275), who died long before 1176. The *Histoire littéraire* (ix. 70, 191) makes the
author William of Soissons. As the archbishop is called legate and not cardinal, the
dedication cannot be later than 1179.

[57] Infra, Chapter XI, n. 4; Chapter XVIII, n. 55; Grabmann, *Aristotelesübersetzungen*, pp. 170–174.

[58] iii. 188–193; and in *Revue de philosophie*, 1909, pp. 163–178.

[59] Supra, Chapter II, n. 93.

[60] Migne, clxxii. 48–55; cf. Baumgartner, in *Beiträge*, ii, no. 4, p. 50. Cf. also
Adelard on the elements: *Questiones*, cc. 1–4.

[61] Migne, clxxii. 95.

[62] *Ibid.*, coll. 50, 93; Duhem, iii. 88 f.

the universe and the elements to be found in the manuscripts of this period.

Let us take as an illustration a group of such works in Cotton MS. Galba E. iv of the British Museum, written in different hands of about the year 1200.[63] First comes the earlier part of an anonymous work on natural philosophy,[64] beginning 'Sciendum est quid sit philosophia,' but coming shortly to the four elements as the main topic, with applications to meteorology. The author, who knows a few Greek words [65] and seems to live in southern Europe,[66] quotes Seneca, Macrobius, and the Latin poets. He accepts the Platonic doctrine of ideas eternally in the mind of the Creator,[67] and quotes the *Timaeus* on motion as the origin of the elements; [68] but his definitions of *phisis* and the three species of fire, as well as the dictum of the earth's immobility, are cited specifically from Aristotle's *Physics*.[69] The treatise breaks off abruptly after six pages of the manuscript.

The lacuna in the codex is likewise responsible for the loss of the beginning of the next treatise, a dialogue in two books between master and pupil, entitled *Liber Marii*.[70] The first book

[63] Ff. 187–204 v. Formerly at Bury St. Edmunds (M. R. James, *On the Abbey of S. Edmund*, Cambridge, 1895, p. 66, no. 154; id., *List of MSS. formerly owned by Dr. John Dee*, Oxford, 1921, p. 29, no. 144). As an indication of date, note that the *e* with cedilla appears throughout these treatises, which are followed by the so-called *Prenon phisicon* (i. e., Nemesius, infra, Chapter VIII, n. 5) and the *Questiones* of Adelard of Bath.

[64] Ff. 187–189 v.

[65] 'Quę en noian dicuntur, id est in mente' (f. 187). 'Anastronica, id est sine stellis' (f. 189 v).

[66] On f. 189 he argues that clouds come from the west and south because the ocean is nearer in that direction.

[67] F. 187.

[68] 'Ut dicit Plato in fine nostri translationis inducens similitudinem pistorii instrumenti': f. 187 v; cf. *Timaeus*, c. 52 E.

[69] 'Ut dicit Aristotiles in phisica (ph'ica) sua, Phisis est naturalis motus alicuius elementi ex se' (f. 187). 'Dicit Aristotiles in phisica sua ignis esse tres species' (f. 188). 'Quod legitur in phi., Terra est immobilis' (f. 188). Cf. *De physico auditu*, 3, 1, 1; 8, 3, 3; 3, 5, 17. The reference on the three species of fire should be to the *Topica*, 5, 5, 11. A more exact quotation of the *Physics* (3, 1, 1) is found on f. 187 v, but without citation of source: 'Phisis proprie est principium motus ex se.'

[70] Ff. 190–200. Inc. of first page 'aque que est.' F. 194 v: 'et ego subtilius potero respondere. Explicit liber primus. Incipit secundus. D. Iam igitur mihi vellem dari argumenta quod animalia atque virentia et ea quę vocant Sarraceni con-

considers the four elements and their qualities, the second treats of their compounds in the form of "animals, plants, and those things which the Saracens call *congelata,* such as quicksilver, sulphur, and all metals." The compounds include odors and complexions as well as the six metals compounded of quicksilver and sulphur, an interesting early example of the standard alchemical doctrine.[71] The author has travelled widely;[72] he has written a *Liber de humano proficuo* and promises a succeeding one on the five senses.[73] He cites 'ancient books' and other philosophers,[74] Plato,[75] and especially the pseudo-Aristotelian *De elementis,* which seems to have been first translated by Gerard of Cremona.[76]

The next treatise, a brief one, is entitled simply *Liber de elementis.*[77] It cites the opinions of various Greek philosophers, mentions especially Aristotle's fifth element, the ether, and relies on the dictum of Hippocrates that man cannot consist of a single element. The series closes with a *Liber de aere et aquis,* a piece of humoral speculation on climatic conditions, designed especially for physicians,[78] for whom astronomy is also indicated as useful. The author, evidently a dweller near the Mediterranean in the twelfth century, contrasts particularly the inhabitants of Europe and Asia, with most detail concerning the Turks.[79]

gelata sicut vivum argentum, sulphur, et metalla cuncta . . . ideoque a philosophis minor mundus nuncupatus est. Qui ipsum super huius seculi universa composita sullimavit sit benedictus in secula seculorum amen' (f. 200).

[71] F. 198. Cf. the *Liber de congelatis* translated by Alfred of Sareshel: infra, Chapter VI, n. 47; Thorndike, ii. 250; Pelzer, in *Archivum Franciscanum historicum,* xii. 49 f. (1919). On flavors cf. the fragment "De saporibus" edited by F. Hartmann, *Die Literatur von Früh- und Hoch-Salerno* (Leipzig diss., 1919), p. 55.

[72] F. 199.   [73] F. 200.

[74] 'Legi quoque in antiquis voluminibus de elementis' (f. 190 v). Cf. ff. 192, 199.

[75] Ff. 190 v, 199.

[76] Ff. 192 v, 193 v. On the translations of this work see Steinschneider, *H. U.,* pp. 232 f.

[77] Ff. 200 v–201 v: 'Elementum in mundo tocius est corporis minima pars . . . alterum ab altero nasci videbis. Explicit.'

[78] Ff. 201 v–204 v: 'Quisquis ad medicinę studium accedere curat . . . et non errabis a veritate. Explicit liber.'

[79] F. 204.

The occurrence of these treatises in the same manuscript does not, of course, show any inherent connection, but the internal evidence refers them to the same general age and milieu. Anterior to ca. 1200, they belong to the epoch when Aristotelian science was coming in through Arabic channels but had not yet been fully absorbed. The authors are more interested in physics than in astronomy, at least one of them also cares for medicine, and there are traces of Greek as well as Arabic learning. All this points to southern Italy and Sicily rather than any other part of Latin Europe.

More specifically astronomical is an anonymous treatise of which the first twenty-five chapters are preserved in MS. Lat. 15015 of the Bibliothèque Nationale.[80] According to the table of contents its forty chapters covered the four elements and their motion, earthquakes and tides and other matters of meteorology, and the motions of the planets. There are many diagrams, but there is nothing very striking in the text.[81] The author quotes authorities sparingly, as in one instance, 'philosophi in libro de rerum natura';[82] if he does not specifically cite Aristotle's *Meteorology*, he refers to Ptolemy 'in codice de sperarum compositione'[83] on the size of the sun, and thus brings us to the close of the twelfth century.

A similar transition to the science of the thirteenth century is seen in a brief tract in the Biblioteca Casanatense at Rome,[84] perhaps also referable to southern Italy because of its allusions to

[80] Ff. 200–223 v, of the early thirteenth century: 'Gratia Deo primo sine principio . . . [chapter headings]. Postquam capitula singulatim computavimus, ad unumquodque explanandum ordine accedamus. Primum quod firmamentum est creatum et gubernatum . . .'

[81] F. 203: 'In toto enim mundo non est locus vacuus.' Cf. supra, Chapter II, n. 97.

[82] F. 206 v.

[83] F. 214 v. See *Almagest*, 5, 16. The *Introduction* of Geminus, translated by Gerard of Cremona, seems to have been current under a similar title (Steinschneider, *E. U.*, no. 46(37); Manitius, *Gemini Elementa*, 1898, pp. xviii f.), but I do not find this passage in the edition.

[84] MS. 2052, ff. 17–18 b (of the early thirteenth century): 'Videndum etiam quid sit philosophia, que eius partes, que sunt partium partes, deinde partium et subpartium executiones . . . ille tamen transeundo per terre venas colantur et sic dulces eunt.'

hot baths and sulphurous flames. After a classification of the
sciences, the author takes up the elements, their qualities, and
their *passiones*. Dionysius the Areopagite is cited, as well as
Lucan, Macrobius, and Seneca's *Quaestiones naturales*. Aristotle is cited once via Macrobius,[85] once specifically in the
Metaphysics.[86]

THE MARSEILLES TABLES

In the diffusion of Arabic learning north of the Pyrenees an
important part was taken by the cities of southern France. We
have already seen Hermann of Carinthia at Béziers and Toulouse,[87] while Jewish scholars like Abraham ibn Ezra at Narbonne
prepared the way for the numerous translations from Arabic into
Hebrew made for the Jewish communities of Provence and Lan-
guedoc.[88] Montpellier was a well known centre of astronomy by
the thirteenth century, while Marseilles appears at the very outset of the new movement.

One of the earliest attempts to adapt the astronomy of Spain
to places north of the Pyrenees is found in the planetary tables
drawn up by a certain Raymond of Marseilles in 1140: *Liber
cursuum planetarum capitisque draconis a Raymundo Massiliensi
super Massiliam factus*.[89] The introduction to this work, it is

---

[85] 'Huiusmodi questiones [salt and fresh waters] Macrobius de sñtialibus [*sic*]
movet et solvit secundum Aristotilem' (f. 18). Cf. Macrobius, *Saturnalia*, 7, 13, 19
(ed. Eyssenhardt, p. 448), where the reference is apparently to the *Problemata*.

[86] 'De actionibus elementorum de quibus Aristotiles in metaphisicis egit nunc
inspiciendum est' (f. 17 v).

[87] *Supra*, Chapter III.

[88] Renan, in *Histoire littéraire*, xxvii. 571–623; Steinschneider, *H. U.*, *passim*;
id., in *Abhandlungen zur Geschichte der Mathematik*, iii. 57–128; Duhem, iii. 298 ff.

[89] This is the title in MS. 243 of Corpus Christi College, Oxford, ff. 53–62 (**saec.**
xv), which lacks the tables and begins with 166 verses:

> O qui stelligeri cursus moderaris Olimpi
> Sideribus septem contra labentibus orbem
>
> . . . .
>
> Ergo lectorem prius hoc novisse iubemus
> In media quod principium sit nocte diei
> Atque quod in simili sit finis parte sequentis,
> Et domini nostri Ihesu Christi super annos
> Massiliamque super nos hunc componere librum

true, bears the date 1111, or 1106, but that this is a scribe's error for 1140 (MCXI for MCXL) appears in a specific reference to a debate of 27 October 1139 [90] as well as in the content of the tables themselves.[91] Their purpose is to adjust the tables of Toledo to the use of Latins in general and the author's own Marseilles in particular:[92]

Cum multos Indorum seu Caldeorum atque Arabum quos in astronomia plurimum valuisse cognovimus [93] cursuum planetarum [94] libros aut super Arin civitatem, que in medio mundi rectissime fore constructa memoratur, aut super Meseram et super annos mundi seu Grecorum aut Gezdaheirt edidisse vidissemus, novissime autem quendam Toletanum hac in arte [95] perspicuum, qui a quibusdam Azarhel vel Albatheni nuncupatur, super annos Arabum et super Toletum, que a nostra civitate, id est Massilia, per horam et alterius partem decimam distat, cursuum similiter librum fecisse comperissemus; non indignum esse [96] credidimus super annos domini Ihesu et super prefatam civitatem nostram librum constituere, et quoniam [97] nos primi Latinorum fuimus [98] ad quos post Arabum translationem hec scientia pervenerat et [99] aliquid utilitatis ex nostro labore cunctis Latinis administrare haud absurdum videbatur, opus presens aggressi sumus [atque predictum Toletanum in eo immitati sumus].[100] Constituimus ergo in eo radices [101] .vii. planetis capiti atque caude [102] draconis super mediam noctem quam

> Senciat; est illic que nostre gentis origo.
> Natalemque locum nostro de numero clarum
>
> . . . .
>
> Carminibus finem facio; laus omnipotenti
> Sit Domino nostro qui regnat trinus et vivus.
> Amen.

Anonymous and without title and preliminary verses in B. N., MS. Lat. 14704, ff. 110–135 v; fragment at Cambridge, Fitzwilliam Museum, McClean MS. 165, ff. 44–47, a MS. anterior to 1175, in which this treatise is entitled "Liber cursuum planetarum .vii. super Massiliam." See Duhem, iii. 201–216; and, for astrology, Thorndike, ii. 91 f. The fulness of Duhem's discussion makes detail unnecessary, but he knew only the anonymous B. N. MS.

[90] Corpus MS. 243, f. 56; MS. Lat. 14704, f. 111; Duhem, iii. 207. The Corpus MS. (f. 55) has a further corruption in the principal date: 'M°. C°. VI°.'

[91] Duhem, iii. 203 f.

[92] MS. McClean 165, f. 44, which begins here; MS. Lat. 14704, f. 116; Duhem, iii. 211 f. The Corpus MS. stops just before this passage.

[93] MS. Lat. *cognovissemus.*   [97] MS. Lat. inserts *quia.*

[94] MS. Lat. om.   [98] MS. Lat. *fueramus.*

[95] MS. Lat. *doctrina.*   [99] MS. Lat. om.

[96] MS. Lat. om.

[100] MS. McClean omits the words in brackets.

[101] MS. McClean *radicem.*   [102] MS. Lat. *atque capiti draconis.*

sequebatur vii^a feria kalendarum ianuarii qua ingressus est annus Latinorum
in quo Dominus incarnatus est, et super Massiliam que ab Arin, cuius tam
latitudo quam longitudo nulla est, trium horarum spatio distat.

The author, whose piety is evident, is at some pains to justify
the study of the stars and their influence on human affairs by
reference to the Bible and by copious quotations from Lucan. He
quotes the Fathers — Ambrose, Augustine, and Gregory — Hippocrates and Galen, Ovid, Priscian, and Boethius, giving evidence
of a considerable Latin culture in the astronomical portion as well
as in the mythology and geography of the introductory verses. If
he does not cite by name Plato and Macrobius, he discusses
briefly the world-soul. Besides al-Zarkali and al-Battani, his
Arabic authorities are the astrologers Albumasar, Alcabitius, and
Messehalla, from whom he promises extracts not found in our
MSS. He inveighs against certain incorrect planetary tables and
apocryphal works ascribed to Ptolemy, and he has himself written
a treatise on the astrolabe to which he makes frequent reference.
His own attainments are respectable. "He is," says Duhem,[103]
"an astronomer not only because he is abreast of the most deli-
cate and most recent discoveries, such as al-Zarkali's discovery of
the proper movement of the sun's apogee, but he also shows himself an astronomer by his sound ideas concerning the methods of
observation and the corrections which they require."

Raymond declares himself the first of the Latins to acquire the
science of the Arabs, in evident ignorance of the work of Adelard
of Bath and Plato of Tivoli. Moreover he describes his debate of
1139 with two masters who possessed incorrect tables,[104] but he
does not say that these were Latins or where the debate took
place.[105] In 1194 Maimonides addresses his treatise on astrology
to certain Jews of Marseilles.

A CRITIC OF MACROBIUS

The decline of the older astronomy can be seen from another
angle in a treatise on the planetary spheres contained in a manu-

---

[103] iii. 208.   [104] MS. Lat. 14704, f. 111; MS. Corpus 243, f. 56.

[105] For later astronomers at Marseilles see Steinschneider, in *Bullettino*, xvii.
775 f., xx. 575–579; P. Tannery, in *Notices et extraits des MSS.*, xxxv, 2, pp. 561–
640; Duhem, iii. 287–291; Thorndike, ii. 92 f., 206, 211, 485–487.

script of Cambrai.[106]   This codex, of the later twelfth century, begins as follows, as if it were a translation of Maimonides:

Incipit liber Mamonis in astronomia a Stephano philosopho translatus

Quoniam in canonem astronomię quas proposueramus regularum exsequto tractatu promissum exsolvimus, secundum hoc opus licet arduum et sub- tilissimo ac multiplici naturę celatum archano non inconsulta aut impu- denti temeritate sed frequenti et animi et utilitatis ammonitione aggredior. Sic enim licet magnorum super his gravissimorumque disputatio philoso- phorum, tamen mediocres persepe maxima quemadmodum maiores curant minora. Illud quoque attendendum est plurimum quod, cum omnis a Deo fit sapientia, ea autem verior et sine scrupulo fallacię concessa sit nemo no- verit. Unde et qui graves habentur philosophi sepe extra se maximis in rebus eorundemque verius et perspicacius alios qui nec philosophiam adepti essent nec ad eam aliquando posse pertingere existimaverent [*sic*] de divini mu- neris larga benignitate hausisse noticiam comperimur. Testes sunt Plato et Aristotiles quos omnium liberalium artium fere magistros habemus. Quorum Plato in multis a veritate dissonat, Aristotilis mundum non esse a Deo con- ditum de nichilo sed cum eo sicut nunc est tamquam cum corpore umbram processisse et condidit et argumentis fallacibus conatur asserere, eo nimirum in loco intellectus et animi et oculorum privatus officio qui fidelium simplici- tati divina nascitur misericordia. Idem ipse in hac de qua proposita est dis- putatio questione cum de celestibus speris dissereret, octo positis de nona non, ut quidam arbitrantur, consulto tacuit sed se ad eius noticiam nequa- quam pervenisse manifestum nobis reliquit testimonium. Quod nullatenus arroganter dictum cuipiam videri velim et quod tantę gravitatis et scientię et ex eisdem auctoritatis adepte philosophus ignorasse dicitur me non latuit. Nam etsi inter maximos locum non obtineam, ad eosdem tamen aspirans mediocrium invasi disciplinam. Habet enim ille sua quibus plurima con- sumpta opera perpetuitatis dum philosophantes vixerint nomen adeptus est quorum tamen pluraque a maioribus omnia autem a Deo preter obfuscata falsitatis errore accepit. Quare nobis quoque, qui nichil aliis derogamus, si quidem idem omnium ditissimus Deus annuat invideri dedecet, cum ab eo accepta alios docere quam ignavie silentię tegere malumus. Hęc autem ideo quia nisi tanta foret obtrectantium multitudo ferociores habuisset latinitas auctores fertiliorque apud nos philosophię seges pullularet. Cum etenim plu- rimi essent exercitus detrahentium pauci qui benigne susciperent, pauciores certe artium scriptores magis exterrebantur multitudinis immanitate quam adunarentur aliquorum benigno studio. Unde factum est ut que fere pleni- tudinem posset habere artium nunc ceteris gentibus Europa videatur humi- lior, quippe que quos educat contra fontem scientię sepius oblatrantes sentit sibi ipsis rebelles nunc hęc nunc illa nunquam consona ruminantes. Quę res

---

[106] MS. 930 (829), 49 folios, formerly belonging to the cathedral. It breaks off before the end of the treatise, but evidently not long before. On the fly-leaf a hand of the fifteenth century has written. 'Quidam tractatus de astronomia .xxii.'

tantum attulit litteralis scientię odium ut a quibus summe venerari debuerat rerum [107] rectoribus summe odiretur. . . .

(f. 2 v) Quoniam autem in canonis regulis multa tetigimus que in hoc opere explicari desiderant, promissum preterire consilium non fuit, ut quod illic dubietatis scrupulus fastidium generaverit hoc [108] operis beneficio sopiatur. Atque hec est ratio que me maxime ad hoc opus coegit ne autem anxium [109] lectorem a studio repulsum iri paterer nostratumque utilitati quoad posse consulerem neve quod pollicitus fueram aut ignorasse aut inertia neglexisse arguerer. Placet igitur celestium sperarum circulos numerum ordinem quo verius potero quantumque humana patitur ratio aperire, ut qui a Ptholomeo in sua sinthasi disponuntur circuli in speris etiam quo modo possint inveniri laborantibus in hac arte via teratur. In quo nichil enim perfectum mihi vel cuiquam ad explicandum concessum arbitror, siquid pretermissum superflueve positum fuerit sapientium arbitrio corrigendum relinquo.

Mundus nomen est ad placitum per quod omnia fere que condita sunt designantur, forma eius rotunda atque speralis . . .

Starting from the solid sphere of the earth as the centre of the spherical universe, the author explains that the earth is immovable and the heavens revolve about it. He knows nothing of the surrounding sea [110] but argues briefly concerning the source of the Nile; whence he passes quickly to the nine heavens which constitute his main theme. The greater heat of the sun in summer is due to its nearness, not, as the Aristotelians think, to the angle of its rays.[111] He has himself tested the effect of the full moon on the weather.[112] Throughout the first book, as for example, on the zodiac, there is a running criticism of Macrobius, concerning whom he thus expresses himself in the preface to book ii: [113]

In astronomie mihi suscepta disputatione laboranti, de qua pauca certe habet latinitas eorumque pleraque erroris obfuscata caligine, obici fortassis animus doctis poterit arrogans in invidia quod in Macrobium inter philosophantes non mediocrem tociens acrius invehar, eoque amplius quod usque ad hec tempora omni caruerit obtrectationis livore. Quibus vellem satis

[107] *p* apparently erased after *rerum*.   [108] *h'* for *huius?*   [109] MS. *anexium*.

[110] 'Et de mari quidem quod quo ambitus quibusve locis terram circumfluat incertum habeo preter id quod septentrionales norunt habitatores, de quibus quoniam apud illos sepe dictum est taceamus' (f. 5).

[111] 'Nam et hos qui more solis super terram causam imponunt plurimum errasse et Aristotelicos qui motui radii tantum a veritate deviasse videmus' (f. 10 v).

[112] 'Nam in estate, quod ego id compertum habeo, plenilunialem noctem humidiorem esse et frigidiorem, sinodalem vero diem minus calidum et siccum' (f. 12 v).

[113] Ff. 15–15 v. On the influence of Macrobius prior to the reception of Arabic astronomy, see Duhem, iii, ch. 3.

esset mea cognita voluntas intelligantque me latine tradere facultati nostratum incognita auribus archana, que cum frequentibus vigiliis diuturnis
cogitationum recessibus exquisita comparaverim quorum Macrobium aut
inscium fuisse video aut intellecta perversa depravasse exponere.[114] Horum
alterum cum ad filium suum, quem sapientia sua sapientiorem fieri vellet,
scriberet fuisse dicendum non est, nemo enim dilectum sciens perverse in-
struit. Non igitur intellecta veraciter depravasse sed non intellexisse potius
et ignorasse iudicandum est. Quam ob rem non mihi in huius artis peritia
philosopho sed cum inscio contencio est. . . . In Macrobium igitur nostra
idcirca maior est animadversio quoniam apud nostratum opinionem ceteris
ipsum copiosiorem in astronomia et sentio et relatum per quam plurimos est.

In the second book we find the usual division into climates, and
the common view that the habitable globe lies between fixed
parallels.[115] The third book takes us further into the subject by
discussing the spheres of moon and sun and their eclipses; the
fourth, 'De retrogradacione,' considers the spheres of the planets
as well as the eighth sphere of the fixed stars and the ninth which
he calls *aplanos*. Naturally the author does not accept the
Macrobian theory of the rotation of Venus and Mercury about
the sun. He loves geometry, especially geometrical proofs 'unknown to the Latin world,' [116] and these are accompanied by
good diagrams.[117] Ptolemy is always cited with respect,[118] the
Almagest specifically as the *Sintaxis*.[119] The author calls himself
a Peripatetic, but disagrees frequently with the Aristotelians.[120]
The other authorities are not cited by name,[121] save the much
criticized Macrobius, but the author's indebtedness to an unnamed
Arab writer is mentioned in the preface to the fourth book: [122]

[114] Probably for 'expositione.' [115] Ff. 5 v, 22, 31 v.

[116] Infra, p. 102. 'Deprehensum est enim a quodam sollertissimo et astronomie
scientie peritissimo philosopho geometricali argumentatione' (f. 37 v). 'Id ita esse
ut aiunt verissimis ostenditur in libro geometrie rationibus' (f. 7 v).

[117] Ff. 4 v, 26 v, 27 v, 30 v, 31 v, 32, 35, 36, 38 v, 43 v, 45 v, 47 v, 48.

[118] 'Tholomeus in astronomia magnificus . . . hec in libro quem de habitatione
dixit scripta sunt. Michi vero tametsi difficillimum videatur, credendum tamen
estimo eius philosophice traditioni quam et multarum constat rerum experientiam
habuisse et antiquorum scriptis et sui temporis hominum relatione multa que nobis
incognita sunt certo cognovisse' (f. 22 v). See also ff. 27 v, 29 v, 30, 49 v.

[119] Ff. 29 v, 49 v; cf. the preface above.

[120] 'Neque enim Epicurum aliquando dogma audivimus sed peripatetice potius
accedimus claritati' (26 v). On the Aristotelians see f. 10, 10 v.

[121] Cf. n. 116 and the preface to book iv, below.

[122] F. 38–38 v.

Quartus hic laboris nostri decursus de .*e.* planetarum speris et circulis et octava denique nona spera disserens transcurso maris alto funere anchore portus tranquillo attinget. Verum cum in aliis Arabem quendam plurimum secuti sumus, in hoc quoque per multum sequimur, licet quedam de sperarum numero et rotunditatum invenerimus et de circulis quidem et inclinationibus planetarum vera perstrinxit a quibus sperarum numerus dissonat. Hoc autem suis in locis aperte monstrabitur. . . . Non enim parva apud Latinos diutius inquievit questio quonam modo erraticorum .*e.* globi quorum natura indictus cursus in orientem est fiant retrogradi et ab oriente relabantur in occiduas partes. Et hec quidem, ut verum fateamur, questio digna est et proponi et solvi sed a nemine tamen eorum absoluta. Nec hoc mirum ducimus, cum occulta sit res et geometricalibus exquisita et aprobata argumentis quorum latinitas inscia indivulgato diu multumque volutatur errore. . . .

There are, however, no Arabic terms of any sort, while words like *extasis*,[123] *sintaxis*, and *panselinus* [124] point to some use of Greek sources or works derived therefrom. Diagrams are lettered *abcdef*, not *abgdez* aş in the case of mechanical transfers from the Greek, but there is a curious system of numerals by which the letters of the Roman alphabet are given a numerical value in succession like the Greek.[125] Thus in the extract printed above *e* is used for the five planets, and we likewise find *g* for the seven climates [126] and *ld* for twenty-four hours.[127] The higher numbers have caused some confusion to the copyist, but the following may serve as an example, in which $k = 10$, $l = 20$, $t = 100$, $u = 200$, etc.: [128]

Que spatia cuius sint proportionis ita videbimus: Inter terram et lunam .*a* gradus esse concedamus. Duplum a terra usque solem, id est .*b.* Triplum huius a terra ad Venerem, id est .*f.* Quadruplum autem huius ad Mercurium, id est .*l.d.* Novies .*l.d.* usque Martem que sunt *u.k.f.* Octies autem ducenta *k.f.* usque Iovem, scilicet mille septingenti .*l.h.* Qui vicesies septies multiplicati spatium a terra usque Saturnum reddunt *qcy.y.g.*[129] et de quibus sublatis *t.u.p.a.* scilicet spatio a terra usque Iovem remanet a Iove usque Saturnum

---

[123] F. 3.   [124] 'Paranselinio (*sic*) quod nos plenilunium dicimus' (f. 34).

[125] I have not found this system elsewhere, unless it is the one found by Friedlein in MS. Erfurt 1127: *Die Zahlzeichen und das elementare Rechnen der Griechen und Römer* (Erlangen, 1869), p. 20.

[126] F. 22 v.   [127] F. 23.

[128] Ff. 27 v–28. So (ff. 8, 17) .*xp.* is used for the 360° of a circle, half of which is .*tr.* For the distances in the passage here printed cf. Macrobius, *In Somnium Scipionis*, 2, 3, 13 (ed. Eyssenhardt, p. 584).

[129] What I have represented by *y* resembles rather the early western form of 5, and the *et* may also be a numeral. The numbers from this point up I must leave to some one else to interpret. As far as 1728 the system is clear.

q.*q*.*q*.*p*.*a*. Sublato de spatio Iovis spatium Martis restat a Marte *i*. *g*.*ip* usque Iovem, sic de reliquis. Cum igitur spatium a luna usque solem *.a.* gradus, a sole ad Venerem *.d.*, ad Mercurium *k.h.*, a Mercurio ad Martem *t.s.b.*, a Marte ad Iovem *i.g.i. p.*, a Iove ad Saturnum *q*.*q*.*q*.*p*.*a.*, cui reliquis omnibus spatiis iunctis prior surgit numerus. Que spatiorum assignatio multis rationibus improbatur. Si enim est ut idem dicit una eademque omnium celeritas, duplo temporis sol suum peragraret circulum quo luna suum circuit, duorum et enim circulorum si alterius diametrum duplum sit diametro alterius et circuli sic se habent. Peragraret igitur, si vere essent assignata spatia eademque citatio, sol *b* mensibus totum zodiacum, Venus *.f.*, Mercurius *.l.d.*, Mars *.u.k.f.*, Iupiter *t.n.d.* annis, Saturnus *.ψ .a.a.d.* annis et *.b.* mensibus. Que cum ita sint, aut falsa est sperarum assignatio aut celeritas non erit eadem.[120]

It is by this time plain that what we have is no translation of Maimonides or any one else, but an independent work using authorities but following consistently its own line of argument. The author has already written *Regule canonis*. His present purpose is to introduce a more correct astronomy into the Latin world, which is still in fog and darkness. He is plainly a Latin, citing Lucan and Cicero,[131] with bits of classical lore like the story of Solon's travels,[132] mentions of Caesar and Constantine,[133] and references to the Epicureans and Peripatetics.[134] As his doctrines are thought new and Macrobius is his chief enemy, he still belongs to the period of the first reception of the new astronomy, when Platonism is still in the ascendant and Arabic learning is just arriving. Whether the name Stephen in the title has any more value than the reference to Maimonides, must remain an open question. The combination of Greek and Arabic influences points toward Sicily, though Stephen of Antioch is also a possibility.[135]

TRANSLATIONS OF PTOLEMY

With the translation of Ptolemy's *Almagest* into Latin the fulness of Greek astronomy reached western Europe. The Μαθηματικὴ Σύνταξις of Ptolemy was for all subsequent times the most

---

[130] On f. 48 v we have: 'Completur enim Saturni lati motus in *.l.i.* annis, *.e.d.* diebus, *.k.e.* horis, horarum seX. *.l.d.*; Iovis autem annis *.k.a.*, x.k.e diebus, horis *.k.d.*, horarum seX *.l.i.*; Martis vero anno uno, diebus *.x.l.b.*, horis *.l.d.*; ac Veneris et Mercurii anno uno, horis *.e.*, seX. horarum *.n.i.*'

[131] Ff. 5 v, 13 v, 27.   [133] F. 27.   [135] Infra, Chapter VII.
[132] F. 15 v.   [134] F. 26 v.

important work of ancient astronomy, summing up, as it did, the labors of Ptolemy and his Alexandrine predecessors in systematic and comprehensive form, and in the Middle Ages it possessed supreme authority as the source of all higher astronomical knowledge. In 827 it was translated into Arabic, and among the Saracens it passed as a divine and preëminent book, about which there grew up a large body of explanatory literature.[136] Indeed the name by which it was generally known, *Almagest*, has been explained as a superlative title, al μεγίστη, though recent writers are inclined to make it a corruption of μεγάλη σύνταξις.[137] In the Latin Europe of the twelfth century Ptolemy's results became known at first indirectly, in the compends of al-Fargani and al-Battani; and even after his great work was translated, an abridgment, the so-called *Almagestum parvum* of ca. 1175–1250, replaced it for many readers.[138]

The first Latin version of the *Almagest* itself has commonly been placed in 1175, the date attached to the translation from the Arabic made in Toledo by Gerard of Cremona.[139] It is now known, however, that a rendering was made from the Greek in

---

[136] On the place of the *Almagest* in the history of astronomy and mathematics, see R. Wolf, *Geschichte der Astronomie* (Munich, 1877), pp. 60–63; Cantor, i. 414–422; P. Tannery, *Recherches sur l'histoire de l'astronomie ancienne* (Paris, 1893); Steinschneider, "Die arabischen Bearbeiter des Almagest," in *B. M.*, 1892, pp. 53–62; and in *Zeitschrift der deutschen morgenländischen Gesellschaft*, l. 199–207 (1896); Manitius, introduction to his German translation (Leipzig, 1912); Duhem, i. 466 ff.

[137] Brockelmann, *Geschichte der arabischen Litteratur* (Weimar, 1898), i. 203.

[138] On the date of the *Almagestum parvum*, see Nallino, *al-Battani*, p. xxviii; Birkenmajer, *Bibljoteka Ryszarda de Fournival* (Cracow, 1922), pp. 29–34. For citations from the *Almagest* in 1143 by Hermann of Carinthia, see Chapter III, n. 61. The extract in MS. Chartres 214 is in a later hand: Suter, *al-Khwarizmi*, p. 16.

[139] On Gerard see Chapter I, supra. The evidence for this date is found on the last folio of a thirteenth-century MS. of Gerard's translation in the Laurentian (MS. lxxxix. sup. 45; cf. Bandini, *Catalogus*, iii, col. 312): 'Finit liber Ptholomaei Pheludensis qui grece megaziti, arabice almagesti, latine vocatur vigil, cura magistri Thadei Ungari anno domini millesimo .c.lxxv°. Toleti consummatis [*sic*], anno autem Arabum quingentessimo .lxx°. [then a blank of about the space of six letters] mensis octavi .xi°. die translatus a magistro Girardo Cremonensi de arabico in latinum.' The two computations agree, and the date has been generally accepted (Wüstenfeld, p. 64; Rose, in *Hermes*, viii. 334; Cantor, i. 907; Steinschneider, *H. U.*, p. 522), but Steinschneider in his latest reference to it inserts an interrogation point (*E. U.*, no. 46 (36)).

Sicily about 1160; first discovered in 1909, this is described at some length in a later chapter.[140] Moreover, while the version of Gerard of Cremona was the one to pass into general circulation, other translations, or partial translations, of the *Almagest* were made, although in each case the date and author are unknown.[141] For purposes of comparison let us begin with Gerard's rendering. First comes certain prefatory matter peculiar to the Arabic text: the biography and maxims of Ptolemy ('Quidam princeps [142] . . .'), and the account of the translation into Arabic under al-Mamun. Then the first book begins: [143]

Capitulum primum. In quo huius scientię ad alias excellentiam et finem eius utilitatis dicam.

Capitulum secundum. De ordinibus modorum huius scientię.

Capitulum iii. Quomodo scitur quod motus celi sit spericus.

Capitulum iiii. De eo quod indicat quod etiam terra sit sperica.

Capitulum v. De eo quod indicat quod terra sit in medio cęli.

Capitulum vi. De eo quod indicat quod terra sit sicut punctum apud celum.

Capitulum vii. Quod terra localem motum non habeat.

Capitulum viii. Quod primi motus qui sunt in celo sunt duo primi motus.

Capitulum ix. De scientia partium cordarum circuli.

Capitulum x. De modo quo tabule arcuum circuli et cordarum eius fiunt.

Capitulum xi. De positione arcuum et cordarum eorum in tabulis.

Capitulum xii. De arte instrumenti quo scitur quantitas arcus qui est inter duos tropicos.

Capitulum xiii. De scientia quantitatis arcuum qui sunt inter orbem equationis diei et inter orbem medii signorum qui est declinatio.

Capitulum xiiii. De scientia quantitatis arcuum equationis diei qui elevatur in spera directa cum arcubus orbis signorum datis.

Ecce ubi initium primi capituli prime distinctionis dedit:

Bonum, Scire, fuit quod sapientibus non deviantibus visum est cum partem speculationis a parte operationis diviserunt, que sunt duę sapientię partes. Licet enim contingat ut operatione sit speculatio prius, inter eas tamen non parva existit differentia, non solum, et si quorundam morum honestatem possibile sit pluribus hominum inesse absque doctrina, tamen non tocius scientiam absque doctrina comprehendere est possibile, verum etiam quia plurimum utilitatis consistit aut in opere propter plurimam perseverantiam agendi in rebus aut in scientia propter augmentum in scientia.

---

[140] Infra, Chapter IX.

[141] Manitius, pp. xii ff., is unsatisfactory on the mediaeval versions.

[142] See Boncompagni, *Gherardo*, p. 400; cf. the description of MS. Vat. lat. 2057 in the printed catalogue.

[143] MS. lat. 14738, f. 1. For the version of the preface from the Greek, see below, p. 163.

Qua propter nobis visum est expedire nobis ut sciamus metiri operationem cum doctrina principiorum eius que reperiuntur in imaginatione et intellectu, ne quid desit ex inquisicione tocius pulchre rei decentis forme secundum mensurationis bonitatem neque in minimis rebus neque in vilibus, et ut expendamus plurimum nostri ocii et plurimum nostri studii in disciplina scientię magne et excelsę et precipue que nominatur scientia. O quam bonum fuit quod Aristotiles divisit theoricam, cum eam in tria prima genera distribuit, in naturale, doctrinale, theologicum! Generatio namque omnis generati ex materia est et forma et motu, neque est possibile ut in aliquo noto horum trium solum per se singillatim stans absque alio videatur, possibile tamen est ut unum absque alio intelligatur. Quod siquis scire querit que sit prima causa primi motus, affirmabitur ei cum illud secundum ordines suos fuerit declaratum quod est Deus invisibilis et immobilis. Species autem theorice qua inquiritur perscrutatio qua scitur quod est in suprema altitudine ordinum mundi nominatur theologica, et hoc quidem intelligitur separatum esse a substantiis sensibilibus. . . .

The most interesting body of evidence respecting other versions is found in a manuscript in the Landesbibliothek at Wolfenbüttel, MS. Gud. lat. 147.[144] This codex, of the thirteenth century, contains first, after the fly-leaf, without heading (f. 2) the preface to the Sicilian version of the *Almagest* from the Greek. On f. 3 we have the Ptolemaic maxims ('Conveniens est intelligenti . . .') which ordinarily accompany the biographical material ('Quidam princeps . . .') in the version of Gerard of Cremona, followed by Gerard's version of Ptolemy's preface, headed 'Alia translatio primi capituli.' On f. 3 v comes the biography as translated by Gerard and his rendering of the chapter headings of the first book. Then on f. 4 begins a quite different version of the preface from the Arabic as follows:

Bonum quidem fecerunt illi qui perscrutati sunt scientiam philosophie, Iekirie,[145] in hoc quod partiti sunt partem philosophie speculativam ab activa. Sed, quamvis activa antequam sit activa est speculativa, tamen quod inter eas de diversitate reperitur est magnum. Non propterea[146] quod quasdam bonas virtutes animales possibile est esse in multis hominum sine doctrina, sed ad scientiam omnium rerum speculativarum non est possibile aliquem pervenire absque doctrina,[147] sed tantum propterea quod perducens

[144] The description in the printed catalogues is too meagre to be of service: F. A. Ebert, *Bibliothecae Guelferbytanae codices Graeci et Latini classici* (Leipzig, 1827), no. 733; O. von Heinemann, *Die Hss. der herzöglichen Bibliothek zu Wolfenbüttel* (Wolfenbüttel, 1913), ix. 163. I have examined the codex by specimen photographs.

[145] 'Id est, O domine Frire' above the line.

[146] MS. *propterea" quod.*   [147] MS. *doctrina" tantum.*

ad finem quesitam in parte quidem activa est multitudo assiduationis super operationem et in parte quidem speculativa additio speculationis. Et propter illud vidimus quod oportet ut sit rectificatio operationis illud quod credimus per mentes nostras ut non recedamus nec in pauco ex rebus a consideratione perducente ad dispositionem pulchram ordinatam et ponamus plurimum nostre occupationis in inquisitione scientie rerum speculativarum propter multitudinem earum et superfluam bonitatem ipsarum et proprie in rebus quibus proprium est ut nominentur doctrinales. O quam bonum quod divisit Aristotiles partem speculativam cum divisit eam in tria prima genera, naturale, disciplinale, et divinum! Quoniam essentia omnium rerum ex materia est et forma et motu, et non est possibile ut sit una rerum trium secundum singularitatem inventa actu, et est iam possibile ut intelligatur unaqueque earum absque alia. Causa igitur prima motui totali primo quando cogitamus motum simplicem videmus quod est Deus qui non videtur neque movetur, et nominabimus hanc speciem inquisitionem de Deo nostro. Et hanc quidem intelligentiam intelligimus in altiore altitudine rerum tantum seiunctam penitus a substantiis sensatis . . . quod primi non comprehenderunt nec consecuti sunt ex eius comprehensione quod oportet.

Then with the second chapter this version is abandoned for Gerard's, which seems to be used thereafter. The beginning of book iv, which I have compared, has the ordinary text of Gerard, and the manuscript closes on f. 161 with Gerard's version (13, 11):[148]

Quia igitur iam consummavimus has intentiones et perfecimus omnia ad quorum scientiam necessarium est invenire in hoc libro, secundum quantitatem status nostre scientie et summe nostri consilii preter extranea eorum, secundum quantitatem qua adiuvit nos tempus quod pervenit ad nos ad inveniendum id cuius est inventio necessaria ex illo et premittendum id cuius est necessaria premissio et verificatio eius ex eo, et secundum quod sit quod scripsimus inde conferens in hac scientia preter quod inquiramus per ipsum prolongationem et abbreviationem, tunc iam sequitur et honestum est ut ponamus hoc finem libri.

Finit liber Ptolomei Pheludensis qui grece megasin. arabice ALMAGESTI latine maior perfectus appellatur.

This, however, is not the whole story, for there are frequent marginal notes containing extracts from a version, or paraphrase, out of the Arabic which is not that of Gerard. Thus at the beginning of book iv the text has:[149]

Iam narravimus et demonstravimus in dictione que est ante hanc totum quod contingit in motu solis, et postquam illud incipere volumus secundum

[148] Cf. Boncompagni, *Gherardo*, p. 401.

[149] F. 38 v (= MS. lat. 14738, f. 55). For the Sicilian version, see Chapter IX, n. 9; for the Greek rendering of the Dresden MS., *Hermes*, xlvi. 216.

quod sequitur loqui de motu lune videmus quod primum per quod oportet
nos illud inquirere est ex considerationibus. . . .

This is Gerard's version, but the margin reads:

Quia in tractatu qui est ante hunc pervenimus super omnia que inveniuntur comitari in motu solis, assumpsimus in hoc tractatu in eis que sequuntur
illud et coniunguntur illi ex sermone in luna primum ergo quod videmus
oportere ut ab eo inciperemus loqui in hoc. . . .

We thus see that the scribe of the Wolfenbüttel manuscript had
before him in the thirteenth century not only Gerard's version
and the preface at least of the Sicilian version, but a third version
of at least the prefatory chapter and, apparently, of the passages
which he inserts in the margin. This third form of the preface is
also found in a manuscript of Gerard's version at Madrid;[150] and
as this codex is of the early thirteenth century and comes from
the cathedral library of Toledo, we can infer that the third version
is anterior to this date and probably of Spanish origin. The preface also occurs at the close of a copy of Gerard's version in the
Vatican, MS. Vat. lat. 2057, also of the thirteenth century.[151]
There is as yet no clue to the translator. The statement that this
version was made from the Arabic under Frederick II [152] seems to
have arisen from a combination of the misunderstood Sicilian
preface with certain notes of the year 1230 in another hand on
the fly-leaf of the Wolfenbüttel manuscript. No translation
under Frederick II is known save that into Hebrew by Jacob
Anatoli.[153]

If we thus have a second version from the Arabic, there is also

---

[150] Biblioteca Nacional, MS. 10113 (Hh. 89), where we have, after the 'Quidam
princeps,' the Wolfenbüttel preface (f. 1 v: 'Bonum quidem') followed by Gerard's
preface and version complete and ascribed to him. Cf. Octavio de Toledo, *Catálogo
de la librería del cabildo Toledano* (Madrid, 1903), no. 335 (469). The 'Bonum
quidem' preface also appears on the last folio of the *Notule almagesti* in the library
of the Academia de la Historia, Est. 11 gr. 1ª, MS. 22 (saec. xiii).

[151] See Nogara's printed catalogue.

[152] Manitius, i, pp. xii f., 459, citing a note of von Zach in 1813 which I have not
seen. Birkenmajer, *Vermischte Untersuchungen* (*Beiträge*, xx, no. 5), p. 21, saw that
this MS. contained the Sicilian version but did not know that this was confined to
the preface.

[153] Steinschneider, *H. U.*, p. 523. The statement that a version was made under
Frederick II is found as early as 1741 (Boncompagni, *Gherardo*, p. 402) and became
widely current (Steinschneider, *E. U.*, no. 177).

evidence of a second version from the Greek,[154] for a manuscript of ca. 1300 in Dresden, formerly the property of the Dominicans of Cologne, contains a quite different rendering of the first four books of the *Almagest*. That it was based ultimately upon the Greek appears from the general character of the text, as well as from the carrying over of specific words and the appearance of the Greek form Hipparchus instead of the corruption Abrachis, as in the versions from the Arabic. It is, however, not a close rendering like the Sicilian, and contains none of the tables so carefully preserved by the other translators, while the numbers are often inaccurate.[155] No other copy has been found, nor did this form of the text deserve a wide circulation. The title 'Phylophonia Wuttoniensis (or Wintoniensis) Ebdelmessie,' which appears at the close of each book, is obviously a corruption, but I cannot guess of what, nor is there any evidence of date other than the age of the manuscript, which begins:

De prologo.
De ordine eorum que sunt in hoc libro.
Quia celum est sperale et suus motus speralis motus.
Quia figura est terre etiam speralis.
Quia terra est in medio celi.
Quia terra ad celum est quasi punctus.
Quia terra non habet motum.
Quia primi motus qui sunt celi sunt duo.
De mensuris cordarum et arcuum qui cadunt in circulo.
De faciendis tabulis arcuum circulorum et suarum cordarum.
De posicione tabularum arcuum et suarum cordarum.
De scienda inclinacione.
De proposicione racionum speralis sciencie.

Preclare fecerunt qui corrigentes scienciam philosophie, O Syre, diviserunt theoricam partem philosophie a practica. Nam si pars practice antequam sit praxis est theorica, sed diversitas inter eas est magna, non propter hoc quod aretius [156] morum anime possit esse in pluribus sine doctrina, omnis autem rei theorice non potest aliquis habere sine doctrina scienciam, sed propter hoc qui ducit ad utilitatem que est acquisicio in parte praxis usus facti et in parte theorice crementum sciencie. Ideo igitur perscrutantes speculati

---

[154] MS. Db. 87, ff. 1–71. I know this from specimen photographs secured in 1910 and from Heiberg's description, *Hermes*, xlvi. 215 f. It was first indicated by Björnbo, in *Archiv für die Geschichte der Naturwissenschaften*, i. 392 (1909).

[155] There is also a confusing form of numerals: b = β = 2, etc. Cf. supra, n. 128.

[156] Gk. ἀρετῶν.

sumus qui debet esse emendacionem nostram in praxis pro sua speculacione
ad nostram ymaginacionem, propter hoc enim non mutabimus re parva spe-
culacionis que nos ducit ad ordinacionem pulcri operis, igitur ponemus maius
de nostro labore in inquirendis theoricis scienciis, nam multe sunt et pul-
criores sunt et maxime in rebus que nominantur mathematice. O quam
pulcra est particio Aristotilis de theorica parte in tria prima genera, phisialoi-
cam, mathematicam, theoloycam! Nam esse omnium rerum ex materia est
et forma et motu, nec potest inveniri unum illorum trium tantum in actu,
potest tamen quique eorum subintelligi unum sine alio. Prima ergo causa
primi motus universi cum ymaginati fuerimus motum per se intelligemus esse
Deum qui nec movetur nec videtur. Nominavimus autem locucionem de
eo theologicam et illud facere intelligemus in alta altitudine mundi tantum
et divisum ab omni sensibili substantia. . . .

The copy closes with the fourth book as follows:

Igitur est manifesta ex hoc quod diximus causa illius discordie est et con-
firmata fides nostra ex hoc quod ostendimus de conputacione discordie que
erit in tempore pansilini et synodi et invenimus illas eclipses quas commemo-
ravimus concordes fundamenti.

Phylophonia Wuttoniensis Ebdelmessie.

Explicit quartus liber [157] sermo libri mathematice Ptholomei qui prenomi-
natur megalixintaxis sive astronomie translacione dictaminis.

What the *Almagest* of Ptolemy was for ancient astronomy, his
Tetrabiblos or *Quadripartitum* was for astrology.[158]   Its authenticity, which was long doubted because of modern unwillingness to
believe that Ptolemy was an astrologer, has been established by
Franz Boll,[159] from whom a critical edition is expected.   Early
translated into Arabic, it was widely popular among the Saracens
and was soon the subject of commentary by Ali ibn Ridhwan and
others.   Naturally it was one of the earliest works to be turned
into Latin, the version of Plato of Tivoli being dated 1138.   Another version was made for Alfonso X by Egidius de Thebaldis of
Parma.[160]   Midway between these two in point of time is a third
version from the Arabic dated 29 August 1206 and preserved in
the manuscript of Wolfenbüttel which we have just been examin-

[157] *Liber* cancelled.

[158] For the translation of Ptolemy's *Planisphere* in 1143, see Chapter III, no. 5;
for the *Optics*, see infra, Chapter IX, n. 70; for the pseudo-Ptolemaic *Centiloquium*,
see Chapter IV, n. 10; and for the *Canons* ascribed to Ptolemy, cf. the present
chapter, n. 53.

[159] Boll, *Studien über Claudius Ptolemäus* (Leipzig, 1894), pp. 111–188.

[160] Steinschneider, *H. U.*, pp. 525 f.; *E. U.*, nos. 9, 98 *h*.

ing, as well as at Parma.[161]   No author is indicated in the text, which begins and ends as follows:

Prolixitatis exosa latinitas artium principia prescriptione quadam insignire sollicita est ut sequens negotium gratiosius elucescat. In huius igitur initio iuxta expositionem .7. sunt que consideranda premittuntur: auctoris intentio, operis utilitas, titulus libri, nomen auctoris, ordo librorum in disciplina, cui parti scientie tractatus innitatur, et operis partitio. Intentio quidem est suscepti operis dilucida consummatio, et utilitas est diligentius intuentis compubescens instructio. . . .

Ex stellarum habitudine prescientie perfectio consecuta, Iezuri, tamquam partes maiores et sublimiores in duo consistit distributa: pars quidem prima in ordine et fortitudine est scientia figurarum solis et lune planetarumque .5. consecutiva, que figure mediantibus motibus stellis eisdem accidunt collatione eorum adinvicem et ad terram observata; pars vero secunda alterationes et operationes investigat que a figuris revolutioni stellarum propriarum et naturalium in rebus quas continent accidunt et perficiuntur. . . .

Quoniam ergo iuxta propositum nostrum in astrorum iudicia viam universalem tradidimus, congruum est ut huic tractatui nostro finem imponamus. Perfecta est huius libri translatio .29. die augusti anno Domini .1206. et 23 die almuharam [162] anno Arabum 603. Et Deus melius novit. Explicit Quadripertitum Ptholomei in iudicia astrorum secundum accidentia editum.

Two other versions of the *Quadripartitum* were discovered by Björnbo at Oxford [163] but have not been specially studied: one, ascribed to the Englishman Simon de Bredon ca. 1305 and preserved in marginal extracts, the other made directly from the Greek. The latter, which seems to be cited by Henri Bate in 1281, begins as follows,[164] after the chapter headings of the first book:

Hiis qui instituunt per astronomiam pronosticum finem, O Sire, cum duo insint maxima et principalissima, unum quidem quod et primum est ordine

---

[161] MS. Gud. lat. 147, ff. 162–194; Parma, Biblioteca Palatina, MS. 719, ff. 311–343 v (saec. xiii).   Also at Florence (S. Marco 200 = J. II. 10): *B. M.*, xii. 197.

[162] MS. Parma 719, f. 343 v has *almihatan*.

[163] *Archiv für die Geschichte der Naturwissenschaften*, i. 391 f.   Another *incipit* appears in B.N., MS. lat. 7432, ff. 5–125 v: 'Res, O Mizor, quibus pronosticationes accepte de astronomia maiores et nobiliores due sunt . . . finem in hoc loco huic libro conveniens existimamus' (with commentary of Conrad Heingarter dedicated to John, duke of Bourbon and Auvergne). What may be still another version of the *Quadripartitum* is found at Madrid, MS. 10053, ff. 89–110: 'Iuxta providam philosophorum assertionem . . .'   See also MS. Chigi, F. iv. 48, f. 23.

[164] MS. Digby 179, ff. 171–208 v.   On Henri Bate see now Birkenmajer, "Henri Bate de Malines," in *La Pologne au Congrès international de Bruxelles*, and separately (Cracow, 1923).

[et] virtute per quod motuum solis et lune et astrorum factas semper adinvicem figuraciones comprehendimus, secundum autem per quod per naturalem proprietatem figuracionum ipsorum inclitas permutaciones contentorum consideramus. Primum quidem propriam et propter se eligibilem habens theoriam, etiam si finis qui est ex connexione si non concludatur, in propria compilacione ut maxime inerat demonstrative tibi traditum est. . . .

The treatise ends:

Consummata iam geneatici sermonis opinione [165] summatim, unde utique habebit huic tractatui convenientem inponere [166] finem. Explicit liber Ptho[lemei]. Que sequuntur in greco exemplari subiniuncta reperi quo mense morietur quis in omni nativitate . . .

There is no indication of date or translator, but the extreme literalness [167] is characteristic of the versions made in Italy in the twelfth and thirteenth centuries.

We have now reached and passed beyond the close of the twelfth century in our examination of the anonymous writers and translators who exemplify its tendencies in the field of astronomy. In the next chapter we shall traverse the same period in a series of datable works by known authors in a single country, England.

[165] MS. *opē*.   [166] MS. *tempore* cancelled before *finem*.
[167] μέν = quidem.   δέ = autem.   ἄν = utique.

CHAPTER VI

THE INTRODUCTION OF ARABIC SCIENCE INTO ENGLAND [1]

IN the diffusion of the science of the Saracens throughout western Europe in the twelfth century England occupies a position of considerable importance. An English scholar, Adelard of Bath, seems to have been the chief pioneer in this movement of study and translation,[2] while the existence of a certain number of dated treatises of his contemporaries and successors makes it possible to follow the spread of the new learning in England with greater definiteness than has so far been attempted elsewhere. At the beginning of the century we have a group of abacists and computists who have in nowise been affected by Arabic influence: the abacists, such as Thurkil and Adelard in his *Regule abaci*, follow the schools of Lorraine and Laon,[3] while in astronomy the older Latin tradition is found in full vigor as late as 1119, when Philip de Thaon wrote his *Cumpoz* with the help of Bede, Helperic, Gerland, a lost treatise of Thurkil on this subject, and the work of the so-called Nimrod, which in its present form probably dates from the Carolingian period.[4] In the following year, however, the new movement begins to make itself felt in Walcher, prior of Malvern, who had possessed one element of the Arabic astronomy, the astrolabe, as early as 1092, and who now begins to utilize the teaching of a converted Spanish Jew, Petrus Alphonsi.

---

[1] Revised from *E. H. R.*, xxx. 56–69 (1915).   [2] Supra, Chapter II.

[3] Poole, *The Exchequer in the Twelfth Century*, pp. 47 ff.; infra, Chapter XV.

[4] Mall, *Li Cumpoz Philipe de Thaün mit einer Einleitung* (Strassburg, 1873); T. Wright, *Popular Treatises on Science* (London, 1841), pp. 20–73; P. Meyer, "Fragment du Comput de Philippe de Thaon," in *Romania*, xl. 70–76. Cf. Langlois, *La connaissance de la nature et du monde au moyen âge* (Paris, 1911), pp. 2, 3, 11; Hamilton, in *Romanic Review*, iii. 314, who suggests the identity of Turkils and Turchillus compotista, but overlooks the fact that the treatise in three books cited by Philip cannot be the *Reguncule super abacum*, which contains nothing on the subjects treated in the *Cumpoz*. I have discussed Philip's sources in Chapter XVI, and the computists in Chapter V.

Of Lotharingian origin, Walcher had come to England by 1091, and at his death, in 1135, had acquired a reputation as mathematician and astronomer [5] which is confirmed by two treatises preserved in the Bodleian MS. Auct. F. 1. 9 (ff. 86–99), a manuscript of the twelfth century in which they precede the Khorasmian tables of Adelard of Bath.[6] The first of these, anonymous in the manuscript, was written between 1108 and 1112,[7] and consists of a set of lunar tables, with explanations, which comprise a cycle of seventy-six years ending in 1112 and are calculated from an eclipse observed in 1092. In 1091, while travelling in Italy, the author saw the eclipse of 30 October but had no means of determining the exact time, save to note that it differed considerably from the hour reported on his return to England by a brother monk, whence he comments on the considerable difference in time between the two countries. In the following year, however, he had the good fortune to observe the eclipse of 18 October and fix it accurately by means of the astrolabe, which he mentions with the Arabic names of three of its points as something well known to his readers.[8] His account reads (f. 90):

De experientia scriptoris

Quod vero ipse expertus sum quodque de his et de ceteris supradictis inquirere et colligere potui non silere curavi, ut his quibus defectus solis et lunę non est visus aut querendi modo supradicto facultas vel otium vel diligentia non famulantur certior faciliorque ad naturalem cuiusque lunationis originem pateat aditus. Anno ab incarnatione domini iuxta Dionisium ᴍ°xc°ɪ° contigit me esse in Italia in parte orientali ab urbe Romona (sic) itinere diei et dimidii ubi defectum lune .x°iiii°. vidi .iii. kal. novembris ad occidentalem plagam ante aurorę exortum, sed nec horologium tunc habui quo plenilunii horam deprehenderem nec ipsa luna conspicue densis obstantibus nebulis

---

[5] See his epitaph in *Monasticon*, iii. 442; and cf. William of Malmesbury, *Gesta regum*, ii. 346. The visit to Italy is known only from the text printed below.

[6] Tanner (*Bibliotheca*, p. 745) gives Walcher a bare mention on the basis of this manuscript (= Bernard, no. 4137). Walcher's authorship of the first treatise is not only an inference from its contents and its occurrence with the second, but is confirmed by cross references, e. g., f. 97 v to f. 94 v.

[7] It refers (f. 95 v) to the eclipses of 11 January and 31 December 1107, and is obviously anterior to the close of the lunar cycle in 1112.

[8] F. 90, col. 2: 'Quia de astrolabio scientibus loquor, primam partem Tauri eidem altitudini superposui in parte Almagrip . . . notato loco quem designabat Almeri, reduxi gradum solis usque ad ultimum Almucantaraz.'

apparebat. Memini me vidisse eam corniculatam in modum .V. sed quando deficere incepit vel quando rursus plenitudinem sui luminis recuperavit vehementius densatis nebulis videre non potui. Reversus itaque in Angliam cum quesissem a quibusdam siquis eo tempore vidisset eclypsin, narravit mihi frater quidam ea die tota quę noctem illam precesserat diurno tractandę causę negotio se occupatum plurima iam noctis parte transacta domum venisse, postea cenasse, post cenam parumper sedisse, et quendam de familia egressum attonitum regredi dicentem horribile prodigium in luna monstrari, quod ipse dum exisset vidit et agnovit diu ante mediam noctem, multum enim adhuc a plaga meridiana distabat quam semper luna plena nocte tenet media. Iamque inter Italiam et hanc nostram Anglię insulam non modicam horarum animadvertebam distantiam, cum illic paulo ante auroram defecerit iam vergens ad occasum, hic vero diu ante mediam noctem adhuc ab ortu ascendens. Sed cum nil certum haberem neque de illa neque de hac terra unde quod in voluntate habebam cyclum texere inciperem, grave ferebam et in instantia querendi permanebam. Et ecce anno sequenti eiusdem mensis lunatio tanquam meis occurrens studiis ut me reficeret iterum defecit et .xv. kal. novembris obscurata me illuminavit, quia ignorantię meę tenebras ipsa lumine privata depulit. Mox enim ego apprehenso astrolapsu horam qua totam nigredo caliginosa lunam absorbuerat diligenter inspexi, et .xiᵃ. noctis agebatur hora .iii. puncto peracto. . . . Modum autem huius inquisitionis si alios non piget legere, me non piget scribere, et credo quia omnino non deerunt quibus placeat. . . .

This clear bit of evidence is of some importance as confirming specifically, what we know in general from the treatises on the astrolabe commonly ascribed to Gerbert and Hermannus Contractus and containing numerous Arabic words,[9] that an acquaintance with this instrument had in some unknown way passed into Latin Europe in the course of the eleventh century, thus preceding considerably the arrival of the Arabian astronomy as a whole. The tables of Walcher's first treatise are worked out by the clumsy methods of Roman fractions, but in the second, written in 1120, he uses the degrees, minutes, and seconds, and the more exact observations which he has learned, evidently in England, from Petrus Anfusi (f. 96):[10]

[9] Bubnov, *Gerberti opera mathematica*, pp. 109-147; Migne, *Patrologia Latina*, cxliii. 379-412; supra, Chapter I, nn. 20, 21.
[10] Professor Thorndike has called my attention to a copy at Erfurt, MS. Q. 351, ff. 17 v-23: 'Alfoncius de dracone.'

Sententia Petri Ebrei cognomento Anphus de dracone quam dominus
Walcerus prior Malvernensis ęcclesię in latinam transtulit linguam

Inter .vii^em. planetas per zodiacum circumeuntes discurrit etiam draco
sed contrario motu . . . Ecce vides si de eclypsi aliquid volumus prescire
quam sit necessarium scire in quibus signis vel signorum gradibus inveniri
vel sibi opponi debeant sol et luna caput et cauda draconis omni tempore.
Ad quod investigandum prius videnda est via per quam discurrunt, quę est
in zodiaco circulo sed non iuxta usum nostrum priorem. Nos enim, quia
traditum a prioribus tenebamus auctoribus unum esse gradum spatium illud
quod sol in zodiaco in una die et nocte peragit, ipsum zodiacum in computa-
tionibus nostris per .ccc^os.lxv^e. gradus et quadrantem dividere soliti sumus
propter totidem anni dies et .vi^ex. horas, ut unusquisque dies suum habeat
gradum et .vi^ex. horę, quę sunt diei unius quadrans, unius gradus quadrantem. In tali divisione unumquodque signum plusquam .xxx^ta. gradus habet
quia solem .xxx^ta. diebus et .x^cem. horis cum dimidia retinet. In presenti
autem negotio magister noster hac divisione non utebatur sed illa quę unumquodque signum in .xxx^ta. gradus equaliter dividit et totum zodiacum .ccc^tis.
.lx^ta. gradibus claudit secundum quam sol in die unum gradum non perficit.
Unde cum de solis inter ipsos gradus progressione queritur cum difficultate
.ccc^tos.lx^ta. gradus per .ccc^tos.lx^ta v^e. dies et quadrantem quibus sol totum
perficit zodiacum dividuntur, quia minorem numerum per maiorem dividi
natura non patitur. Oportet itaque hanc divisionem per minutias fieri, sed
magister noster minutiarum quibus utuntur Latini usum non habens tali
utebatur divisione: Zodiacum totum sicut et nos in .xii^eim. signa unumquodque signum in .xxx^ta. gradus unumquenque gradum in .lx^ta. punctos unumquenque punctum in .lx^ta. minutias unamquamque minutiam in .lx^ta. minutias minutiarum dividebat, et per harum particularum collectiones ubi sol
vel luna vel caput seu cauda draconis inveniri possent quacunque die vellet
vel hora diei vel horę particula investigabat. Et ad hęc investiganda tale
nobis posuit fundamentum:

Anno ab incarnatione domini .M^illesimo. C^o.XX°. kal. aprilis feria V^ta.
hora diei VI^ta.plena fecerat sol in Ariete VII^em.gradus et XVIIII^em.punctos
et LVII^em. minutias; luna vero in eodem signo XX^tiIII^es. gradus et XXX.
punctos et LI. minutias; caput draconis erat in primo gradu Scorpionis in
primo puncto in prima minutia. Nimirum miraris sicut et nos mirati sumus
quod solem kal. aprilis in .VII°. gradu Arietis esse dixerit, cum omnium
Latinorum, non dico modo aliorum, auctoritas habeat ipsum solem ipsa die
XV^mum. gradum eiusdem signi tenere. Unde et interrogatus a nobis respondit dicens, Tunc quod dixi de die et sole et gradu signi verum esse scietis cum
per hoc eclypsim futuram inveneritis. . . . Nos autem tantummodo videamus ubi ponat initia vel fines signorum et in hac supputatione in qua ipsum
magistrum habemus sic eius institutionem teneamus ut nostram in aliis non
relinquamus.

Questioned respecting the diurnal motion of the sun and the
moon, the master says (f. 96 v), after giving the median motion
of the moon:

Habet et ipsa motum maiorem et minorem quorum diversitatem ad purum in promptu se non habere dicebat et codices suos in quibus de his et de aliis pluribus omnia certa habebat se trans mare tunc temporis reliquisse. . . . Ecce totum quod dixit nobis de investigatione futurę eclypsis. Unam siquidem id est solis in convenientia ipsius solis et lunę et capitis sive caudę draconis fieri dixit, alteram id est lunę in oppositione ipsorum ut dictum est. Indicavit etiam loca diem et horam unde initium investigandi debeamus assumere et cursum siderum per quem ad finem inquisitionis debeamus pervenire. Quod amplius est prudentię calculatoris relinquitur.

Peter explains the discrepancies in tables by the retardation of the sun in the zodiac. Walcher then works out the motion of sun, moon, and nodes for groups of days and months, in the course of which he says (f. 97 v):

De luna vero, quia accensionem eius et plenilunium sequitur solis eclypsis et lunę, nil melius ad presens dicere possumus quam supra dictum est ubi de naturali accensione eius tractavimus, quanvis ad certam illius horam propter diversos eius motus pervenire non valeamus. Quam diversitatem et nos in ipso tractatu deprehendimus et testimonio Petri Anfusi confirmatum est dicentis eam habere .iii^{es}. motus ut supradiximus.

The statement that Walcher 'translated' Petrus must plainly be taken in the general sense of a paraphrase rather than as meaning a version which would require knowledge of Arabic on Walcher's part.

Further evidence of the astronomical labours of Petrus Anfusi is contained in a treatise preserved in MS. 283 of Corpus Christi College, Oxford.[11] Here we have first a set of chronological tables of the sort usual in treatises based on the Arabic, including a concordance of eras for the year 1115,[12] then a series of tables for the various planets, and finally an explanation of the use of the chronological tables covering four pages and beginning as follows:[13]

Dixit Petrus Anfulsus servus Ihesu Christi translatorque huius libri: Gratias Deo omnipotenti et domino nostro qui creavit mundum sua sapiencia et disposuit suo intellectu omnia. . . . Hec autem trina cognitio

---

[11] Ff. 113–144, saec. xii. exeuntis. Cf. Coxe, *Catalogus*, p. 122; supra, Chapter II, n. 17.

[12] F. 113: 'Tabula ad cognoscendum quantum temporis secundum omnes subscriptos terminos restat usque ad principium huius operis.' This table is also found for the same year in the *Liber ysagogarum Alchoarismi ad totum quadrivium* (Ambrosian MS. A. 3 sup., f. 18; B.N., MS. lat. 16208, f. 70), so that there may be some relation between the two treatises.

[13] F. 142 v. Cf. Steinschneider, *H. U.*, p. 985.

vocatur stellarum scientia que in tres partes dividitur in cogitacione mirabiles et in rerum significatione notabiles et in experimento approbabiles. Quarum prima est scientia qualitatis et quantitatis circulorum firmamenti cum his que in eo sunt, ad quam vivacitas humani ingenii pervenit geometrali figura numero et mensura; secunda est scientia motuum firmamenti circulorum et stellarum que per numerum sciri potest; tercia vero est scientia nature circulorum et stellarum et significationes eorum in rebus terrenis que contingunt eorum ex nature virtute et suorum motuum diversitate que experimento cognoscuntur. Fuit etiam ex animi mei sententia ut inde librum ederem et ut per ipsius noticiam eiusdem utilitas cognosceretur, scilicet numerus et motus circulorum et stellarum pertinentibusque cum ipsis annis videlicet et mensibus diebus horis ipsarumque punctis, itaque primum necessarium est quota feria annus vel mensis incipiat nosse. Hoc autem opus magno labore desudatum et summo studio ab Arabicis Persicis Egipciacis translatum Latinis benigne impertiri volui, et quia volo ut hic liber predictis omnibus clareat, ideo sub eorumdem numero intitulavi et prout in ordine in eorum lingua repperi sic seriatim in latinam linguam digessi.

Evidently we have not this pretentious work in its original and full form, for the chronological tables seem out of place with reference to the explanation of them, while the planetary tables are notable, so far as they extend, for their close agreement with the Khorasmian tables as translated by Adelard of Bath, in the earlier form of his text preserved in the Bodleian.[14] There can be no question of two independent versions, for in the explanatory portions the verbal coincidence is exact. As there is no specific reference to the planetary tables in Peter's preface, their insertion here may be due to a copyist, but their occurrence raises interesting questions respecting the relations of the two contemporaries and their work. Conceivably Adelard may have used Peter as an interpreter, after the fashion of the later translators from the Arabic; his own authorship of the *Liber ezic* is positively asserted by Adelard, but we find others engaged on the Khorasmian tables in some form.[15]

The only known Petrus Anfusi, or Alphonsi, is the author of the *Disciplina clericalis* and the *Dialogi cum Iudeo*, who was baptized at Huesca in 1106 with the name of his godfather, Alfonso I of Aragon. Nothing is known of his biography save that he was then in his forty-fourth year, the common assertion that he died

---

[14] Tables, ff. 113–140 v = (in most respects) Suter, pp. 111–167; text, ff. 141 v–142 v = Suter, pp. 7–14. See Chapter II, no. 3.

[15] Below, n. 31; supra, Chapters II, no. 3; and III, no. 2.

in 1110 being based apparently upon a misunderstanding of Oudin.[16] There is no reason why he may not have journeyed to England, leaving his books *trans mare*, and as a matter of fact we find in a Cambridge manuscript of the *Disciplina clericalis* this heading, in language exactly parallel to the passage in the astronomical treatise: *Dixit Petrus Amphulsus servus Christi Ihesu Henrici primi regis Anglorum medicus compositor huius libri*.[17] The statement that Peter was Henry I's physician I have not found corroborated, but it fits in chronologically with the dates in the astronomical writings, and while there is no necessary connection between their author and the author of the *Disciplina clericalis*, it is more natural to assume identity than to suppose that there were at the same time two converted Spanish Jews of this name, both occupied with translation from the Arabic. In any case it is to a Petrus Alphonsi that we must ascribe a certain share in the introduction of the Arabic astronomy into England before 1120.

Whatever further investigation may discover in the way of predecessors or collaborators, the work of Adelard of Bath remains comprehensive and fundamental, alike with reference to mathematics, astronomy, astrology, philosophy, and his advocacy of the experimental method, but it yields few specific dates. We know that his version of the Khorasmian tables dates from 1126 and that he was in England in 1130 and probably well on into the reign of Stephen; but his earlier life was spent chiefly on the Continent and in the East, and we cannot say when the results of his labours first reached England or affected English learning. John of Worcester knew the translation of the tables probably for the first time in 1138.[18]

[16] Antonio, *Bibliotheca Hispana vitus*, ii. 10 f.; Oudin, *De scriptoribus ecclesiae*, ii. 992; Migne, clvii. 527–706. Oudin says merely, 'Claruit circa annum 1110.'

[17] University of Cambridge, MS. Ii. vi. 11, f. 95. Cf. *Catalogue of MSS.*, iii. 508; Bernard, *Catalogi*, ii. 390, no. 65 (Moore MSS.); Tanner, *Bibliotheca*, p. 40. The latest editors of the *Disciplina clericalis*, Hilka and Söderhjelm, in *Acta Societatis Fennicae* (1911), xxxviii, no. 4, pp. xi, xix, who are unacquainted with the astronomical evidence, consider the statement due to a confusion with some one else. For another astronomical treatise of Petrus, see Thorndike, ii. 70 f.

[18] Ed. Weaver, p. 53.

Adelard's younger contemporary, variously known as Robert of Ketene, Robertus Retinensis, and Robert of Chester,[19] is likewise of interest for the history of Arabic learning in England. In his case the connection with Spain clearly appears. An Englishman by birth, Robert's life is unknown to us until 1141, when, already familiar with Arabic and engaged in the pursuit of astrology, he and his associate, Hermann of Carinthia, were discovered in the region of the Ebro by Peter the Venerable, abbot of Cluny, who engaged them upon a translation of the Koran and upon various controversial pamphlets directed against Mohammedanism. For these facts we have both Peter's correspondence and Robert's prefaces. The version of the Koran was completed in 1143, when Peter tells us that Robert had become archdeacon of Pamplona,[20] and when the dedication of Hermann's *De essentiis* celebrates the reunion of the two friends;[21] but the assumption of the older bibliographers that Robert spent the rest of his life in Navarre disappears if we admit the probability of his identity with Robert of Chester, who is found at Segovia in 1145 and in London in 1147 and 1150. The preface to the Koran tells us,[22]

---

[19] On Robert, see Steinschneider, *E. U.*, nos. 101, 102, whose results have been employed, with some use of English manuscripts, by Archer, in the *Dictionary of National Biography*, xlviii. 362–364; and Karpinski's edition of the *Algebra*. The form 'Retinensis,' which has led some writers to surmise a connection with Reading, is not sufficiently supported by the manuscripts, 'Ketenensis' being found in most of the copies of the translation of the Koran and in the preface of Hermann of Carinthia to his translation of the *Planisphere* (Heiberg, *Ptolemaei Opera astronomica minora*, p. clxxxvi), while the Cotton MS. of the *Iudicia* has 'de Ketene.' The place is probably to be identified with Ketton (in Rutland), which appears as Ketene in charters of the twelfth century: Round, *Calendar of Documents in France*, nos. 530, 532; *Index of Charters and Rolls in the British Museum*, i, s. v. The later works (nos. 2–6) have regularly 'Robertus Cestrensis,' who has sometimes been treated as a different person. The coincidence, however, of time, subjects, English birth, and residence in Spain, tells strongly against the assumption of two distinct Roberts, although the connection with Chester still remains to be explained, unless there is a scribe's confusion of 'Kestrensis' and 'Ketenensis' (Langlois, in *Journal des savants*, 1919, p. 70).

[20] Migne, clxxxix. 650; supra, Chapter III, no. 4.

[21] Dated at Béziers 1143 and subsequent to 1 June, the date of the *Planisphere*, which refers to it as unfinished. See Chapter III, nos. 3, 6, where it appears that the dedication of Albumasar to Robert may be of 1140.

[22] Migne, clxxix. 659.

INTRODUCTION OF ARABIC SCIENCE INTO ENGLAND 121

what we also learn from his other works and from the prefaces of Hermann of Carinthia,[23] that Robert's real interest lay in the study of geometry and astronomy, which he had interrupted for this undertaking, and that his chief ambition was to produce a comprehensive treatise on astronomy. In the field of mathematics and natural science he has left the following works:

1. A translation of the *Iudicia* of al-Kindi. See Steinschneider, *E. U.*, no. 101; and for other manuscripts, Nagy, in *Rendiconti dei Lincei*, 5th series, iv. 160 f.; and Thorndike, i. 648. This has been attributed to another Robert, because of the date 1272 which has slipped into certain manuscripts, probably from the date of a copy, but the authorship of Robert is formally asserted in the Cotton MS. App. VI, and is clear from the preface which is there addressed to Hermann:[24]

Incipiunt iudicia Alkindi astrologi Rodberti de Ketene translatio [25]

Quamquam post Euclidem Theodosii cosmometrie libroque proportionum [26] libencius insudarem, unde commodior ad Almaiesti quo precipuum nostrum aspirat studium pateret accessus, tamen ne per meam segniciem nostra surdesceret amicicia, vestris nutibus nil preter equum postulantibus, mi Hermanne, nulli Latinorum huius nostri temporis astronomico sedere [27] penitus parare paratus, eum quem commodissimum et veracissimum inter astrologos indicem vestra quam sepe notavit diligentia voto vestro serviens transtuli, non minus amicicie quam pericie facultatibus innisus. In quo tum vobis tum ceteris huius scientie studiosis placere plurimum studens, enodato verborum vultu rerum seriem et effectum atque summam stellarium effectuum pronosticationisque quorumlibet eventuum latine brevitati diligenter inclusi. Cuius examen vestram manum postremo postulans non indigne vobis laudis meritum, si quod assit, communiter autem fructus pariat mihi-

---

[23] Preface to the *De essentiis*, supra, Chapter III; preface to the *Introductorium* of abu Ma'aschar, *ibid.*; preface to translation of the *Planisphere*, in Heiberg, *Ptolemaei opera astronomica minora*, pp. clxxxvi f.

[24] F. 109 (156).

[25] The heading is from the Cotton MS. App. vi, f. 109 (156), which contains a corrupt form of the text, here printed from Ashmole MS. 369, f. 85. The *Dictionary of National Biography*, under 'Robert the Englishman,' is in error in inferring from the tract of abu Ali, which follows in the Cotton MS., a connection between Robert and Plato of Tivoli.

[26] On the basis of this passage Steinschneider, no. 101, assigns to Robert, whom he makes a distinct Robertus Anglicus, an anonymous *Liber proportionum* found in several manuscripts.

[27] *sedem?*

que non segne res arduas aggrediendi calcar adhibeat, si nostri laboris munus amplexu favoris elucescat. Sed ne proemium lectori tedium lectionique moram faciat vel affetat, illius prolixitate supersedendo rem propositam secundum nature tramitem a toto generalique natis exordiis texamus, prius tamen libri tocius capitulis enumeratis ad rerum evidenciam suorumque locorum repertum facilem.

2. A translation of Morienus, *De compositione alchemie*, completed 11 February 1144 (era 1182). This is "one of the earliest treatises of alchemy translated from Arabic into Latin." See Steinschneider, *E. U.*, no. 102 *c*; Thorndike, ii. 83, 215–217. The Basel edition of 1559 contains the preface; there is an English version in the British Museum, Sloane MS. 3697. Robert may also have had something to do with a version of the *Mappe clavicula*: Steinschneider, *E. U.*, no. 102 *d*; di Marzo, *I MSS. della Biblioteca comunale di Palermo*, iii. 239.

3. A translation of the *Algebra* of al-Khwarizmi, dated Segovia, 1145 (era 1183). The first Latin version of this fundamental treatise, through which the name as well as the processes of algebra first penetrated to Latin Europe. See now Karpinski, *Robert of Chester's Latin Translation of the Algebra of al-Khowarizmi* (New York, 1915), in the *University of Michigan Studies*; and, for the Arabic work, J. Ruska, in Heidelberg *Sitzungsberichte*, phil.-hist. Kl., 1917, no. 2.

4. A treatise on the astrolabe, dated London 1147 (era 1185). See Steinschneider, *E. U.*, no. 102 *f*.; and in *Z. M. Ph.*, xvi. 393. There are differences in the various manuscripts (e. g., Digby MS. 40, which has the date and place, but a different *incipit*, and no mention of Robert), and there was evidently a revision after 1150, as the tables of that year are cited (see the next paragraph).[28]

5. A set of astronomical tables for the meridian of London in 1149–50, based upon the tables of al-Zarkali and al-Battâni and probably adapted from a translation of the *Opus astronomicum* of the latter by Robert, to which Hermann of Carinthia refers in 1143 in the preface to his *Planisphere* but which is otherwise unknown. See Steinschneider, *E. U.*, no. 102 *b*; Nallino, *al-Battâni*, pp. xxxiv f., xlix f. The London tables formed the second part of a work of which the first part

---

[28] The Ambrosian MS. H. 109 sup., to which reference has heretofore been made on the authority of Muratori, has (f. 11) clearly 'Robertum Cestrensem'; the treatise is followed on f. 17 v by an anonymous *Canon super chilindrum*, beginning, 'Accepturus horas.'

was calculated for the year 1149 [29] and the meridian of Toledo. Both are cited in Robert's treatise on the astrolabe: [30]

De ratione coequationis .xii. domorum in libro canonum quem super Toletum et civitatem Londoniarum edidimus, prout tractatus exposcebat ratio, tractavimus.

6. A revision, likewise for the meridian of London, of Adelard's version of the tables of al-Khwarizmi. Madrid, Biblioteca Nacional, MS. 10016, f. 8: 'Incipit liber Ezeig id est chanonum Alghoarizmi per Adelardum Bathoniensem ex arabico sumptus et per Rodbertum Cestrensem ordine digestus.' F. 14: 'He autem adiectiones omnes iuxta civitatem Londonie in hoc libro computantur et mediis cursibus planetarum adiciuntur.' [31] There are numerous differences from Adelard's version of 1126 as preserved in the Bodleian MS. Auct. F. 1. 9, where the tables are based upon Cordova, and where various Arabic words are retained which the later text omits or turns into Latin. The word 'sine' first appears here. The text of the Madrid MS. corresponds in general with that of the Chartres MS. 214 and of the extracts in MS. 3642 of the Bibliothèque Mazarine. See Suter's edition, pp. xi–xiii, 69.

How far Robert's labours were carried in the works of Euclid, Theodosius, and Ptolemy, we cannot say, for we have only his statement in the preface to al-Kindi, but in his work upon the tables of al-Battâni and al-Khwarizmi he continued worthily the tradition of Adelard of Bath, and in the fields of algebra and alchemy he broke new ground for Latin Europe.

The Madrid manuscript [32] which preserves Robert's revision of the Khorasmian tables also contains various tables for the meri-

[29] Not 1169, as is generally stated on the basis of Ashmole MS. 361, f. 24 (Black, *Catalogue*, col. 277). The correct statement is found in Savile MS. 21, f. 88 v: 'Ea namque eius pars que ad meridiem civitatis Toleti constituitur a .1149. anno domini incipit et ab eodem termino annos domini per .28. colligens lineas annorum collectorum in mediis planetarum cursibus in tempus futurum extendit, altera vero eius pars cuius videlicet ratio ad meridiem urbis Londoniarum contexitur ab anno domini .1150. sumpsit exordium.'

[30] Canonici MS. Misc. 61, f. 22 v.

[31] On this manuscript, which is of English origin, see the following note and cf. Chapter II, n. 15.

[32] The manuscript, no. 10016, containing 85 leaves, is of the early thirteenth century. It belonged originally to an English Cluniac monastery, as appears from the calendar on ff. 5–7 v in the same hand as certain of the tables, but had reached Spain, perhaps via Italy (Suter, p. xi), by 1439, when a Spanish notary, Juan de

dian of Hereford, which are obviously the work of another English astronomer of the twelfth century, Roger of Hereford.[33] We have from him the following:

1. *Compotus*, in five books, comprising in all twenty-six chapters: Digby MS. 40, ff. 21–50 v; cf. Macray, *Catalogue*, col. 37. The author criticizes the errors of Gerland and the Latin computists generally, and compares their reckoning with that of the Hebrews and Chaldeans. In the preface, the beginning of which is printed by Wright, *Biographia literaria*, ii. 90 f., he says that although still 'iuvenis' he has given many years to the 'regimen scholarum.' The date of the work is exactly given as 9 September 1176 (f. 48): 'Ut exempli gratia circa tempus huius compositionis huius tractatus anno scilicet Domini .m. c.lxx.vi° cicli decemnovenalis .xviii. que in vulgari compoto dicitur accensa .vᵃ. feria anni illius nona die septembris.'[34] The author is not specifically named in the body of the treatise, but appears in the acrostic of the table of chapters, GILLEBERTO ROGERUS SALUTES H[IC?] D[ICIT?], where Gilbert is probably Gilbert Foliot, who had been bishop of Hereford till 1163, and one of whose documents is attested in 1173–74 by Rogerus de Herefordia.[35] The heading in the manuscript reads, 'Prefatio magistri Rogeri Infantis in compotum,' whence the treatise has been assigned to an otherwise unknown Roger Infans, or, as Le-

---

Ornos, began to use the margins for family memoranda; until 1869 it was in the cathedral library at Toledo. Ff. 1 v–2 contain astronomical diagrams with astrological notes. F. 2 v, explanation of calculation of eclipses. F. 3, spera de morte vel vita. F. 4, tabula eclipsis tam solis quam lune. F. 4 v, Easter cycle, beginning 1063. Ff. 5–7 v, calendar. Ff. 8–72 v, Liber Ezeig. Ff. 73–83 v, with heading 'Herefordie,' tabule medii motus solis super mediam noctem Herefordie secundum annos domini, the cycles beginning 1120, 1148, 1176, etc., followed by tables for the moon and planets. F. 84, scienciam latitudinum quinque planetarum erraticorum. F. 85, in same hand as f. 4, ortus signorum super Hereford' latitudo .li. gr. et .xxx. minutorum, longitudo .xxiiii. grad. F. 85 v, letter of Petosiris to Nechepso (cf. *Philologus*, suppl. vi. 382; Wickersheimer, in *Seventeenth International Congress of Medicine*, section xxiii, pp. 315–318; Spiegelberg, in Heidelberg *Sitzungsberichte*, 1922, no. 3).

[33] Roger has been a source of confusion to bibliographers, who have made of him two or even three distinct persons: see Bale's *Index*, ed. Poole and Bateson, pp. 401 f.; Tanner, pp. 641, 788; Wright, *Biographia literaria*, ii. 89–91, 218 f.; *Dictionary of National Biography*, xlix. 106 f. Cf. Thorndike, ii. 181–187, to whom I owe two minor corrections.

[34] Cf. f. 49 v, printed by Macray, who, however, misreads mclxxvi as mclxxvii by mistaking the final punctuation for a unit.

[35] *Epistolae*, no. 210 (Migne, cxc. 913).

land called him, Yonge, to whom Wright, followed by the *Dictionary of National Biography*, gave the date 1124, which is found on f. 50 and indicated in a marginal gloss as the date of the work. This year, however, is used only in the course of a calculation of discrepancies, and the date 1176 appears clearly in two other passages. Inasmuch as the astronomical tables of Roger of Hereford belong to 1178 and no other contemporary astronomer of the name is known, we are justified in assigning the *Compotus* to him. The 'Infantis' of the title may be a corruption of 'h'efort,' or an inference from the 'iuvenis' of the preface; the gloss on Alfred de Sereshel (see below) calls him 'Rogerus Puer.'[36]

2. Astronomical tables for the meridian of Hereford in 1178, based upon tables for Toledo and Marseilles: Madrid, MS. 10016, ff. 4, 73–83 v, 85; British Museum, Arundel MS. 377, ff. 86 v–87: 'Anni collecti omnium planetarum compositi a magistro Rogero super annos domini ad mediam noctem Herefordie anno ab incarnatione domini .m°.c°. lxx°.viii°. post eclipsim que contigit Hereford eodem anno' (13 September). There is only one page of tables under Roger's name in the Arundel MS., but he is probably the author of those which precede (ff. 77–85), and which are calculated for the meridian of Toledo and the year 1176.

3. (?) *Theorica planetarum.* An explanation in thirty-two chapters of the use of astronomical tables: 'Diversi (*al.* Universi) astrologi secundum diversos annos tabulas et computaciones faciunt . . . per modum foraminis rotundi.' Bodley MS. 300 (Bernard, no. 2474), ff. 1–19 v; Digby MS. 168, ff. 69 v–83 v; Savile MS. 21, f. 42 (37), where it is attributed to Robert of Northampton. The treatise refers to 'tabulas ad Londonias factas.' There was a copy at Peterhouse in 1418 (James, *Catalogue*, p. 15), and according to Bale and Leland one at Clare College (James, *Catalogue*, pp. vii, viii). Other MSS. are cited by Duhem, iii. 499–523, who urges that the ascription to Roger of Hereford is the error of a copyist, since this treatise cites the London tables of 1232; Duhem conjectures that the treatise may be by Roger Bacon. There were, however, London tables in the twelfth century.[37]

4. *Tractatus de ortu et occasione signorum.* 'Orizon rectus est circulus magnus . . . maiora erit ut poterit apparere.' Bodley MS. 300, ff. 84–90. According to Bale's *Index*, p. 402, there was formerly a copy at Clare College.

[36] A. Thomas, in *Bulletin hispanique*, vi. 25.

[37] Supra, p. 123. Note also the meridian of Angers and Winchester in Arundel MS. 377, f. 56 v.

5. One or more astrological works: '*Liber de quatuor partibus
iudiciorum astronomie.* Quoniam circa tria sit omnis astronomica con-
sideratio . . . si non respiciens tertia.' Bibliothèque Nationale,
MS. Lat. 7434, ff. 76–79; Limoges, MS. 9, ff. 124 v–128 v; Dijon,
MS. 1045, ff. 172 v–180. A treatise beginning, 'Quoniam regulas
astronomie,' seems to be part of the same work: Digby MS. 149, f. 189
(cf. Macray, *Catalogue*, col. 149); Selden MS. supra 76, f. 3 (Bernard,
no. 3464); MS. e Musaeo 181 (Bernard, no. 3556); University of
Cambridge, MS. Gg. vi. 3, f. 139, MS. Ii. 1. 1, ff. 40–59; Trinity
College, Dublin, MS. 369; Berlin, Staatsbibliothek, MS. 964 (Rose,
Verzeichnis, ii. 1210); Erfurt, MS. O. 84, ff. 39–52. Brief extracts
in Digby MS. 57, f. 145; Ashmole MS. 369, f. 32; Laud MS. Misc.
594, f. 136. The *Iudicia Herefordensis* in Ashmole MS. 192 and Royal
MS. 12 F. 17 of the British Museum consist probably of extracts from
this work (cf. also James, *Ancient Libraries of Canterbury and Dover*,
p. 322, no. 1135). There is also an astrology in four books in MS.
10271 of the Bibliothèque Nationale, ff. 179–201 v: *Liber de divisione
astronomie atque de eius quatuor partibus compositus per dominum*
(MS. *datum*) *Rogerium Herfort astrologum*, beginning, 'Quoniam
principium huic arti dignum duximus.'

6. *De rebus metallicis.* Seen by Leland at Peterhouse (Tanner,
p. 641), but not since identified; *Expositiones Alphidii* are also cited
by Tanner.

Roger of Hereford, accordingly, was a teacher and writer on
astronomical and astrological subjects, who was still a young
man in 1176, and who, two years later, adapted astronomical
tables of Arabic origin to the use of Hereford. How much longer
his activity continued we cannot say, unless he is the Roger,
clerk of Hereford, who acted as itinerant justice with Walter Map
in 1185,[38] nor do we know whether he travelled in Spain or what
were his relations with Robert of Chester.

In the case of Roger's contemporary, Daniel of Morley, the
dependence upon the schools of Spain is clearly indicated.[39]

[38] Pipe Roll, 31 Henry II, p. 146. Master Roger of Hereford attests a York
charter of 1154–63: Farrer, *Yorkshire Charters* (Edinburgh, 1914), no. 158. A
Roger, vice-dean of Hereford, was the owner of three manuscripts of the twelfth
century (MSS. 66, 105, 106) in the library of Jesus College, Oxford: Coxe, *Catalogus*, pp. 23, 35.

[39] Until recently the fundamental study on Daniel was that of Rose, "Ptolemäus

Finding Paris dominated by law and pretentious ignorance, he hastened, he tells us, to Toledo, as the most famous centre of Arabic science, in order to hear the wiser philosophers of the world. One of his masters there was Gerard of Cremona, the indefatigable translator of the later twelfth century, who had been drawn to Spain by the love of that which he could not find among the Latins, Ptolemy's *Almagest*; and it is likely that the *pretiosa multitudo librorum* with which Daniel returned to England included certain of the mathematical and astronomical treatises which Gerard had turned into Latin.[40] Certainly the *Philosophia*, or *Liber de naturis inferiorum et superiorum*, our sole source of information respecting Daniel, [41] was written to explain the teaching of Toledo to Bishop John of Norwich (1175–1200); its astronomical chapters are based upon al-Fargani and other Arabic authorities, although its philosophy is still tinged by the *Timaeus* and its astrology by Firmicus Maternus.

Could we but follow them, there were doubtless other Englishmen who frequented the schools of Spain in this period, and other learned Jews who visited England. Thus John of Seville composes a treatise on the conversion of Arabic years into Roman at the request of two Englishmen, Gauco and William.[42] Anglo-Norman horoscopes of ca. 1150 have been preserved.[43] We find a William Stafford, archdeacon of Madrid, attesting a Toledo charter of 1154,[44] and the much-travelled mathematician and astrologer, Abraham ibn Ezra, a native of Toledo, spending some

---

und die Schule von Toledo," in *Hermes*, viii. 327–349 (1874), who prints the introduction and conclusion of his *Philosophia*, with a brief analysis, from Arundel MS. 377. Briefer extracts were given by Wright, *Biographia literaria*, ii. 227–230; and by Holland, in Oxford Hist. Soc., *Collectanea*, ii. 171 f. The best account is now Thorndike, ii. 171–181; cf. *E. H. R.*, xxxvii. 540–544 (1922); and the general article of Charles Singer, *Isis*, iii. 263–269. The *Philosophia* has now been edited in full by Sudhoff, in *Archiv für die Geschichte der Naturwissenschaften*, viii. 1–40; see Birkenmajer, *ibid.*, ix. 45–51.

[40] On Gerard's translations see supra, Chapter I, n. 43.

[41] Save for an entry in the pipe rolls under Norfolk and Suffolk for the years 1184–1187; see the index to the printed rolls for 31–33 Henry II.

[42] Oxford, St. John's College, MS. 188, f. 99 v. See supra, Chapter I, n. 39.

[43] Royal MS. App. 85.

[44] Printed by Fita, in *Boletín de la Academia de la Historia*, viii. 63 (1886); cf. Bonilla y San Martin, *Historia de la filosofía*, i. 367.

time in London in 1158–59.[45]   The diffusion of the Arabic astrology is well illustrated by the predictions for the year 1186, which occupy considerable space in the English chroniclers, William the astrologer, clerk of the constable of Chester, being specifically named as one of the authors.[46]

The natural philosophy and metaphysics of Aristotle, cited in part but little utilized by Alexander Neckam, first appear to come to their own in England in the writings of Alfred of 'Sereshel' or Alfred the Englishman, a contemporary of Roger of Hereford, to whom he dedicates his version of the Pseudo-Aristotelian treatise *De vegetabilibus*.[47]   In the accompanying commentary he cites the *De anima*, the *De generatione et corruptione*, and a *Liber de congelatis* which he had translated from the Arabic as an appendix of three chapters to the *Meteorology*.   A still wider acquaintance with Aristotle appears in a subsequent work, the *De motu cordis*, where he refers to the *Physics*, *Metaphysics*, and *Nicomachean Ethics*;[48]   in a commentary on the *Meteorology* used by Roger Bacon;[49]   and in a lost commentary on the *Parva naturalia*.[50]

----

[45] Steinschneider, in *Z. M. Ph.*, xxv, sup., pp. 57–128; Jacobs, *Jews of Angevin England*, pp. 29–38.

[46] Roger of Hoveden, ii. 290–298; Benedict of Peterborough, i. 324–328.

[47] Jourdain, pp. 106, 430.   A copy in the library of the University of Barcelona (MS. 7–2–6) reads: 'Incipit liber de plantis quem Alveredus de arabico transtulit in latinum mittens ipsum magistro Rogero de Herfodia.'

[48] Baeumker, *Die Stellung des Alfred von Sareshel (Alfredus Anglicus) und seiner Schrift* De motu cordis *in der Wissenschaft des beginnenden XIII. Jahrhunderts*, in Munich *Sitzungsberichte*, 1913, no. 9, especially pp. 33–48; and his recently published edition of the *De motu cordis* in *Beiträge*, xxiii, nos. 1–2 (1923).   Extracts from the *De motu cordis* were published by Barach (Innsbruck, 1878), and it is discussed by Hauréau in *Mémoires de l'Académie des Inscriptions*, xxviii, 2, pp. 317–334.

[49] A. Pelzer, "Une source inconnue de Roger Bacon," in *Archivum Franciscanum historicum*, xii. 44–67 (1919).

[50] The library of Beauvais cathedral possessed in the seventeenth century 'Alfredus Anglicus in Aristotelem de mundo et celo, de generatione et corruptione, de anima, de somno et vigilantia, de morte et vita, de colore celi'.   Omont, "Recherches sur la bibliothèque de l'église cathédrale de Beauvais," in *Mémoires de l'Académie des Inscriptions*, xl (Paris, 1914), p. 48, no. 143.   Other treatises attributed to Alfred by the older bibliographers (Tanner, pp. 37 f.) have not been confirmed by recent studies.   Steinschneider, *E. U.*, nos. 13, 23, does not identify the translator of the appendix to the *Meteorologica*, whom he calls, after certain manuscripts, Aurelius.

Being dedicated to Neckam, the *De motu cordis* cannot be later than his death in 1217, and as Neckam himself seems to have been acquainted several years earlier with the *Metaphysics*, *De anima*, and *De generatione et corruptione*,[51] it may go back to the beginning of the century. Even if we assign the latest possible limit to the treatise, it shows a wealth of Aristotelian citation such as we cannot find in any other Latin author of its time,[52] and its philosophy, based partly upon western Platonism and partly upon the older Arabic tradition, is singularly free from theological pre-possessions. While Alfred's knowledge of Aristotle was derived in part from versions made from the Greek,[53] we know from Roger Bacon and from internal evidence that he visited Spain,[54] and he must be placed in the series of intermediaries between Arabic and western learning. With him, however, the movement passes from its mathematical and astronomical phase to that which occupied itself primarily with natural philosophy and metaphysics, and we are thus brought into the philosophical currents of the thirteenth century.

[51] *Infra*, Chapter XVIII. [52] Baeumker, *Die Stellung*, p. 33.
[53] *Ibid.*, pp. 36–41.
[54] *Opus majus*, ed. Bridges, i. 67; *Compendium studii*, ed. Brewer, p. 471; Baeumker, *op. cit.*, p. 23; *Bulletin hispanique*, vi. 25.

CHAPTER VII

TRANSLATORS IN SYRIA DURING THE CRUSADES

THE influence of the Crusades upon the intellectual life of Europe has been variously judged. Once considered the great channel for the westward flow of Arabic culture, the estimate of their importance has greatly diminished with the clearer apprehension of the manifold contacts established with the East through Spain, Africa, Sicily, and the Byzantine Empire. It has even been denied that the Crusades had any direct effect upon the diffusion of Arabic learning, and it is certainly surprising that even in so practical a field as geography the writers of the thirteenth century should continue to draw upon the classical Latin authors rather than upon the fresher and more direct knowledge of Arabian explorers.[1] Plainly the Crusaders were men of action rather than men of learning, and there was little occasion for western scholars to seek by long journeys to Syria that which they could find nearer home in Spain. Nevertheless, intellectual relations with the Arabs of Syria were not wholly lacking. Early in the twelfth century Adelard of Bath is known to have visited Antioch and Tarsus, though it is not clear to what extent his acquaintance with Arabic science was gained there;[2] while toward the close of the Crusading epoch Frederick II included the East in the distribution of his questionnaires, and when in Syria came into direct relations with Mohammedan philosophers and scientists, while his 'philosopher' Theodore hailed from Antioch.[3] In the intervening hundred years or more our information is but fragmentary, yet it includes, in the twelfth century, translations of the great medical work of Ali-ben-Abbas and a treatise on divina-

[1] On the slow diffusion of Arabic geography, see J. K. Wright, *Geographical Lore of the Time of the Crusades* (New York, 1924).

[2] Supra, Chapter II.

[3] Infra, Chapter XII. On an alleged translation of the so-called *Theology* of Aristotle at Damascus by a Jew of Cyprus, Moses Arovas, see Steinschneider, *H. U.*, p. 244; *E. U.*, nos. 85, 92.

tion, and in the following century the transmission of one of the most famous of mediaeval books, the *Secretum secretorum* ascribed to Aristotle.

STEPHEN OF ANTIOCH

Of these translators who are definitely known to have worked in the East the first is a Pisan, Stephen, trained apparently in the schools of Salerno and Sicily, who followed his countrymen to Antioch, where he appears in 1127 as translating the medical writings of Ali-ben-Abbas and planning further versions from the Arabic. Moreover, his work makes clearer the significance of the Pisan contribution to the learning of the twelfth century, already attested by a medical translation from the Arabic in 1114 [4] and by the versions from the Greek made at Constantinople half a century later by the Pisan scholars Leo Tuscus, Hugo Eterianus, and especially Burgundio.[5]

Ali-ben-Abbas, one of the outstanding Arabic writers of the tenth century, planned his *al-Malaki*, or *Regalis dispositio*, as a comprehensive treatise on medicine intermediate between the enormous *Continens* of Rhazes and the concise *Liber medicinalis* of the same writer, and succeeded in formulating clearly therein the best medical knowledge of his time.[6] Stephen's translation of the *Liber regalis* is found in numerous manuscripts and in two early editions printed at Venice in 1492 [7] and at Lyons in 1523. The editions and two of the manuscripts comprise two parts, each in ten books, the *Theorica*, of which I know only these manuscripts,[8] and the *Practica*, much more common.[9] The printed

[4] See below.

[5] See Chapter X. Note also the astronomical tables of Abraham ben Ezra for the meridian of Pisa: *B. M.*, vi. 232; Birkenmajer, *Ryszarda de Fournival*, pp. 35–42; cf. Arundel MS. 377, ff. 56 v–68 v.

[6] Neuburger, *Geschichte der Medizin* (Stuttgart, 1911), ii, 1, pp. 176, 210.

[7] Hain 8350*. I have used the copy in the Surgeon General's Library and the copy belonging to Dr. E. C. Streeter of Boston. For the edition of 1523 I have used the copy in the Bibliothèque Nationale.

[8] Vatican, MS. Urb. lat. 234; MS. Vat. lat. 2429. These MSS. include the *Practica* as well.

[9] Berlin, Cod. elect. 898; Erfurt, MS. F. 250; Basel, MS. D. ii. 18; Cesena, Plut. xxvi, Cod. iv; Worcester Cathedral, MS. F. 40; Cambrai, MS. 911 (incom-

text lacks at the close a glossary of the technical terms of Dioscorides, first noted by Valentin Rose in his description of the Berlin manuscript, where the essential facts regarding Stephen's translation are first brought together.[10] The *Theorica* had previously been translated into Latin under the title *Pantegni* by Constantine the African, who likewise translated the beginning and the first half of the ninth *particula* of the *Practica*, also found separately as *De chirurgia*.[11] The second half of this *particula* was turned into Latin by Constantine's pupil John the Saracen, or Johannes Afflacius,[12] and a Pisan physician named Rusticus at the time of the great expedition against Majorca in 1114.[13] Stephen, according to his preface, having come upon Ali's book in Arabic, found there was no complete Latin version, while what had been translated suffered from omissions and transpositions. He accordingly decided to prepare an entirely new version, which appears upon collation to be quite different, every book being signed by the translator to emphasize his work.[14]

At the close of the *Regalis dispositio* Stephen adds a glossary of the technical terms in Dioscorides, *Medicaminum omnium breviarium*, which in more or less complete form appears in the manuscripts and is cited as Stephen's *Synonyms* by later writers. In its full form this is an alphabetical list, Greek, Arabic, and Latin in three parallel columns. Readers who have difficulty with the Latin terms càn thus consult experts, "for in Sicily and at Salerno,

plete); University of Leipzig, MS. 1131, dated 1179 (Arndt-Tangl, *Schrifttafeln*,[4] no. 23): Bibliothèque Nationale, MS. lat. 6914. See also Delisle, *Cabinet des MSS.*, ii. 534, §§ 149, 151, 152. "Aly Stephanon Phlebotomia," in MS. Vienna 1634, ff. 94 v–97 v, is probably an extract. I have used the Basel and Paris MSS. and extracts from the Roman.

[10] Cod. elect. 898, *Verzeichniss der lateinischen Hss. der königlichen Bibliothek*, ii. 1059–1065 (1905). Steinschneider (Virchow's *Archiv*, xxxvii. 356 ff., xxxix. 333–335, lii. 479; *E. U.*, no. 111) had seen only the incomplete printed text.

[11] Ed. by Pagel, in *Archiv für klinische Chirurgie*, lxxxi, i, pp. 735–786 (1906); cf. Sudhoff, *Die Chirurgie im Mittelalter* (Leipzig, 1914–16), ii. 95.

[12] This point is overlooked by Friedrich Hartmann, *Die Litteratur von Früh- und Hoch-Salerno* (Leipzig diss., 1919), p. 20.

[13] On the presence of Pisan physicians with this expedition, cf. *Liber Maiolichinus*, ed. Calisse (Rome, 1904), lines 2375 ff.

[14] Cf. Steinschneider, in Virchow's *Archiv*, xxxix. 333 f.; Rose, *l. c.*

where students of such matters are chiefly to be found, there are both Greeks and men familiar with Arabic." [15]

That Antioch was the place of Stephen's work admits of no doubt, for the explicit has *scriptusque eius manu Antiochie*. Steinschneider once suggested as more probable a small place of this name in Spain; [16] but Stephen speaks definitely of the East and in his concluding paragraph of Syria.[17] There are indications of date at the end of certain of the books, as follows:

I. 5. Finitur sermo quintus prime partis libri completi artis medicine que dicitur regalis dispositio Hali filii Abbas discipuli Abimeher Moysi filii Seyar translatio Stephani phylosophie discipuli de arabico in latinum, et Deo sicut est dignus laus et gloria. Scriptus novembris die vicesima octava feria secunda anno a passione Salvatoris millesimo .c. vicesimo septimo Alduini manu, expletus manu Panci vi° diebus existente mense aprilis .M°.C°. XXVII.[18]

I. 10. Translatio Stephani de arabico in latinum die octubris septima feria tercia anno a passione Domini millesimo centesimo vicesimo vii°, Deo gratias, Alduini manu.[19]

II. 3. Scriptus vicesimo septimo et centesimo M. anno.[20]

II. 7. Finitur sermo septimus . . . translatio Stephani phylosophie discipuli de arabico in latinum scripsitque ipse et complevit anno a passione Domini millesimo centesimo vicesimo .vii. mense novembris die .iii. feria septima apud Antiochiam. Deo gratias rerum principio et fini.

Incipit sermo .viii. . . . . scripsitque ipse et complevit anno a passione Domini M°.C.XXVII°. mense novembris die tertio feria .vi. apud Antiochiam.[21]

---

[15] See the preface in Rose, p. 1063, and cf. Stornajolo, *Codices Urbinates Latini*, i. 227. The Basel MS. omits the synonyms; the Paris codex has, ff. 147–156, a different list of Arabic and Latin terms only, without the concluding paragraph. The *Glose magistri Stephani* of this period noted by Traube (Wölfflin's *Archiv*, vi. 265) appear to be different. On medico-botanical glossaries see *Anecdota Oxoniensia*, i (1882–1887); and Götz in *Corpus glossariorum Latinorum*, i. 227–236; and for the related material in prescriptions, H. E. Sigerist, *Studien und Texte zur frühmittelalterlichen Rezeptliteratur* (Leipzig, 1923).

[16] Virchow's *Archiv*, xxxix. 333; *Serapeum*, xxxi. 292.

[17] Rose, p. 1063.

[18] MS. Vat. lat. 2429, f. 41 v; MS. Urb. lat. 234, f. 78 v, which gives the final date as ' vii. diebus ext mense aprilis M°.C°C°. xxxvii.° '

[19] MS. Vat. lat. 2429, f. 86 v; MS. Urb. lat. 234, f. 162; Venice edition, f. 78 v; Lyons edition, f. 134 v.

[20] Berlin, 898, f. 116 v.

[21] MS. Vat. lat. 2429, f. 168 v; MS. Urb. lat. 234, f. 307, omitting the incipit of book viii; the Venice edition reads 'sunt vi' for 'feria vi.'

II. 10. Scriptusque eius manu Antiochie a passione Domini millesimo centesimo vicesimo septimo mense ianuario vicesimo septimo die feria quarta.[22]

These dates are hopelessly inconsistent with one another, and most of them are also inconsistent with 1127, and no simple emendation or adjustment of chronological styles will harmonize them all. The one element in all is the year 1127, with which the explicit of i. 5 also agrees, while Rose brings the concluding date (ii, 10) into harmony by emending January 26, with the Venice edition, because of possible confusion with the twenty-seven of the year. In that case the translation of the *Practica* would antedate that of the *Theorica*. In any event we may conclude that some part of Stephen's version was made in 1127, the exact dates having been confused by errors of copyists or by the attempt to reduce all dates to this single year.

Of the translator Stephen the preface and epilogue tell us but little. He is a Latin, who quotes Boethius and follows the advice of Solomon to get wisdom. He has studied Arabic in order to mount to the fountain head of learning, and he has evidently some knowledge of Greek. He knows, probably from personal acquaintance, of the scholars of Salerno and Sicily. Matthew of Ferrara adds that Stephen was a Pisan, who went to Saracen lands, learned Arabic, and made a complete translation of Ali, later called *Practica pantegni et Stephanonis*.[23] As Stephanonus he is cited by Platearius.[24] That Stephen should be a Pisan is not surprising, for the Pisans had had a special quarter in Antioch since 1108,[25] and Pisan activity in medical translation has already been noted.

Stephen's interest in Arabic literature was not limited to medicine. He expressly tells us that the version of the *Regalis dis-*

<hr />

[22] MS. lat. 6914, f. 147; MS. Basel D. ii. 18, f. 255 v; and the Lyons edition. The Berlin and Cesena MSS. have 1107. The Venice edition has a paraphrase: 'Ipsum autem ex arabico in latinum ornatissime traduxit sermonem Stephanus philosophie discipulus in Anthiochia. Anno dominice passionis .M°.C°. xxvii. xxvi. ianuarii feria quarta.'

[23] Gloss printed by Rose, p. 1060. · Ganszyniec, in *Archiv für Geschichte der Medizin*, xiv. 110, claims Stephen as the author of a *De modo medendi*.

[24] Rose, p. 1059.

[25] Röhricht, *Regesta Regni Hierosolymitani* (Innsbruck, 1893), no. 53.

positio was his first work, but he hopes to translate something out of "all the secrets of philosophy which lie hidden in the Arabic tongue," passing thus from those things which concern the body to the far higher things of the mind.[26] This obviously suggests philosophy, and a search among the treatises of the period may show traces of his work in this field. In any such inquiry the name 'Stephen the philosopher' is a source of confusion, denoting, as it also may, Stephen of Alexandria in the seventh century;[27] a Greek writer on astrology in the following century;[28] and the alleged translator of an astronomical treatise of Maimonides from the late twelfth century;[29] not to mention the Stephen of Provins who was commissioned by Gregory IX in 1231 to revise the natural philosophy of Aristotle, and who was in scientific relations with Michael Scot.[30] The supposed treatise of Maimonides turns out not to be his, but the work of a Latin writer of the twelfth century who had some knowledge of Greek terms and of Arabic astronomy; if the name 'Stephen the philosopher' does not fall with the ascription of the tract to Maimonides, it is conceivable, though hardly probable from internal evidence, that the author was Stephen of Antioch.

'BERNARD SILVESTER'

Associated in certain manuscripts with the *Experimentarius* of Bernard Silvester is a brief bit of oriental divination whose origin is narrated as follows:[31]

[26] 'His igitur in libris nostri primum consumere laboris proposuimus operam, tametsi alia his preclariora lingua habeat apud se arabica, recondita omnia scilicet philosophie archana, quibus deinceps si divina dederit benignitas exercitatum dabimus transferendis ingenium; leviora enim hec preferimus ut ad difficilia via nobis sit et que corporibus necessaria sunt tempore preponimus, ut his sanitate preposita arte medicine que ad animi attinent excellentiam longe altiora subsequantur.' Edition of 1522, f. 5.

[27] Usener, *De Stephano Alexandrino* (Bonn, 1880); Krumbacher, p. 621.

[28] Cumont, in *Catalogus codicum astrologicorum Graecorum*, ii. 181 ff.

[29] Cambrai, MS. 930. See above, Chapter V, where it is shown that the treatise which appears with this title is not a translation but an original Latin work.

[30] On the various men who bore the name Étienne de Provins in the first half of the thirteenth century, see my paper "Two Roman Formularies in Philadelphia," in the *Miscellanea Francesco Ehrle* (Rome, 1924).

[31] Bodleian, Digby MS. 46, f. 3-3 v; Savile MS. 21, f. 182.

Quidam invictissimi ac benignissimi regis Amalrici medicus hoc opus .xx. et .viii. questionum super fata secundum .xx. et .viii. mansiones in quibus sol in toto anno moratur naturam et potestatem .vii. planetarum considerans instituit. Hoc autem ad regis laudem et gestorum eius memoriam et maxime triumphi nuper domiti Syraconis, qui dux Persarum, Turcorum, Turcomanorum, Cordiorum, Agarenorum, et Arabum et multarum diversarum gentium cum omnibus viribus suis totam Egyptum violenter invaserat preter quandam municionem quam Cassarum vocant; dominus [32] Egipciorum et cum eo inclusi ad regem miserunt et auxilium postulantes ab eo impetraverunt. Rex autem Amalricus cum paucis per deserta transiens in civitatem quandam munitissimam Siraconem perteritum fugavit suique multitudinem exercitus intrare coegit ibique eum diucius expugnando, quod omnibus mirum fuit, divina adiutus [33] potentia cum marte potenter domuit ac de toto Egipto expulit et facti sunt Egipcii Amalrico regi tributarii in eternum. Post quod gestum prefatus regis medicus predictum opus secundum planetarum ordinem [sicut] infra in serie apparet ordinavit et regi domino Francorum .v.^{to} in Ierusalem feliciter Deo protegente regnanti.

The reference is either to the events of 1164, when Shirko drove the Egyptians into Cairo and was in turn defeated by Amaury I, king of Jerusalem, and shut up in Bilbais, or, more probably because of the mention of permanent tribute, to the Egyptian campaign against Shirko in 1167.[34] The date of the translation is not given, but Amaury (†1173) is apparently thought of as still alive, and in any case the writer knows nothing of the second king of that name who came to the throne in 1197.[35] Amaury's physician, the original compiler, is not named, but the treatise sometimes appears as part of the *Experimentarius* of Bernardus, or Bernardinus, Silvester, who is in one manuscript called a translator from the Arabic.[36] There is, however, no reason for ascribing any knowledge of Arabic to Bernard Silvester of Tours, a well known figure in the literary history of the twelfth century, nor can he be traced beyond the middle of the century.[37] More prob-

[32] MS. *dominum*.   [33] MS. *adintus*.

[34] Röhricht, *Geschichte des Königreichs Jerusalem* (Innsbruck, 1898), pp. 314–330.

[35] William of Tyre says in 1184 that he wrote a history of events in the East from Arabic materials furnished by King Amaury: Steinschneider, *E. U.*, no. 123.

[36] 'Titulus talis est, Experimentarius Bernardini Silvestris, non quia inventor fuit sed fidelis ab arabico in latinum interpres': MS. Ashmole 304, f. 2; cf. MS. Digby 46, f. 1.

[37] On Bernard, see particularly Cousin, *Fragments philosophiques* (1840), pp. 336 ff.; Hauréau, *Philosophie scholastique* (1872), i, ch. 16; id., in *Mémoires de*

ably, as Thorndike suggests,[38] the similarity of subject-matter led to the early association of such treatises on divination, whence it is but a step to the ascription of a common oriental origin.

PHILIP OF TRIPOLI

Directly connected with Syria is the transmission to Europe of the *Secret of Secrets* ascribed to Aristotle, one of the most widely popular books in the whole of the later Middle Ages and the sixteenth century, more than two hundred manuscripts being known of the Latin version, besides early imprints and translations into most of the European languages. Purporting to have been written by Aristotle for the guidance of Alexander the Great, it seemed to contain the distilled essence of practical wisdom and occult science for every reader, as well as the secret maxims of government for the use of princes.[39] There had been a translation of the medical portion by John of Seville in the first half of the twelfth century, but the first and the standard version of the whole was due to a certain Philip, clerk of Tripoli, and dedicated to Guido de Vere of Valence, or Valencia, who appears in different manuscripts as bishop of Tripoli, or as archbishop of an unnamed see or of Naples.[40] The original had been found when Philip and this prelate were at Antioch, and it is by the patron's command that it was turned into Latin, "sometimes literally and sometimes according to the sense, for the Arabs have one idiom and the Latins another." A philosophic pearl of such great price, dealing

---

l'Académie des Inscriptions, xxxi, 2, pp. 77 ff.; id., in *Histoire littéraire*, xxix. 569 f.; Langlois, "Maître Bernard," in *B. E. C.*, liv. 225–250; Clerval, *Les écoles de Chartres*, pp. 158–163; Duhem, iii. 68, 117; R. L. Poole, "The Masters of the Schools at Paris and at Chartres in John of Salisbury's Time," in *E. H. R.*, xxxv. 326–331 (1920); Thorndike, ii, ch. 39.

[38] ii. 115.

[39] Of the vast literature on the *Secretum secretorum*, see particularly R. Förster, *De Aristotelis quae feruntur Secretis secretorum commentatio* (Kiel, 1888); and "Handschriften und Ausgaben des pseudo-aristotelischen Secretum Secretorum," in *Centralblatt für Bibliothekswesen*, vi. 1–22, 57–76, 218 (1889); Steinschneider, *H. U.*, pp. 249 ff.; R. Steele, introduction to Roger Bacon's edition, *Opera hactenus inedita*, v (Oxford, 1920, with an English version from the Arabic); Thorndike, ii. 267–278; and in *Journal of English and Germanic Philology*, xxi. 248–258 (1922).

[40] 'Tripolis,' 'metropolis,' 'Napolis.'

with every kind of knowledge, was deemed a worthy gift to a prelate so learned in letters, law, and theology.

As to date, Philip's translation is subsequent to the version of John of Seville, which it utilizes, and anterior to the commentary of Roger Bacon, written between 1243 and 1254, probably about 1247. If, which seems to me doubtful, this translation rather than the Arabic original was used in Michael Scot's *Physiognomy*, it was anterior to 1236 and probably to 1228.[41] No manuscripts have been noted earlier than the thirteenth century. Guido de Vere of Valence, or Valencia, is unknown in the East or as archbishop of Naples; but there are many gaps in our lists of this period, and many unconfirmed elections. There are, for example, gaps in Tripoli between 1145 and 1170 and between 1209 and 1217 or later.[42] There was a Philip, chanter of Tripoli, in 1126.[43] In 1177 Alexander III uses a certain Philip, his own physician, as an intermediary with Prester John,[44] but, in spite of recent assumptions, there seems nothing to connect him specifically with Tripoli. A Master Philip of Tripoli appears in a fictitious attribution of "1212."[45] More probable, and much better known, is Philip, canon of Tripoli, who meets us in the papal registers from 1227 to 1251. 17 May 1227, as Master Philip, clerk of Foligno, he received from Gregory IX a canonry at Tripoli in recognition of his services to the patriarch and church of Antioch and his loss of property in such service, but in the face of opposition from the bishop and chapter.[46] This opposition appears to have been for a time successful, for he received a reappointment at the beginning

[41] Infra, Chapter XIII.

[42] See Röhricht, "Syria sacra," in *Zeitschrift des deutschen Palästina-Vereins*, x. 1–48 (1887); *Regesta*, no. 800. Some connection of Valence with Tripoli appears in John of Valentia, canon of St. Michael of Tripoli in 1244: Berger, *Registres*, no. 737. Cf. also Gerald, bishop of Valence, who became patriarch of Jerusalem in 1226.

[43] Röhricht, *Regesta*, nos. 117, 1274.

[44] *Ibid.*, no. 544; Jaffé-Löwenfeld, no. 12942; Thorndike, ii. 244.

[45] Brown, *Michael Scot*, p. 20; Steinschneider, *H. U.*, p. 793; Thorndike (ii. 271) vainly attempts to save this date by assuming the Spanish era.

[46] Auvray, *Registres de Grégoire IX*, nos. 118, 119. 'Philippus subdiaconus noster nepos bone memorie R. Antiocheni patriarche canonicus Antiochenus,' who appears in a bull of Honorius III, 25 September 1225 (Pressutti, *Regesta*, no. 5660), would seem to be a different person.

of the next pontificate,[47] when, as canon of Byblos, he also complains to the Pope of his bishop's ignorance of Donatus and Cato, in the course of a controversy between them which had begun before 1236.[48] 11 September 1245 he witnesses an act at Genoa.[49] In 1247, as plain Philip of Tripoli, he is at Lyons with the Pope, representing the patriarch of Jerusalem, who is ordered to give him an additional ecclesiastical appointment in that province because of his qualities of character and his knowledge of letters.[50] In 1248, chaplain of Hugh, cardinal priest of St. Sabina, he resigns his prebend at Byblos in favor of his nephew, and is confirmed by the Pope in his prebend at Tripoli, conferred upon him by Innocent IV five years before but a subject of protracted litigation with the bishop.[51] Canon of Tyre in the same year,[52] he declines a disputed election to the see of Tyre in 1250 and succeeds the archbishop elect as chanter of Tripoli, meanwhile retaining his cathedral prebends in Tyre and Sidon.[53] In 1251 he is also chaplain of the Pope.[54] At this point Philip disappears from the printed papal registers, but local documents show him as chanter of Tripoli in 1257 and 1259.[55] In all this history of pluralities and controversy and steady support from Rome there is no word of Philip's literary labors save the Pope's special mention of his *scientia litterarum* in 1247, but there is ample evidence of his sojourn in the East and his journeys westward. Moreover the chronological difficulty which appeared to exist when he had not been traced back of 1243 vanishes with the discovery of the documents of 1227 which show him to have been already at Antioch. There is every reason to believe that this canon is the Philip of

---

[47] Berger, *Registres d'Innocent IV*, no. 4394.

[48] *Ibid.*, nos. 57, 2403.

[49] Document cited by A. Ferretto, in *Giornale storico della Liguria*, i. 362, n. (1900). I know of no foundation for this author's assertion that Philip was a Florentine.

[50] Berger, no. 3138.

[51] *Ibid.*, nos. 4354 f., 4394.

[52] *Ibid.*, no. 4355.

[53] *Ibid.*, nos. 5048, 5390.

[54] *Ibid.*, no. 5178.

[55] Röhricht, *Regesta*, nos. 1258 b, 1274 a; Delaville Le Roulx, *Cartulaire de S. Jean de Jérusalem*, nos. 2875, 2921.

Tripoli who made the translation of the *Secretum secretorum* some time in the first half of the thirteenth century.

Philip's version went through a revising and standardizing process which may explain the Gallicisms that have been found in the text.[56] The translation has been pronounced remarkably close and accurate.

[56] Förster, p. 28.

CHAPTER VIII

THE GREEK ELEMENT IN THE RENAISSANCE OF
THE TWELFTH CENTURY [1]

THE renaissance of the twelfth century consisted in part of a revival of the Latin classics and the Roman law, whence the movement has sometimes been called a 'Roman renaissance,' in part of a rapid widening of the field of knowledge by the introduction of the science and philosophy of the ancient Greeks into western Europe. This Greek learning came in large measure through Arabic intermediaries, with some additions in the process, so that the influence of the Saracen scholars of Spain and the East is well understood. [2] It is not always sufficiently realized that there was also a notable amount of direct contact with Greek sources, both in Italy and in the East, and that translations made directly from Greek originals were an important, as well as a more direct and faithful, vehicle for the transmission of ancient learning. Less considerable in the aggregate than what came through the Arabs, the Greek element was nevertheless significant for the later Middle Ages, while it is further interesting as a direct antecedent of the Greek revival of the Quattrocento. No general study has yet been made of this movement, but detailed investigation has advanced sufficiently to permit of a brief survey of the present state of our knowledge.

The most important meeting-point of Greek and Latin culture in the twelfth century was the Norman kingdom of southern Italy and Sicily. [3] Long a part of the Byzantine Empire, this region still retained Greek traditions and a numerous Greek-speaking population, and it had not lost contact with the East. In the eleventh century the merchants of Amalfi maintained an active commerce with Constantinople and Syria; Byzantine craftsmen

---

[1] Revised from the *American Historical Review*, xxv. 603–615 (1920). Cf. *Isis,* iv. 582.

[2] Supra, Chapter I.   [3] See Chapters IX, XII.

wrought great bronze doors for the churches and palaces of the south,[4] and travelling monks brought back fragments of Greek legend and theology to be turned into Latin.[5] Libraries of Greek origin, chiefly of Biblical and theological writings, were gathered into the Basilian monasteries,[6] and more comprehensive collections were formed at the Norman capital. Only in the Norman kingdom did Greek, Latin, and Arabic civilization live side by side in peace and toleration. These three languages were in current use in the royal charters and registers, as well as in many-tongued Palermo, so that knowledge of more than one of them was a necessity for the officials of the royal court, to which men of distinction from every land were welcomed. The production of translations was inevitable in such a cosmopolitan atmosphere, and it was directly encouraged by the Sicilian kings, from Roger to Frederick II and Manfred, as part of their efforts to foster learning. While Roger commanded a history of the five patriarchates from a Greek monk, Nilus Doxopatres, and a comprehensive Arabic treatise on geography from the Saracen Edrisi, translation appears to have been more actively furthered during the brief reign of his successor. Under William I a Latin rendering of Gregory Nazianzen was undertaken by the king's orders, and a version of Diogenes Laertius was requested by his chief minister Maio. Indeed the two principal translators were members of the royal administration, Henricus Aristippus and Eugene the Emir, both of whom have left eulogies of the king which cele-

---

[4] A. Schaube, *Handelsgeschichte der romanischen Völker* (Munich, 1906), pp. 34-37; F. Novati, *Le origini*, in the coöperative *Storia letteraria d'Italia*, pp. 312 ff.

[5] The principal examples are Nemesius, *De natura hominis*, translated by Alfano, bishop of Salerno (ed. Burkhardt, Leipzig, 1917); and a collection of miracles put into Latin by the monk John of Amalfi. On Alfano, see particularly C. Baeumker, in *Wochenschrift für klassische Philologie*, xiii. 1095-1102 (1896); and G. Falco, in *Archivio della Società romana di storia patria*, xxxv. 439-481 (1912); and in *Bullettino dell' Istituto storico italiano*, no. 32, pp. 1-6 (1912); *Neues Archiv*, xxxviii. 667; Manitius, *Lateinische Litteratur*, ii. 618-637. On John, M. Huber, *Iohannes Monachus, Liber de Miraculis* (Heidelberg, 1913); Hofmeister, in *Münchner Museum*, iv. 129-153 (1923); Manitius, ii. 422-424.

[6] F. Lo Parco, "Scolario-Saba," in *Atti della R. Accademia di Archeologia di Napoli*, n. s., i, pt. ii, pp. 207-286 (1910), with Heiberg's criticism in *B. Z.*, xxii. 160-162.

brate his philosophic mind and wide-ranging tastes and the attractions of his court for scholars.[7]

Archdeacon of Catania in 1156, when he worked at his Plato in the army before Benevento, Aristippus was the principal officer of the Sicilian *curia* from 1160 to 1162, when his dismissal was soon followed by his death. Besides the versions of Gregory Nazianzen and Diogenes, which, if completed, have not reached us, Aristippus was the first translator of the *Meno* and *Phaedo* of Plato and of the fourth book of Aristotle's *Meteorology*, and his Latin rendering remained in current use during the Middle Ages and the early Renaissance. An observer of natural phenomena on his own account, he was also instrumental in bringing manuscripts to Sicily from the library of the Emperor Manuel at Constantinople. One of these possesses special importance, a beautiful codex of Ptolemy's *Almagest*, from which the first Latin version was made by a visiting scholar about 1160. The translator tells us that he was much aided by Eugene the Emir, "a man most learned in Greek and Arabic and not ignorant of Latin," who likewise translated Ptolemy's *Optics* from the Arabic. The scientific and mathematical bent of the Sicilian school is seen in still other works which were probably first turned into Latin here: the *Data*, *Optica*, and *Catoptrica* of Euclid, the *De motu* of Proclus, and the *Pneumatica* of Hero of Alexandria. A poet of some importance in his native Greek, Eugene is likewise associated with the transmission to the West of two curious bits of Oriental literature, the prophecy of the Erythraean Sibyl and the Sanskrit fable of Kalila and Dimna. If it be added that the new versions of Aristotle's *Logic* were in circulation at the court of William I, and that an important group of New Testament manuscripts can be traced to the scribes of King Roger's court, we get some further measure of the intellectual interests of twelfth-century Sicily, while the medical school of Salerno must not be forgotten as a centre of attraction and diffusion for scientific knowledge.

Italy had no other royal court to serve as a centre of the new learning, and no other region where East and West met in such constant and fruitful intercourse. In other parts of the peninsula

[7] *Hermes*, i. 388; *B. Z.*, xi. 451.

we must look less for resident Greeks than for Latins who learned their Greek at Constantinople, as travellers, as diplomats, or as members of the not inconsiderable Latin colony made up chiefly from the great commercial republics of Venice and Pisa.[8]

Among the various theological disputations held at Constantinople in the course of the twelfth century, Anselm of Havelberg has left us an account of one before John Comnenus in 1136, at which "there were present not a few Latins, among them three wise men skilled in the two languages and most learned in letters, namely James a Venetian, Burgundio a Pisan, and the third, most famous among Greeks and Latins above all others for his knowledge of both literatures, Moses by name, an Italian from the city of Bergamo, and he was chosen by all to be a faithful interpreter for both sides."[9] Each of these Italian scholars is known to us from other sources, and they stand out as the principal translators of the age, beyond the limits of the Sicilian kingdom.

Under the year 1128 we read in the chronicle of Robert of Torigni, abbot of Mont-Saint-Michel, and well informed respecting literary matters in Italy, that "James, a clerk of Venice, translated from Greek into Latin certain books of Aristotle and commented on them, namely the *Topics*, the *Prior* and *Posterior Analytics*, and the *Elenchi*, although there was an older version of these books."[10] Long the subject of doubt and discussion, this passage has recently been confirmed from an independent source,[11] so that James can be singled out as the first scholar of the twelfth century who brought the *New Logic* of Aristotle afresh to the attention of Latin Europe. What part his version had in the Aristotelian revival, and what its fate was as compared with

[8] On the north-Italian translators, see below, Chapter X; and in general, G. Gradenigo, *Lettera intorno agli Italiani che seppero di greco* (Venice, 1743). Sandys, *History of Classical Scholarship*,[3] i. 557 ff., touches the matter very briefly.

[9] L. d'Achery, *Spicilegium* (Paris, 1723), i. 172; Migne, clxxxviii. 1163; infra, p. 197.

[10] Robert of Torigni, *Chronique*, ed. Delisle, i. 177; *M. G. H., Scriptores*, vi. 489. In the eleventh century, St. Anastasius, a Venetian monk of Mont-Saint-Michel, is said to have known Greek: *Acta Sanctorum*, October, vii. 1125–1140; Paul Fournier, in Baudrillart, *Dictionnaire d'histoire et de géographie*, ii. 1469.

[11] Infra, Chapter XI.

the traditional rendering of Boethius, are questions which for our present purpose it is unnecessary to examine.

Moses of Bergamo evidently found his eastern connections by way of Venice. He is the author of an important metrical description of Bergamo, and kept up relations with his native city through letters to his brother and through benefactions to various churches, but his messengers pass through Venice, and he lives in the Venetian quarter at Constantinople. Here he is found in the emperor's service in 1130, when he has lost by fire a precious collection of Greek manuscripts, brought together by long effort at the price of three pounds of gold. He tells us that he learned Greek for the special purpose of turning into Latin works not previously known in the West, but the only specimen which has been identified is a translation of an uninteresting theological compilation. He has also left grammatical *opuscula*, including a commentary on the Greek words in St. Jerome's prefaces, which attest his familiarity with the language and with the writings of the Greek grammarians. Apparently what we have left are only the fragmentary remains of a many-sided activity, as grammarian, translator, poet, and collector of manuscripts,[12] which justifies us in considering him a prototype of the men who "settled *hoti's* business" in the fifteenth century.

Burgundio the Pisan [13] is a well known figure in the public life of his native city who made several visits to Constantinople. Although translation from the Greek seems to have been the occupation of his leisure moments only, his output was more considerable than that of any of his Latin contemporaries. Much of it was theology, including works of Basil and Chrysostom and John of Damascus which exerted a distinct influence on Latin thought. Philosophy was represented by Nemesius, law by the Greek quotations in the *Digest*, agriculture by an extract from the *Geoponica*. He was perhaps best known as the author of the current translations of the *Aphorisms* of Hippocrates and ten works of that Galen whom another Pisan, Stephen of Antioch, helped bring in from the Arabic.[14] His epitaph celebrates the universal learning of this *optimus interpres*:

[12] Infra, pp, 197–206.　　[13] See below, Chapter X.　　[14] Supra, Chapter VII.

Omne quod est natum terris sub sole locatum
Hic plene scivit scibile quicquid erat.

Less noteworthy than Burgundio, two other members of the Pisan colony should also be mentioned, Hugo Eterianus and his brother Leo, generally known as Leo Tuscus.[15] Hugo, though master of both tongues, was not so much a translator as an active advocate of Latin doctrine in controversy with Greek theologians, a polemic career which was crowned with a cardinal's hat by Lucius III. Leo, an interpreter in the emperor's household, translated the mass of St. Chrysostom and a dream-book (*Oneirocriticon*) of Ahmed ben Sirin. The interest in signs and wonders which prevailed at Manuel's court is further illustrated by one Paschal the Roman, who compiled another dream-book at Constantinople in 1165 and is probably the author of the version of Kiranides made there in 1169; as well as by other occult works which found their way westward about this time, perhaps in part from the imperial library. Indeed the relations, formal and informal, between the Greek empire on the one hand, and the Papacy and the Western empire on the other, offered many occasions for literary intercourse; and while we hear most of the resultant disputes between Greek and Latin theologians, it is altogether likely that other materials came west in ways which have so far escaped detection.

North of the Alps there is little to record in the way of translation, although it is probable that certain of the anonymous translators who worked in Italy came from other lands. In Germany we have the *Dialogi* with the Greeks written down by Anselm of Havelberg about 1150, and the *De diversitate persone et nature* which another emissary of the Western Empire brought back in 1179. Before the middle of the century a monk in Hungary, Cerbanus, translated the *Ekatontades* of Maximus the Confessor and perhaps also a treatise of John of Damascus.[16] In 1167 a certain William the Physician, originally from Gap in Provence,

---

[15] See Chapter X.

[16] See below, Chapter X; and for Cerbanus, Ghellinck in *B. Z.*, xxi. 453–457 (1913). On ignorance of Greek in mediaeval Germany, see Pendzig, in *Neue Jahrbücher*, xlii. 213–227 (1918).

brought back Greek manuscripts from Constantinople to the monastery of Saint-Denis at Paris,[17] where he later became abbot (1172–86). Sent out originally by Abbot Odo, he was evidently specially charged with securing the works attributed to Dionysius the Areopagite, who was confused with the patron saint of the monastery and of France, and a volume of these which he brought back is still preserved among the Greek codices of the Bibliothè-que Nationale.[18] He also brought with him and translated the text of the *Vita Secundi*, a philosophical text of the second cen-tury,[19] and summaries (*hypotheses*) of the Pauline epistles, while still other manuscripts may have been included in the *opes atticas et orientales* mentioned by one of his fellow-monks. This monk, also named William and sometimes confused with the physician, translated the eulogy of Dionysius by Michael Syncellus, but the writings which occupy the remainder of the Dionysian volume — *De caelesti hierarchia, De ecclesiastica hierarchia, De divinis nomini-bus, De mystica theologia*, and ten epistles — were rendered into Latin by John Sarrazin.[20] This John had himself visited the Greek East, where he had sought in vain the *Symbolica theologia* of Dionysius, as we learn from one of his prefaces.[21] In spite of the crudeness of his translations, his learning was valued by John of Salisbury, who turns to him on a point of Greek which Latin masters cannot explain, and who even expresses a desire to sit at Sarrazin's feet.[22]

[17] The material relating to William the Physician is conveniently given by De-lisle, in *Journal des savants*, 1900, pp. 725–739.

[18] MS. Gr. 933.

[19] Delisle, in *Journal des savants*, p. 728. The version is critically edited, and its use by French writers traced, by A. Hilka in *88. Jahresbericht der schlesischen Ge-sellschaft für vaterländische Cultur* (Breslau, 1910), iv. Abt., c. 1. See further F. Pfister, in *Wochenschrift für klassische Philologie*, 1911, coll. 539–548. On the popu-larity of the Latin version, see Manitius, *Geschichte der lateinischen Litteratur im Mittelalter*, i. 285; Thorndike, ii. 487.

[20] Delisle, pp. 726 ff.; *Histoire littéraire de la France*, xiv. 191–193. MSS. of these translations, with the prefaces, are common, e. g., Bibliothèque de l'Arsenal, MS. 529; Chartres, MS. 131; Vatican, MS. Vat. Lat. 175; Madrid, Biblioteca Na-cional, MS. 523 (A. 90); Munich, MSS. 380, 435. On the influence of Sarrazin, who also wrote a commentary on the *Celestial Hierarchy*, see now Grabmann, in *Festgabe Albert Ehrhard* (Bonn, 1922), pp. 180–199; and G. Théry, in *Revue des sciences philosophiques et théologiques*, xi. 72–81 (1922).

[21] Delisle, p. 727.   [22] *Epistolae*, no. 169; cf. also nos. 147, 149, 223, 229, 230

The dependence of the leading classicist of the age upon a man like Sarrazin shows the general ignorance of Greek. "The most learned man of his time," John of Salisbury made no less than ten journeys to Italy, in the course of which he visited Benevento and made the acquaintance of the Sicilian chancellor; he knew Burgundio, whom he cites on a point in the history of philosophy; [23] he studied with a Greek interpreter of Santa Severina, to whom he may have owed his early familiarity with the *New Logic*; yet his culture remained essentially Latin.[24] "He never quotes from any Greek author unless that author exists in a Latin translation." [25] So the theologian whom John considers his most learned contemporary, Gilbert de la Porrée, though he knows something of the Greek Fathers, is quite ignorant of that language.[26] Greek could be learned only in southern Italy or the East, and few there were who learned it, as one can see from the sorry list of Greek references which have been culled from the whole seventy volumes of the Latin *Patrologia* for the twelfth century.[27] The Hellenism of the Middle Ages was a Hellenism of translations — and so, in large measure, was the Hellenism of the Italian Renaissance.[28]

Finally there remain to be mentioned the anonymous translations, made for the most part doubtless in Italy. Where we are fortunate enough to have the prefaces, these works can be dated approximately and some facts can be determined with respect to their authors, as in the case of the first Latin version of the *Al-*

[23] *Metalogicus*, bk. iv., c. 7.

[24] Schaarschmidt, *Johannes Saresberiensis* (Leipzig, 1862); Poole, in *Dictionary of National Biography*; C. C. J. Webb, *Ioannis Saresberiensis Policraticus*, i, introd.

[25] Sandys, *History of Classical Scholarship*[3], i. 540.

[26] *M. G. H., Scriptores*, xx. 522; infra, Chapter X, p. 213.

[27] How sorry this list is, the Abbé A. Tougard does not seem to realize when he has drawn it up: *L'hellénisme dans les écrivains du moyen âge* (Paris, 1886), ch. v. On the reserve necessary in using such citations, cf. Traube, *O Roma Nobilis* (Munich, 1891), p. 65. For a list of theological MSS. of the twelfth century not in the *Patrologia*, see Noyon, in *Revue des bibliothèques*, 1912, pp. 277–333; 1913, pp. 297–319, 385–418. On Greek in the twelfth century, see Sandys, pp. 555–558. Miss Louise R. Loomis, *Medieval Hellenism* (Columbia thesis, 1906), adds nothing on this period.

[28] Loomis, "The Greek Renaissance in Italy," in *American Historical Review*, xiii. 246–258 (1908).

magest, made in Sicily about 1160, and a version of Aristotle's *Posterior Analytics* (1128–59) preserved in a manuscript of the cathedral of Toledo.[29]   In the majority of cases no such evidence has been handed down, and we have no guide beyond the dates of codices and the citations of texts in a form directly derived from the Greek.   Until investigation has proceeded considerably further than at present, the work of the twelfth century in many instances cannot clearly be separated from that of the earlier Middle Ages on the one hand, and on the other from that of the translators of the thirteenth and fourteenth centuries who follow in unbroken succession.   Often we know only that a particular work had been translated from the Greek before the time of the humanists.   The most important body of material with which the twelfth century may have occupied itself anonymously is the writings of Aristotle.[30]   The *Physics*, *Metaphysics*, and briefer works on natural history reach western Europe about 1200; the *Politics*, *Ethics*, *Rhetoric*, and *Economics* only in the course of the next two generations.   In nearly every instance translations are found both from the Greek and from the Arabic, and nearly all are undated.   At present about all that can be said is that by the turn of the century traces are found of versions from the Greek in the case of the *Physics*, *De caelo*, *De anima*, and the *Parva naturalia*, and perhaps of the *Metaphysics*.

On the personal side these Hellenists of the twelfth century have left little of themselves.   James of Venice is only a name; the translator of the *Almagest* is not even that.   Moses of Bergamo we know slightly through the accident which has preserved one of his letters; others survive almost wholly through their prefaces. Characteristic traits or incidents are few — Moses lamenting the loss of his Greek library, and the three pounds of gold it had cost him; the Pisan secretary of Manuel Comnenus trailing after the emperor on the tortuous marches of his Turkish campaigns; Burgundio redeeming his son's soul from purgatory by translating Chrysostom in the leisure moments of his diplomatic journeys; a Salerno student of medicine braving the terrors of Scylla and Charybdis in order to see an astronomical manuscript just ar-

[29] Infra, Chapters IX, XI.   [30] Infra, Chapters XI, XVIII.

rived from Constantinople, and remaining in Sicily until he had mastered its contents and made them available to the Latin world; Aristippus working over Plato in camp and investigating the phenomena of Etna's eruptions in the spirit of the elder Pliny; Eugene the Emir, in prison at the close of his public career, writing Greek verse in praise of solitude and books. Little enough all this, but sufficient to show the kinship of these men with "the ancient and universal company of scholars."

So far as we know, these Hellenists produced no grammars like Roger Bacon's or the *Erotemata* of Chrysoloras, though Moses of Bergamo turned into Latin the substance of two chapters of the grammar of Theodosius of Alexandria.[31]   Nor was their knowledge of Greek reflected in Greek dictionaries or in any permanent improvement in lexicography; indeed the Greek of the etymologists grows worse rather than better as the Middle Ages wear on. When, about 1200, the learned Pisan canonist Hugutio, professor at Bologna and bishop of Ferrara, compiles his *Derivationes*, he takes his Greek etymologies chiefly from his predecessors, the Lombard Papias (1053) and the Englishman Osbern, both likewise ignorant of Greek; yet Hugutio was the standard lexicographer of the later Middle Ages and was by Petrarch bracketed with Priscian as the chief of grammarians.[32]   The *Grecismus* of Evrard de Béthune (1212), a favorite grammar in its time, is notable chiefly for its ignorance of Greek.[33]   Some acquaintance with the language was claimed by William of Corbeil, who in the early twelfth century dedicated his *Differentie* to Gilbert de la Porrée.[34]

In all its translations the twelfth century was closely, even painfully literal, in a way that is apt to suggest the stumbling and conscientious school-boy.   Every Greek word had to be repre-

---

[31] Infra, Chapter X, n. 64.

[32] On Hugutio see particularly G. Götz, in Leipzig *Sitzungsberichte*, lv. 121–154 (1903); id., in *Corpus glossariorum Latinorum*, i. ch. 17, who cites 106 MSS. and prints the pompous preface. On Osbern's writings see Miss Bateson, in the *Dictionary of National Biography*. Besides the MSS. there cited (Royal 6 D ix of the British Museum; and 654 of Rouen), I have used the dialogues in MS. 301 at Tours, ff. 76–110.

[33] Ed. Wrobel (Berlin, 1887); cf. Sandys³, i. 667.   [34] Infra, Chapter X, n. 119.

sented by a Latin equivalent, even to μέν and δέ. Sarrazin laments that he cannot render phrases introduced by the article, and even attempts to imitate Greek compounds by running Latin words together.[35] The versions were so slavish that they are useful for establishing the Greek text, particularly where they represent a tradition older than the extant manuscripts. This method, *de verbo ad verbum*, was, however, followed not from ignorance but of set purpose, as Burgundio, for example, is at pains to explain in one of his prefaces.[36] The texts which these scholars rendered

[35] John of Salisbury, *Epistolae*, nos. 149, 230; cf. William the Physician, in *Journal des savants*, 1900, p. 738.

[36] 'Verens igitur ego Burgundio ne, si sentenciam huius sancti patris commentacionis assumens meo eam more dictarem, in aliquo alterutrorum horum duorum sapientissimorum virorum sentenciis profundam mentem mutarem et in tam magna re, cum sint verba fidei, periculum lapsus alicuius alteritatis incurrerem, difficilius iter arripiens, et verba et significationem eandem et stilum et ordinem eundem qui apud Grecos est in hac mea translatione servare disposui. Sed et veteres tam Grecorum quam et Latinorum interpretes hec eadem continue egisse perhibentur,' the Septuagint being an example, though St. Jerome made a new version of Isaiah. 'Sanctus vero Basilius predictum Ysaiam prophetam exponens lxx duorum interpretum editione[m] mirabiliter ad litteram commentatur, eiusque commentacionem ego Burgundio iudex domino tercio Eugenio beate memorie pape de verbo ad verbum transferens ex predicta lxx duorum interpretum editione facta[m?] antiquam nostram translationem in omnibus fere sum prosequtus, cum Sancti Ieronimi novam suam editionem nullatenus ibi expositam invenirem nec eam sequi ullo modo in ea commentacione possem. Psalterium quoque de verbo ad verbum de greco in latinum translatum est sermonem, et diverse ille quoque eius proferuntur apud Latinos edictiones romana < > ex equivocacione grecarum dictionum ortas esse perpendo, interpretibus modo hanc modo illam in eis assumentibus significacionem.' He then passes in review the other literal translations previously made from the Greek — the Twelve Tables, the *Corpus Juris Civilis*, the *Dialogues* of Gregory the Great, Chalcidius's version of the *Timaeus*, Priscian, Boethius, the *Aphorisms* of Hippocrates and the *Tegni* of Galen, John the Scot's version of Dionysius the Areopagite, and the *De urinis* of Theophilus — and concludes: 'Si enim alienam materiam tuam tuique iuris vis esse putari, non verbo verbum, ut ait Oratius, curabis reddere ut fidus interpres, ymo eius materiei sentenciam sumens tui eam dictaminis compagine explicabis, et ita non interpres eris sed ex te tua propria composuisse videberis. Quod et Tullius et Terentius se fecisse testantur. . . . Cum igitur hec mea translatio scriptura sancta sit et in hoc meo labore non gloriam sed peccatorum meorum et filii mei veniam Domini expectavi, merito huic sancto patri nostro Iohanni Crisostomo sui operis gloriam et apud Latinos conservans, verbum ex verbo statui transferendum, deficienciam quidem dictionum intervenientem duabus vel etiam tribus dictionibus adiectis replens, idyoma vero quod barbarismo vel metaplasmo vel scemate vel tropo fit recta et propria sermocinacione retorquens.' Preface to translation of

were authorities in a sense that the modern world has lost, and their words were not to be trifled with. Who was Aristippus that he should omit any of the sacred words of Plato?[37] Better carry over a word like *didascalia* than run any chance of altering the meaning of Aristotle.[38] Burgundio might even be in danger of heresy if he put anything of his own instead of the very words of Chrysostom. It was natural in the fifteenth century to pour contempt on such translating, even as the humanists satirized the Latin of the monks, but the men of the Renaissance did not scruple to make free use of these older versions, to an extent which we are just beginning to realize. Instead of striking out boldly for themselves, the translators of the Quattrocento were apt to take an older version where they could, touching it up to suit current taste. As examples may be cited the humanistic editions of Aristotle's *Logic*, of Chrysostom and John of Damascus, and even of Plato.[39] It has always been easier to ridicule Dryasdust than to dispense with him!

Apart from such unacknowledged use during the Renaissance, the translators of the twelfth century made a solid contribution to the culture of the later Middle Ages. Where they came into competition with translations from the Arabic, it was soon recognized that they were more faithful and trustworthy. At their best the Arabic versions were one remove further from the original and had passed through the refracting medium of a wholly different kind of language,[40] while at their worst they were made in haste and with the aid of ignorant interpreters working through the Spanish vernacular.[41] In large measure the two sets of trans-

Chrysostom's St. John, Vatican, MS. Ottoboni Lat. 227, ff. 1 v–2, a corrupt text respecting which I owe much to the aid of Monsignore Giovanni Mercati. For specimens of Burgundio's method, see Dausend, in *Wiener Studien*, xxxv. 353–369; and cf. the parallel versions studied by Hocedez, in *Musée belge*, xvii. 109–123 (1913).

[37] Even to the point of rendering τε καί by *que et*. *Rassegna bibliografica della letteratura italiana*, xiii. 12.   [38] Infra, pp. 234 f.

[39] Infra, pp. 167, 208, 240 f.; *Wochenschrift für klassische Philologie*, 1896, col. 1097; Minges, in *Philosophisches Jahrbuch*, xxix. 250–263 (1916).

[40] Eugene of Palermo remarks on the difference of Arabic idiom. G. Govi, *L'Ottica di Claudio Tolomeo* (Turin, 1885), p. 3; infra, p. 172.

[41] Cf. Rose in *Hermes*, viii. 335 ff.

lators utilized the same material. Both were interested in philosophy, mathematics, medicine, and natural science; and as most of the Greek works in these fields had been turned into Arabic, any one of these might reach the West by either route. If Plato could be found only in the Greek, Aristotle was available also in Arabic, and for most of his works there exist two or more parallel Latin versions. Theology, liturgy, and hagiography, as well as grammar, naturally came from the Greek alone, while astrology was chiefly Arabic. Nevertheless in the realm of the occult and legendary we have Kiranides and the dream-books, *Kalila and Dimna* and the Sibyl, some alchemy perhaps, and the *Quadripartitum* of Ptolemy and other bits of astrology.[42] In many instances it was more or less a matter of accident whether the version from the Greek or that from the Arabic should pass into general circulation; thus the Sicilian translation of the *Almagest*, though earlier, is known in but four copies, while that made in Spain is found everywhere. The list of works known only through the Greek of the twelfth century is, however, considerable. It comprises the *Meno* and *Phaedo* of Plato, the only other dialogue known to the Middle Ages being the *Timaeus*, in an older version; the advanced works of Euclid; Proclus and Hero; numerous treatises of Galen; Chrysostom, Basil, Nemesius, John of Damascus, and the Pseudo-Dionysius; and a certain amount of scattered material, theological, legendary, liturgical, and occult.[43]

The absence of the classical works of literature and history from the list of translations from the Greek is as significant as it is in the curriculum of the mediaeval universities. We are in the twelfth century, not the fifteenth, and the interest in medicine, mathematics, philosophy, and theology reflects the practical and ecclesiastical preoccupations of the age rather than the wider interests of the humanists. The mediaeval translations "were not regarded as *belles lettres*. They were a means to an end."[44] It is

---

[42] Chapter V, end; Chapter X, end.

[43] Sabbadini, *Le scoperte dei codici: nuove ricerche*, pp. 262–265, gives a list of mediaeval versions from which Euclid, Hero, the *Geoponica*, Nemesius and others are absent.

[44] D. P. Lockwood, in *Proceedings of the American Philological Association*, xlix. 125 (1918).

well, however, to remember that these same authors continue to be read in the Quattrocento, in translations new or old; they are merely crowded into the background by the newer learning. In this sense there is continuity between the two periods. There is also a certain amount of continuity in the materials of scholarship — individual manuscripts of the earlier period gathered into libraries at Venice or Paris, the library of the Sicilian kings probably forming the nucleus of the Greek collections of the Vatican.[45] To what extent there was a continuous influence of Hellenism is a more difficult problem, in view of our fragmentary knowledge of conditions of the south. The Sicilian translators of the twelfth century are followed directly by those at the courts of Frederick II and Manfred, while in the fourteenth century we have to remember the sojourn of Petrarch at the court of Robert of Naples, and the Calabrian Greek who taught Boccaccio. The gap is short, but it cannot yet be bridged.

[45] See the studies of Heiberg, Ehrle, and Birkenmajer cited in Chapter IX, n. 35. Björnbo, "Die mittelalterlichen lateinischen Uebersetzungen aus dem Griechischen," in *Archiv für die Geschichte der Naturwissenschaften*, i. 385–394 (1909), should be consulted for later versions of mathematical works. See also the more general pages of Heiberg, "Les sciences grecques et leur transmission," in *Scientia*, xxxi. 1–10, 97–104 (1922).

CHAPTER IX

THE SICILIAN TRANSLATORS OF THE TWELFTH CENTURY[1]

THE Norman kingdom of southern Italy and Sicily occupies a position of peculiar importance in the history of mediaeval culture.[2] Uniting under their strong rule the Saracens of Sicily, the Greeks of Calabria and Apulia, and the Lombards of the south-Italian principalities, the Norman sovereigns were still far-sighted and tolerant enough to allow each people to keep its own language, religion, and customs, while from each they took the men and the institutions that seemed best adapted for the organization and conduct of their own government. Greek, Arabic, and Latin were in constant use among the people of the capital and in the royal documents;[3] Saracen emirs, Byzantine logothetes, and

---

[1] Based upon *Harvard Studies in Classical Philology*, xxi. 75–102 (1910), xxiii. 155–166 (1912), the first being a joint article with Professor Dean Putnam Lockwood which he kindly permits me to incorporate here. His discovery of MS. Vat. 2056 was the starting-point of the essay. For discussion, see, particularly, Heiberg, "Noch einmal die mittelalterliche Ptolemaios-Uebersetzung," in *Hermes*, xlvi. 207–216; Paul Marc, in *B. Z.*, xix. 568, 569; Bresslau, in *Neues Archiv*, xxxvi. 304, xxxix. 253; and the description of MS. 2056 in the new catalogue of *Codices Vaticani Latini*.

[2] On the culture of southern Italy and Sicily in the twelfth century, see M. Amari, *Storia dei Musulmani di Sicilia* (Florence, 1854–72), iii. 441–464, 655 ff.; V. Rose, "Die Lücke im Diogenes Laërtius und der alte Uebersetzer," in *Hermes* (1866), i. 367–397; E. A. Freeman, *The Normans at Palermo*, in his *Historical Essays*, third series, pp. 437–476; G. B. Siragusa, *Il regno di Guglielmo I in Sicilia* (Palermo, 1885–86), i. 139–148, ii. 101–144; O. Hartwig, "Die Uebersetzungsliteratur Unteritaliens in der normannisch-staufischen Epoche," in *Centralblatt für Bibliothekswesen* (1886), iii. 161–190, 223–225, 505 f.; E. Caspar, *Roger II und die Gründung der normannisch-sicilischen Monarchie* (Innsbruck, 1904), pp. 435–472; F. Chalandon, *Histoire de la domination normande en Italie et en Sicile* (Paris, 1907), ii. 708–742, where the literary side of the subject is treated much too briefly; Haskins, *The Normans in European History* (Boston, 1915), chs. 7, 8. On the Greek element in the South, see also F. Lenormant, *La grande Grèce* (Paris, 1881–84); P. Batiffol, *L'abbaye de Rossano* (Paris, 1891); and the studies on Casule in *Rivista storica calabrese*, vi.

[3] K. A. Kehr, *Die Urkunden der normannisch-sicilischen Könige* (Innsbruck, 1902), pp. 239–243.

Norman justiciars worked side by side in the royal *curia*; and it has been a matter of dispute among scholars whether so fundamental a department of the Sicilian state as finance was derived from the *diwan* of the caliphs, the *fiscus* of the Roman emperors, or the exchequer of the Anglo-Norman kings.[4] King Roger, like his grandson Frederick II, drew to his court men of talent from every land, regardless of speech or faith: an Englishman, Robert of Selby, stood at the head of his chancery, and others from beyond the Alps found employment in his government;[5] a Greek monk, Nilus Doxopatres, wrote at his command the history of the five patriarchates which was directed at the supremacy of the Roman see; a Saracen, Edrisi, prepared under his direction the comprehensive treatise on geography which became celebrated as 'King Roger's Book.' A court where so many different types of culture met and mingled inevitably became a place for the interchange and diffusion of ideas, and particularly for the transmission of eastern learning to the West. Easy of access, the Sicilian capital stood at the centre of Mediterranean civilization, and while the student of Arabic science and philosophy could in many respects find more for his purpose in the schools of Toledo, Palermo had the advantage of direct relations with the Greek East and direct knowledge of works of Greek science and philosophy which were known in Spain only through Arabic translations or compends. Especially was a cosmopolitan court like the Sicilian favorable to the production of translations. Knowledge of more than one language was almost a necessity for the higher officials as well as for the scholars of Sicily, and Latin versions of Greek and

---

[4] R. Pauli, in *Nachrichten* of the Göttingen Academy, 1878, pp. 523–540; Hartwig and Amari, in *Memorie dei Lincei*, third series, ii. 409–438; C. A. Garufi, in *Archivio storico italiano*, fifth series, xxvii. 225–263; O. von Heckel, in *Archiv für Urkundenforschung* (1908), i. 371 ff.; my article on "England and Sicily in the Twelfth Century," in *E. H. R.*, xxvi. 433–447, 641–665 (1911).

[5] Hugo Falcandus, *Liber de regno Sicilie*, ed. Siragusa, p. 6: 'Quoscumque viros aut consiliis utiles aut bello claros compererat, cumulatis eos ad virtutem beneficiis invitabat. Transalpinos maxime, cum ab Normannis originem duceret sciretque Francorum gentem belli gloria ceteris omnibus anteferri, plurimum diligendos elegerat et propensius honorandos.' Cf. Romualdus of Salerno, in *M. G. H.*, *Scriptores*, xix. 426; John of Salisbury, *ibid.*, xx. 538; John of Hexham, *ibid.*, xxvii. 15; ibn-al-Atir, in Amari, *Biblioteca Arabo-Sicula*, i. 450.

Arabic works were sure to be valued by the northern visitors of scholarly tastes who came in considerable numbers to the South and wished to carry back some specimen of that eastern learning whose fame was fast spreading in the lands beyond the Alps.

The achievements of the Sicilian scholars of the twelfth century are in part known, thanks particularly to the studies of Amari and Valentin Rose, but the sources of information are of a very scanty sort, and new material is greatly needed. We can now add Ptolemy's *Almagest* to the list of works known to have been turned into Latin in Sicily, and, with that as our starting-point, bring out additional facts concerning the Sicilian translators and their work.

The mediaeval versions of the *Almagest* we have discussed in another connection.[6] The earliest of those made from the Arabic, that of Gerard of Cremona, was completed in 1175, and three others are known before George Trapezuntius made his version directly from the Greek in 1451.[7] Of these the most interesting is what appears to be the earliest Latin version of all, made in Sicily about 1160 and based directly upon the original Greek. Four manuscripts are known:

A. MS. Vat. Lat. 2056, belonging to the fourteenth or possibly to the very end of the thirteenth century, a well-written parchment codex formerly in the possession of Coluccio Salutati.[8] The translation of the *Almagest* occupies the ninety-four numbered folios,[9] and there are four

---

[6] Chapter V, end.

[7] Voigt, *Die Wiederbelebung des classischen Alterthums*[3], ii. 141.

[8] F. 88 v: 'Liber Colucii.' F. 94 v: 'Liber Colucii Pyeri de Salutatis.'

[9] The *incipit* and *explicit* of each book are given for identification of other possible copies: F. 1–1 v, preface, as printed below, pp. 191–193. Ff. 1 v–9, book i: 'Valde bene qui proprie philosophati sunt, o Sire, videntur michi sequestrasse theoreticum philosophie a practico . . . atque inde manifestum est quoniam et reliquorum taetartimoriorum ordinatio contingit eadem omnibus in unoquoque eisdem contingentibus propter rectam speram, id est equinoctialem, sine declinatione ad orizontem subiacet.' Ff. 9 v–26, book ii: 'Pertranseuntes in primo sintaxeos de totorum positione capitulatim debentia prelibari . . . minutione vero quando occidentalior subiacens.' Ff. 26–33, book iii: 'Assignatas a nobis in ante hoc coordinatis et universaliter debentibus de celo et terra mathematice prelibari . . . piscium gradus .vi .xlv., anomalie vero .iiiᵃ. g[radus] et .viii. ad proximum sexagesima piscium.' Ff. 33–41, book iv: 'In eo quod ante hoc coordinantes quecunque utique quis videat contingentia circa solis motum . . . in coniugationibus lune et ipsis

fly-leaves, partly in blank and partly covered with astronomical notes and symbols in a hand different from the text. The text averages fifty lines to a page, and the written page measures ca. 14.7 by 25.5 centimetres. There are no illustrations in the text, but the outer margins have many geometrical figures, beautifully drawn and often of great intricacy, and lettered in a hand which seems to be that of the original scribe. The text and the titles of chapters which appear at the head of each book are written in a single hand, but the hands of several correctors and annotators appear both in the text and in the tables. This, the only complete MS. so far known of the Sicilian version, was discovered by Professor Lockwood in the spring of 1909 and described in *Harvard Studies in Classical Philology* (xxi. 78 f.) in 1910. See now the new catalogue of the Vatican MSS. (1912); and Heiberg in *Hermes*, xlvi. 207 ff. (1911).

B. Florence, Biblioteca Nazionale, Conventi Soppressi, MS. A. 5. 2654. Written in a southern hand of ca. 1300. Lacks preface and the first twelve chapters of book i. Discovered by Björnbo and indicated in *Archiv für die Geschichte der Naturwissenschaften*, i. 392 (1909); described by Heiberg, "Eine mittelalterliche Uebersetzung der Syntaxis

eclipsibus consonius maxime nostris ypothesibus inventis.' Ff. 41–47 v, book v: 'Causa vero earum que ad solem sinzugiarum et sinodicarum vel panselinicarum . . . periferiam maiorem esse ea que est .zb. habuimus et .aiz. angulum g[radus] .xxxv. et d[imidium], quod propositum erat demonstrando.' Ff. 47 v–55 v, book vi: 'Deinceps ergo contingente eo quod circa eclipticas sinzugias solis et lune negotio . . . universalius recipientes lunarium partes primas et extremas eclipsium et completionum significationes.' Ff. 56–61 v, book vii: 'Pertranseuntes in ante hoc coordinatis, o Sire, et circa rectam et circa inclinatam speram contingentia . . .' [table]. Ff. 62–66 v, book viii: [Table] '. . . spatia sumptis ad solem significationibus et in ipsis in parte lune acclinationibus.' Ff. 66 v–72 v, book ix: 'Igitur quecunque quidem quis et de fixis stellis velut in capitulis commemorat secundum quantum usque nunc apparentia processum conceptionis . . . tantis vero .i. et .vi. superant chelarum g[radus] qui secundum observationem.' Ff. 72 v–76 v, book x: 'Igitur stelle quidem mercurii ypotheses et quantitates anomaliarum, . . . optinebit manifestum quoniam et secundum expositum epochis temporis cancri g[radus] .xvi. .xl.' Ff. 76 v–83 v, book xi: 'Demonstratis circa martis stellam periodicis motibus et anomaliis et epochis . . . et collectum g[raduum] numerum dementes ab eo quod tunc apoguio stelle, in apparentem ipsius progressionem incuremus.' Ff. 83 v–88 v, book xii: 'His demonstratis consequens utique erit et secundum unamquamque quinque erraticarum factas precessiones . . . tertio vero hesperias et rursum quarto eoas et quinto esperias, et est canon huiusmodi:' [table]. Ff. 88 v–94 v, book xiii: 'Delictis autem in eam que de quinque erraticis coordinationem adhuc duobus his et secundum latitudinem . . . et que ad commoditatem solam contemplationis sed non ad ostentationem commemoratio suggerebat, proprium utique nobis hic et commensurabilem recipiat finem presens negotium.'

des Ptolemaios," in *Hermes*, xlv. 57–66 (1910). Neither of these scholars then knew of the existence of A.

C. Vatican, MS. Pal. lat. 1371, ff. 41–97 v; thirteenth century. Complete only as far as 6, 10, including the preface, but offering a text superior to A in accuracy and in the mechanical execution of the illuminations, though omitting some of the tables. The scribe seems to have tried to improve the text, especially in the order of words. Opposite the title an Italian hand of the fourteenth century has written in the margin 'Translatus in urbe Panormi tempore regis Roggerii per Hermannum de greco in latinum.' [10] Discovered by me in June, 1911, and described in *Harvard Studies*, xxiii. 155–166 (1912). Since noted by Monsignore A. Pelzer, in *Archivum Franciscanum historicum*, xii. 60 (1919), who dates it '12ᵉ–13ᵉ siècle,' and the marginal note '13ᵉ siècle.'

D. Wolfenbüttel, MS. Gud. lat. 147, f. 2. Preface only; see above, Chapter V, pp. 106–108.

In the preface, printed at the close of the present chapter, the translator, writing to the teacher of mathematics to whom he dedicates his work, says (lines 23–37) that, as he was laboring over the study of medicine at Salerno, he learned that a copy of Ptolemy's great treatise had been brought from Constantinople to Palermo, as a present from the Greek emperor, by an ambassador of the Sicilian king. This emissary, by name Aristippus, he set out to seek, and braving the terrors of Scylla and Charybdis and the fiery streams of Etna — this last doubtless on the way to Catania, where we know Aristippus was archdeacon — he found him at Pergusa,[11] near the fount, engaged, not without danger, in investigating the marvels of Etna. Our Salernitan scholar's astronomical knowledge was not, however, sufficient to permit his attempting at once the translation of the book which he had

---

[10] See above, Chapter III, no. *b*.

[11] This name gives rise to a difficulty, for the lake of Pergusa, the fabled scene of the rape of Proserpine (Ovid, *Metam.* 5, 386; Claudian, *De raptu Proserpinae*, 2, 112), lies in the vicinity of Castrogiovanni, the ancient Enna, at so considerable a distance from Etna that there would be no possible danger to an observer. Cf. *Hermes*, xlvi. 208, n. The phrase *ethnea miracula* would seem too definite to be interpreted as volcanic phenomena which might occur in the region of Pergusa at a time of disturbance of Etna. Very possibly the author meant some fount in the neighborhood of Etna otherwise unknown to us.

sought, even if there had been no other obstacles in the way, and, already familiar with Greek (*preinstructus*), he applied himself diligently to the preliminary study of the *Data*, *Optica*, and *Catoptrica* of Euclid and the *De motu* of Proclus. When ready to attack the *Almagest* he had the good fortune to find a friendly expositor in Eugene, a man most skilled in Greek and Arabic and not unfamiliar with Latin, and succeeded, contrary to the desire of an ill-tempered man,[12] in turning the work into Latin.

The date of these events can be fixed with some definiteness owing to the mention of Aristippus, who was an important personage in Sicilian history in the reign of William I. Made archdeacon of Catania in 1156, in which year he is found with the king at the siege of Benevento, Henricus Aristippus was in November, 1160, after the murder of the emir of emirs, Maio, advanced to the position of royal *familiaris* and placed in charge of the whole administration of the kingdom; but in the spring of 1162, while on the way to Apulia, he was suddenly seized by the king's order and sent to Palermo to prison, where he shortly afterward died.[13] The meeting at the fount of Pergusa was thus anterior, not only to the events of 1162, but probably also to the promotion of 1160, after which the necessity of constant presence at the *curia* left no time for scientific pursuits. If we follow the diplomatic history of Sicily back to the assumption of the royal title in 1130, we find only three embassies to Constantinople, and the relations of the Greek emperor and the Sicilian king were such during this period that it is quite unlikely that there were others. The first series of

---

[12] 'Contra viri discoli voluntatem.' This may be connected with the unexplained obstacle ('cum occulte quidem alia . . . prohiberent') referred to above, but if the opposition of an unnamed person is meant, we should expect *cuiusdam*, while the mention of Eugene's assistance makes one hesitate to apply the reference to him, as does Heiberg (*Hermes*, xlvi. 209, no. 1). I give Heiberg's interpretation of *preinstructus*, though one would expect *iam instructus* if the knowledge of Greek had been previously acquired.

[13] Except for his prologues to the *Meno* and *Phaedo* of Plato (*Hermes*, i. 386–389) and for the text which we print below, the facts concerning the life of Aristippus are known only from the chronicle of Hugo Falcandus, ed. Siragusa, pp. 44, 55, 69, 81. See Siragusa, *Il regno di Guglielmo I*, i. 144–145; ii. 18, 51–52, 107–112; Kehr, *Die Urkunden der normannisch-sicilischen Könige*, pp. 80 (on the date of the death of Aschettinus, predecessor of Aristippus as archdeacon), 82–83; Chalandon, *Domination normande*, ii. 174, 272, 273, 276, 277, 282, 289.

negotiations falls in 1143 and 1144, when a mission sent to arrange a marriage alliance failed of its purpose because of the death of the Emperor John Comnenus and when a second set of ambassadors was put in prison by his son Manuel.[14] In neither of these instances is it at all probable that the emperor presented a valuable manuscript to King Roger, nor would Aristippus have been a man of sufficient importance to be employed in so responsible a position. For similar reasons he can hardly have been one of the emissaries despatched by William I on his accession in 1154, for these were all bishops and were not well received.[15] By 1158, on the other hand, when peaceful relations were resumed between the two sovereigns, Aristippus occupied a higher position, and the Emperor Manuel, who had not been successful in the preceding campaigns, had every reason to deal generously with the envoys who concluded the peace of that year.[16] If, accordingly, the manuscript of the *Almagest* was brought to Sicily at this time,[17] the meeting with Aristippus can hardly have been much earlier than 1160, and it certainly was not more than two years later. Some time must be allowed for the studies described and for the actual labor of translation, but three or four years would suffice for all this, and we can with reasonable certainty conclude that the translation was completed at least ten years before Gerard of Cremona produced his version in 1175.

Of the name and nationality of the author of this translation nothing is revealed beyond the fact that he is a stranger to southern Italy and Sicily. The statement of the gloss that his name was Hermann we have already had occasion to examine and reject.[18] He calls himself a tardy follower of philosophy (*philoso-*

[14] Caspar, *Roger II*, pp. 362–364; Chalandon, *o. c.*, ii. 127–129.

[15] Cinnamus, 3, 12 (ed. Bonn, p. 119): ἦκον οὖν ἄνδρες ἐπίσκοπον ἕκαστος περικείμενος ἀρχήν. Cf. Chalandon, *Domination normande*, ii. 188 f. Nor does Aristippus in 1156 (*Hermes*, i. 388) mention the *Almagest* in his enumeration of notable books available in Sicily.

[16] Siragusa, *Il regno di Guglielmo I*, i. pp. 74–76; Chalandon, *o. c.*, i. 253 f.

[17] Beyond the fact that there was an eruption before 1162, the chronology of Mount Etna's eruptions in the period preceding 1169 is not known with sufficient fulness and exactness to be of assistance in dating the reference in our text. Cf. Sartorius von Waltershausen, *Der Aetna* (Leipzig, 1880), i. 210–211; Amari, *Biblioteca Arabo-Sicula*, i. 134–135.   [18] Supra, Chapter III, p. 53.

phie tardus assecla) in almost the same words used by Hermannus
Contractus and Adelard of Bath,[19] and seeks to defend the divine
science against the attacks of the profane; but his main interest
is plainly in the studies of the *quadrivium*, in which he has been
instructed by the master to whom his version of the *Almagest* is
dedicated, and which he defends at some length from the criticism
of the religious.[20] He must have been familiar with Euclid's *Elements* before his arrival in Sicily, for he is able to take up the more
advanced applications of geometry contained in Euclid's other
works, and he has made at least a beginning in medicine. He has
picked up an Arab proverb, and can quote Boethius and Remigius of Auxerre, as well as Ovid. He also quotes, though perhaps
not at first hand, Aristotle's *De caelo* from a Greek source,[21] and
his own knowledge of Greek is respectable.[22]

How fully our translator succeeded in mastering the difficult
subject-matter of Ptolemy's treatise is a question that must be
left to specialists in ancient astronomy. Granted, however, that
his work was done with reasonable intelligence, it has an importance for the study of the Greek text far superior to the version of
Gerard of Cremona, who worked from the Arabic with the aid of
a Spanish interpreter.[23] Not only did the author of the Sicilian
translation draw directly from the original Greek, but, like other
mediaeval translators from this language, he made a word-for-
word rendering which, while not so painfully awkward and school-
boyish as the translations of Aristippus,[24] is still very close and
literal.[25] For purposes of textual criticism a translation of this

[19] Migne, cxliii. 381; *Bullettino*, xiv. 91.

[20] Cf. Heiberg, in *Hermes*, xlvi. 210–213.

[21] Line 5: 'earum quas Aristotiles acrivestatas vocat artium doctrina.' The reference is evidently to the *De caelo*, 3, 7: μάχεσθαι ταῖς ἀκριβεστάταις ἐπιστήμαις, i. e.,
αἱ μαθηματικαί. No other mention of the *De caelo* has been found in the West before
the translation which Gerard of Cremona is said to have made from the Arabic. Cf.
Wüstenfeld, p. 67; Steinschneider, *E. U.*, no. 46 (11); id., *Centralblatt für Bibliothekswesen*, Beiheft xii. 55–57 (1893).   [22] Heiberg, in *Hermes*, xlvi. 210.

[23] On Gerard's method see above, Chapter I, n. 57. Yet it has been proposed
(Manitius, in *Deutsche Litteraturzeitung*, 1899, col. 578) to use his translation as an
aid to the establishment of the Greek text.

[24] See the specimen printed below, n. 42.

[25] Generally the number and order of the words in the Latin corresponds exactly
with the Greek, although a genitive absolute in the Greek may be rendered by a

sort is not much inferior to a copy of the Greek text, and as there
are but three existing manuscripts of the Μαθηματικὴ Σύνταξις
anterior to the twelfth century, such a translation would deserve
careful collation and study. Heiberg, however, has shown that
ours is based upon his MS. C, now no. 313 at St. Mark's, apparently the very codex of Aristippus, but through a lost copy
which had probably been emended by Eugene.[26]

However great its merits as a faithful reproduction of the original, it is clear that our translation exerted far less influence than
that of Gerard of Cremona upon the study of mathematical astronomy. Gerard himself was plainly unaware of its existence
when he started for Toledo, although when he came to translate
Aristotle's *Meteorologica* he knew of Aristippus' rendering of a
portion of that work,[27] and the evidence of citations and numerous
surviving copies shows that Gerard's was the version in current
use from the close of the twelfth century to the second half of the

cum-clause in the Latin, or the optative with ἄν be represented by *utique* with the
future indicative or subjunctive; ὅτι regularly becomes *quoniam*. A characteristic
practice is the use of *id quod* when a modifier, other than a simple adjective, stands
in the attributive position in the Greek; e. g., ἡ τῶν ὅλων θεωρία = ea que universorum speculatio. This Grecism occurs in the translator's own composition; see the
preface, l. 18: ad eam que astrorum, which would equal εἰς τὴν τῶν ἄστρων. In the
handling of technical terms the Greek words are often merely transliterated (for an
example see the beginning of book v, printed above, n. 9), but this is not done with
any consistency (e. g., συζύγια is rendered by both *sinzugia* and *coniugatio*, and
σύνταξις may appear as *sintaxis* or as *coordinatio*). The following passage from the
opening chapter of the first book may serve as a more connected specimen of the
translation:

Valde bene qui proprie philosophati sunt, o Sire, videntur michi sequestrasse
theoreticum philosophie a practico. Et enim si accidit (MS. accīt) et practico prius hoc ipsum theoreticum esse, nichilominus utique quis inveniet
magnam existentem in ipsis differentiam; non solum quod moralium quidem virtutum quedam multis et sine disciplina inesse possunt, eam vero que universorum
speccullationem absque doctrina consequi inpossibile, sed et eo quod ibi quidem ex
ea que in ipsis rebus est continua operatione, hic autem ex eo qui in theorematibus
processu, plurima utilitas fiat. Inde nobis ipsis duximus competere actus quidem in
ipsarum imaginationum investigationibus ordinare, ut nec in minimis eius que ad
bonum et bene dispositum statum considerationis obliviscamur. Scole vero dare
plurimum in theorematum multorum et bonorum existentium doctrinam, precipue
vero in eam que eorum que proprie mathematica nominantur. . . .'

[26] *Hermes*, xlv. 60–66, xlvi. 213–215.
[27] See below, n. 48.

fifteenth.[28] On the other hand, while only four manuscripts of the earlier translation have been found, this was not wholly forgotten. These manuscripts are copies, considerably posterior to the date of translation, and as one of them formed part of the library of Coluccio Salutati, the influence of this version can be followed into the period of the early Renaissance. Salutati's correspondence makes no mention of this manuscript, or indeed of the *Almagest*,[29] but it is altogether likely that this was one of the sources of his acquaintance with the opinions of famous astronomers,[30] including Ptolemy.

Of the incidental information furnished by the preface, special interest attaches to the fact that the manuscript of the *Almagest*, probably the very codex now in Venice,[31] was brought to Sicily as a present from the Greek emperor. We know that Manuel Comnenus took a special interest in astronomical and astrological studies,[32] and it is characteristic of the culture of the court of Palermo, as well as of the emperor's own tastes, that the great work of Ptolemy should be thought an appropriate gift to the Sicilian envoys. There is reason for thinking that other manuscripts went at this time from Constantinople to enrich Italian libraries. Certain early treatises on alchemy mention the Emperor Manuel in a way that suggests his reign as the period when

[28] Thus the Bibliothèque Nationale has ten copies of Gerard's translation (MSS· Lat. 7254–60, 14738, 16200, 17864), one of which (MS. Lat. 14738) is of the close of the twelfth century. The use of a version from the Arabic by Roger Bacon can be shown by the appearance in his citation (*Opus majus*, ed. Bridges, i. 231) of the form Abrachis, the Arabic corruption of Hipparchus in *Almagest*, 5, 14. Albertus Magnus uses Gerard's version (Pelzer, in *Revue néo-scolastique*, 1922, pp. 344, 479 f.), as does the *Speculum astronomie* commonly ascribed to him. As late as 1512 a copy of Gerard's version was made at Salamanca: Madrid, Biblioteca del Palacio, MS. 2. L.12. Another version from the Arabic was also current in Spain: see Chapter V, n. 150. Thomas Aquinas, however, knew a translation from the Greek: Jourdain, pp. 397 f.

[29] On the likelihood of its use, see Novati, *Epistolario di Coluccio Salutati*, iv, 1, p. 90, n. 1, who however supposes that Gerard's translation was employed.

[30] *Epp.*, 4, 11; 7, 22; 14, 4, 12, 24 (ed. Novati, i. 280, ii. 348, iv, 1, pp. 12, 86, 226). Cf. Voigt, *Wiederbelebung des classischen Alterthums*[3], i. 204. A copy of the Sicilian translation (not MS. A) was at Bologna in 1451: Sorbelli, *La biblioteca capitolare di Bologna nel secolo xv*, p. 93, no. 36.

[31] Heiberg, in *Hermes*, xlvi. 213.

[32] Chapter X, n. 174.

they were brought to the West,[33] and, as we shall see below, the Latin text of the prophecy of the so-called Erythraean Sibyl expressly states that it was translated from a copy brought from the treasury of the Emperor Manuel (*de aerario Manuelis imperatoris eductum*). Plainly manuscripts from the imperial library must be taken into account, as well as ecclesiastical and commercial influences, in tracing the intellectual connections between the Greek Empire and the West in the century preceding the Fourth Crusade.[34]

It is significant in relation to Latin learning, not only that the Sicilian court brought together an important library of Greek manuscripts, but that this collection probably passed, in part, from Manfred's library to that of the Popes, and thus became the nucleus of the Greek collections of the Vatican. This suggestion, first made by Heiberg, has been confirmed by Ehrle and Birkenmajer,[35] and opens up interesting possibilities of further inquiry.

In mentioning the envoy Aristippus and the expositor Eugene our text introduces us to the two leading figures among the Sicilian translators of this period. That King William's minister Aristippus was a man of learning in Greek and Latin literature had long been known from the chronicle of one of his associates in the royal administration,[36] but it was reserved for Valentin Rose to discover and publish in 1866 the prologues to the translation of the *Meno*

---

[33] J. Wood Brown, *Michael Scot* (Edinburgh, 1897), pp. 83–85. Brown conjectures that alchemical MSS. were brought to Sicily as a result of the Greek campaigns of George of Antioch, but even if the MSS. with which this admiral enriched the church of the Martorana were thus secured, they could not have been obtained from the imperial library, and it is hard to explain the mention of the emperor's name on any other ground than that the treatises had been in his possession.

[34] See the following chapter.

[35] Heiberg, *Les premiers MSS. grecs de la bibliothèque papale*, in *Oversigt* of the Danish Academy, 1891, pp. 315–318; id., in *Hermes*, xlv. 66, xlvi. 215; Ehrle, *Nachträge zur Geschichte der drei ältesten päpstlichen Bibliotheken*, in *Festgabe Anton de Waal* (Rome and Freiburg, 1913), pp. 348–351; Birkenmajer, *Vermischte Untersuchungen* (*Beiträge*, xx, no. 5, 1922), pp. 20–22. The Sicilian library appears also to have suffered losses before Parma in 1248: infra, Chapter XIV, n. 38.

[36] Hugo Falcandus, ed. Siragusa, p. 44: 'mansuetissimi virum ingenii et tam latinis quam grecis litteris eruditum.' That the author of this chronicle was a member of the Sicilian *curia*, very possibly a notary, is shown by Besta, "Il 'Liber de Regno Siciliae' e la storia del diritto siculo," in *Miscellanea di archeologia di storia e di filologia dedicata al Prof. A. Salinas* (Palermo, 1907), pp. 283–306.

and *Phaedo* of Plato which give us an idea of the range of his scholarship and constitute our chief source of information respecting the intellectual life of the Sicilian court.[37]

Dedicating his version of the *Phaedo* to a favorite of fortune (*roborato fortune* [38]) who is returning to his home in England, Aristippus pleads with him to remain in Sicily, where he has at his disposal not only the wisdom of the Latins but a Greek library and the aid of that master of Greek literature, Theoridus of Brindisi,[39] and of Aristippus himself, useful as a whetstone if not as a blade. In Sicily he will have access to the *Mechanics* of Hero, the *Optics* of Euclid, the *Posterior Analytics* of Aristotle, and other philosophical works. Best of all he will have a king whose equal cannot be found — *cuius curia schola comitatus, cuius singula verba philosophica apofthegmata, cuius questiones inextricabiles, cuius solutiones nihil indiscussum, cuius studium nil relinquit intemptatum.* It is, we learn from the prologue to the *Meno*, at the king's order that the archdeacon has begun a translation of Gregory Nazianzen, and at the instance of his chief minister, Maio, and the archbishop of Palermo that he has undertaken to render Diogenes Laertius into Latin. Neither of these, if ever completed, has reached us,[40] but the translations of the *Phaedo*[41] and

---

[37] *Hermes*, i. 386–389. The prologues are reprinted by Hartwig, *Archivio storico per le province napoletane*, viii. 461–464.

[38] See below.

[39] Otherwise unknown; he is not the 'Teuredus noster grammaticus' of John of Salisbury (Rose, *o. c.*, p. 380; Webb in *E. H. R.*, xxx. 658–660). He may possibly have been the ἱερέα καλὸν τῆς Βρενδύσου with whom Eugene the admiral exchanged verses: *B. Z.*, xi. 437–439. In any case this priest should be added to the list of west-Greek poets of the twelfth century.

[40] Unless, as Rose suggests, this translation be the source of the passages which John of Salisbury and others cite from the portion of Diogenes Laertius now lost. Cf. Webb, *Ioannes Saresberiensis Policraticus* (Oxford, 1909), i, pp. xxviii, 223, note. Mr. Webb suggests to me that the citations of Gregory Nazianzen in the *Policraticus* (ii. 91, 167, 170) may be derived from the version of Aristippus.

[41] The *Phaedo* is found at Erfurt, MS. O. 7, ff. 1–18 v (Schum, *Verzeichniss der Amplonianischen Handschriften-Sammlung*, p. 673); at Cues, Spitalbibliothek, MS. 177, ff. 58–89; in the Bibliothèque Nationale, MS. Lat. 6567 A, ff. 6–35, and MS. 16581, ff. 95–162 v (formerly MS. Sorbonne 1771; see Cousin, *Fragments — philosophie scholastique*, Paris, 1840, p. 406); in the Vatican, MS. Vat. lat. 2063, ff. 69–115; at Florence, Biblioteca Nazionale, MS. Palatino 639 (*I codici Palatini della R. Biblioteca Nazionale Centrale di Firenze*, ii. 207); Venice, St. Mark's, Cl. X, MS.

Meno [42] are preserved in several manuscripts and constituted the only medium through which these dialogues were known to Latin Europe until the new translations of the fifteenth century.[43] Men like Petrarch and Salutati were dependent upon a Latin version of the *Phaedo* which was doubtless that of Aristippus,[44] and the author of the translation which ultimately superseded his, Leonardo Bruni Aretino, seems, like more than one humanistic trans-

138 (Valentinelli, *Bibliotheca Ms. ad S. Marci Venetiarum*, iv. 88); University of Leyden, MS. 64 (Rashdall, *Universities of the Middle Ages*, ii. 745); Oxford, Corpus Christi College, MS. 243, ff. 115 v–135 v. For a specimen of the translation see Cousin, *l. c.* (also in his *Oeuvres*, 1847, third series, ii. 325). A marginal note in the Corpus Christi MS. (f. 135 v) comments: 'Hic liber omnium librorum Platonis est agrestissimus, vel quia Socrates in die mortis inornate locutus est et simpliciter, vel quia Plato interitum magistri commemorans pre dolore stilum non ornavit, vel quia etiam Plato quasi fidem et quod omni modo credi voluit hic predicans non obscuro verborum ornatu sed simplici relacione exequtus est.'

[42] The *Meno* is found at Erfurt, in Amplonian MS. O. 7 and MS. Q. 61 of the University; at Cues, Spitalbibliothek, MS. 177, ff. 89 v–100 v; and in Corpus Christi College, MS. 243, ff. 184 v–193 v (Rose, *o. c.*, p. 385). The beginning and end of the text of the Corpus MS. may serve as a specimen of the translation:

'*Menon.* Habes mihi dicere, o Socrate, utrum docile virtus, seu non docibile verum usu et conversacione comparabile, sive neque usu et conversacione comparabile ceterum natura inest hominibus, sive alio aliquo modo. *Socrates.* O Meno, hactenus quidem Tessali laudabiles erant inter Grecos et ammirandi effecti sunt in re equestri (MS. sequestri) et diviciis, nunc autem, ut mihi videtur, etiam in sapientia et non nullatenus tui amatoris Aristippi cives Larissei. Huius rei utique vobis causa est Gorgias. . . . Nunc autem mihi utique hora aliquo ire. Tu autem hec que ipse persuasus es persuade eciam peregrinum istum Anitum uti micior fiat, quia si persuaseris hunc est est [*sic*] quoniam et Atheniensibus proderis etc. Finit Menon Platonis scriptus per Fredericum Naghel de Trajecto anno domini .mcccc. xxiii. dominica infra octavas ascensionis in alma universitate Oxoniensi.'

[43] It would not be strange if the selection of these particular dialogues of Plato was influenced by the fact that they are the only ones which name an Aristippus. On mentions of the *Phaedo* in the Middle Ages see Rose, *o. c.*, p. 374; Delisle, *Cabinet des Mss.*, ii. 530, iii. 87; Roger Bacon, *Opus majus*, ed. Bridges, ii. 274; L. Gaul, *Alberts des Grossen Verhältnis zu Plato* (*Beiträge*, xii, no. 1, 1913), pp. 22–25. Although no other direct source of these citations is known, they are usually not sufficiently specific to enable us to recognize Aristippus' version; but a copy of this was in the library of the Sorbonne at the beginning of the fourteenth century (Delisle, *o. c.*, iii. 87) and is doubtless to be identified with the MS. given to this library by Geroud d'Abbeville which is now MS. Lat. 16581 of the Bibliothèque Nationale (Delisle, ii. 148). Cf. Birkenmajer, *Ryszarda de Fournival*, pp. 70, 73.

[44] Nolhac, *Pétrarque et l'humanisme²*, ii. 140, 141, 241; Novati, *Epistolario di Coluccio Salutati*, ii. 444, 449, iii. 515. MS. Lat. 6567 A belonged (f. 35 v) to 'M. Iacobi Finucii de Castro Aretiñ.' See also the conjectures of F. Lo Parco, *Petrarca e Barlaam* (Reggio, 1905).

lator, to have had at hand a copy of the mediaeval rendering.[45] Both dialogues were copied at Oxford as late as 1423,[46] and both are found in a collection of Latin translations of Plato which was used by Nicholas of Cusa in his Platonic studies.[47] Aristippus was also the author of the standard translation of the fourth book of Aristotle's *Meteorologica*, which passed into circulation so quickly that Gerard of Cremona did not find it necessary to include this book in his version;[48] and the prologue to the *Phaedo* indicates still further literary activity.[49]

To the list of Aristippus's translations our text makes no additions, but it shows him under a new aspect as the intermediary in bringing the *Almagest* and, doubtless, other manuscripts from Constantinople to Sicily. Even more noteworthy is the glimpse it affords of his observations of Mount Etna, for the actual examination of such natural phenomena was a rare thing in mediaeval learning, and the willingness of the translator of the *Meteorologica* to go beyond his authorities, even at some personal risk, reveals a spirit which reminds us less of the schoolmen than of the death of the elder Pliny.

The translation of the *Phaedo* by Aristippus was, as we learn from the prologue, begun at the siege of Benevento, in the spring of 1156, and finished after the author's return to Palermo. It is dedicated to a certain *Roboratus*, or *Roboratus fortune*, who is about to return from Sicily to his home in England, where Aristippus reminds him he will not have at his disposal the scientific and philosophical writings of the Greeks nor the stimulus of the

---

[45] Luiso, "Commento a una lettera di L. Bruni," in *Raccolta di studii critici dedicata ad Alessandro d'Ancona* (Florence, 1901), p. 88. The humanistic version of the *Meno* was the work of Marsiglio Ficino.

[46] Supra, n. 42. Cf. Coxe, *Catalogus*, on this MS.

[47] Kraus, "Die Handschriften-Sammlung des Cardinals Nicolaus v. Cusa," in *Serapeum*, xxvi. 74 (1865), codex K 1; Marx, *Verzeichnis der Handschriften-Sammlung des Hospitals zu Cues* (Trier, 1905), p. 165, MS. 177.

[48] Rose, *o. c.*, p. 385. See now F. H. Fobes, "Medieval Versions of Aristotle's Meteorology," in *Classical Philology*, x. 297–314 (1915); and his edition of the Greek text, Cambridge, 1919; and cf. C. Marchesi, "Di alcuni volgarizzamenti toscani," in *Studi Romanzi*, v. 123–157 (1907). Hammer-Jensen argues that the fourth book is not Aristotelian: *Hermes*, l. 113–136 (1915).

[49] Rose, p. 388: 'atqui theologica, mathematica, meteorologica tibi propono theoremata.'

literary circle which had gathered around King William I. *Roboratus*, as Rose long since pointed out,[50] is probably a play upon *Robertus*, but the further identification with Robert of Selby has been generally rejected, since King Roger's chancellor was not a scholar and is not heard of after he leaves office in 1154.[51] I venture to suggest another Englishman who is known to have been in Sicily at this time, Robert of Cricklade, prior of St. Frideswide's at Oxford from before 1141 until after 1171,[52] and author, not only of a biography of Becket and various theological commentaries, but also of a *Defloratio*, in nine books, of Pliny's *Natural History*, which he dedicates to King Henry II.[53] Contributing in 1171 or early in 1172 to the collection of St. Thomas' miracles which was already in process of formation, he narrates his own miraculous recovery from a disease of the leg which he had contracted while journeying from Catania to Syracuse in the midst of a sirocco more than twelve years before.[54] The visit to Sicily, whose occasion he does not care to set forth,[55] and from which he returned to England by way of Rome, can be placed even more definitely in 1158, when he secured, 26 February, from Adrian IV at the Lateran a detailed confirmation of the possessions of his

[50] *Hermes*, i. 376.

[51] Cf. Hartwig, in *Archivio storico napoletano*, viii, 433; Siragusa, *Guglielmo I*, ii. 111; K. A. Kehr, *Urkunden*, p. 77, n. 6. Rose's identification of Aristippus with the *grecus interpres* of John of Salisbury (cf. *Policraticus*, ed. Webb, i, pp. xxv f.) is also highly conjectural.

[52] He is addressed in a bull of Innocent II of 8 January 1141 (*Cartulary of the Monastery of St. Frideswide*, ed. Wigram, Oxford Historical Society, 1895, i. 20, no. 15), and in a bull of Alexander III which from the Pope's itinerary may belong to 1171, 1172, or 1181 (*ibid.*, ii. 95, no. 792).

[53] Tanner, *Bibliotheca Britannico-Hibernica* (London, 1748), p. 151; Hardy, *Descriptive Catalogue* (Rolls Series), ii. 291; *Oxford Collectanea*, ii. 160-165; *Dictionary of National Biography*, xlviii. 368, 369; Wright, *Biographia Britannica literaria*, ii. 186, 187; Rück, "Das Excerpt der Naturalis Historia des Plinius von Robert von Cricklade," in Munich *Sitzungsberichte*, phil.-hist. Kl., 1902, pp. 195-285.

[54] 'Preteritis iam ferme duodecim annis aut eo amplius cum essem in Sicilia et vellem transire a civitate Catinia usque ad Syracusam, ambulabam secus mare Adriaticum; sic enim se protendebat via.' *Materials for the History of Thomas Becket* (Rolls Series), ii. 97, 98; *M. G. H., Scriptores*, xxvii. 34. Also, somewhat more fully, in *Thómas Saga Erkibyskups* (Rolls Series), ii. 94-97, 284; see the introduction, ii, pp. lxxiv, xcii-xciv.

[55] *Thómas Saga*, ii. 94.

priory.[56] Indeed, as the Italian sojourn would seem to have been a long one,[57] he may also have been present at Benevento, 13 March 1156, when the Pope issued an order in his behalf to the bishop of Lincoln.[58] The coincidence of date, the visit to Catania, where Aristippus was archdeacon, and to Syracuse, whose library Aristippus especially mentions,[59] Robert's reported knowledge of Hebrew,[60] and his interest in natural science,[61] all combine to render it highly probable that he is the translator's English friend. If this be the case, another link is found in the intellectual connections between England and Sicily in the reign of Henry II.[62] Very likely Robert's associations with the South began still earlier than 1156, for personal visits to Rome were probably necessary to secure the confirmation of the monastery's possessions in 1141 [63] and to prosecute its claims against the monks of Oseney ten years later.[64] The prior's interest in secular learning seems to have been a thing of his earlier years,[65] while his theological writings, one of which is posterior to 1170,[66] fall rather in the later period of his

[56] *Cartulary of St. Frideswide's*, i. 27, no. 23. The bull of 27 February *sine anno* (*ibid.*, ii. 327, no. 1125) was doubtless issued at the same time.

[57] The priory lost the island of Medley during his absence. *Ibid.*, i. 33, no. 30.

[58] *Ibid.*, i. 29, no. 24. The year is clear from the Pope's itinerary.

[59] 'Habes in Sicilia Siracusanam et Argolicam bibliothecam.' *Hermes*, i. 388. Lo Parco, *Scolario-Saba*, in *Atti della R. Accademia di Archeologia di Napoli* (1910), new series, i. 241, seeks to identify the *Argolica bibliotheca* with that collected by Scolario-Saba at Bordonaro, near Messina; but see Heiberg, in *B. Z.*, xxii. 160.

[60] Giraldus Cambrensis, *Opera* (Rolls Series), viii. 65.

[61] Cf. his description of the Ionian Sea in *Thómas Saga*, ii. 96. The marginal notes which he tells us (Rück, pp. 213, 266) he added to his excerpts from Pliny might have proved of interest in connection with his Sicilian sojourn, but an examination of the copies at Eton (MS. 134) and in the British Museum (Royal MS. 15. C, xiv) shows that very few of these survive. In one of these (Eton MS., bk. ii, c. 49; Royal MS., bk. ii, c. 51) he shows some spirit of observation when he says, with reference to eels, 'quod et ego expertus sum:'

[62] The eulogy of King William by Aristippus may contain an implied comparison with Henry II: 'verum cum omnia dederis, regemne dabis Willelmum,' etc. Peter of Blois makes an explicit comparison of Henry II and William II: Migne, ccvii. 198.

[63] *Cartulary of St. Frideswide's*, i. 20, no. 15.

[64] 'Eodem anno [1151] perrexit abbas Wigodus Romam provocatus a Roberto priore Sancte Frideswide': *Annales Monastici* (Rolls Series), iv. 27; *M. G. H., Scriptores*, xxvii. 487.

[65] See the preface to his *De conubio Iacobi* in *Oxford Collectanea*, ii. 161.

[66] The preface to his *Speculum fidei* in the library of Corpus Christi College,

life, and the veil which he draws over the occasion of his presence in Sicily may well cover an outgrown interest in things at which religious men then looked askance.[67]

If the interest of Aristippus centred in the philosophical writings of the Greeks, Eugene of Palermo was primarily a student of their mathematics. Of noble birth and nephew of the admiral Basil,[68] he had himself risen to the dignity of admiral, or more accurately emir,[69] in the royal administration, while his intellectual attainments won him also the title of 'the philosopher.' We are indebted to him for a Latin version, made from the Arabic, of a work which would otherwise have been lost, the *Optica* of Ptolemy, the translation having been preserved in a score of manuscripts and having been printed;[70] and it is not surprising to learn that he had at hand the Greek text of Euclid's *Data*, *Optica*, and *Catoptrica*, as well as the treatise of Proclus on mechanics, and was sufficiently familiar with them to give instruction in the difficult matter of the *Almagest*. All of this implies a knowl-

---

Cambridge, MS. 380 (James, *Catalogue*, p. 228), mentions a bull of Alexander III of 28 May 1170 (Jaffé-Löwenfeld, *Regesta*, no. 11806).

[67] See the reference to the *libellus ludicris plenus* in *Oxford Collectanea*, ii. 161; and cf. the remarks of the translator of the *Almagest*, preface, lines 47 ff.; and Heiberg, in *Hermes*, xlvi. 210–212.

[68] *B. Z.*, xi. 449: Στίχοι Εὐγενίου φιλοσόφου, ἀνεψιοῦ Βασιλείου τοῦ ἀμοιρᾶ. *Ibid.*, p. 408: τὸν πανευγενέστατον ἄρχοντα κυρὸν Εὐγένιον. Infra, p. 175: Εὐγενὴς Εὐγένιος.

[69] On the significance of this title at the Sicilian court see Caspar, *Roger II*, p. 301; Chalandon, *Histoire de la domination normande*, ii. 637. The admiral Eugene who appears under Roger I in documents of 1093 and following (Caspar, *o. c.*, n. 7) must have been another person, but the translator was probably the father of Ἰωάννης, υἱὸς τοῦ ἐνδοξοτάτου ἄρχοντος κυρίου Εὐγενίου ἀμηράδος, who sells a garden in Palermo in 1201 (Cusa, *I diplomi greci ed arabi di Sicilia*, p. 89; cf. p. 23). Cf. Hartwig, in *Centralblatt für Bibliothekswesen*, iii. 173.

[70] Described by Boncompagni, "Intorno ad una traduzione latina dell' ottica di Tolomeo," in *Bullettino*, iv. 470–492, vi. 159–170; and edited by Govi, *L'ottica di Claudio Tolomeo da Eugenio ammiraglio di Sicilia ridotta in latino* (Turin, 1885). To the MSS. there enumerated should be added MS. 569 of the University of Cracow (Narducci, in *B. M.*, 1888, p. 98) and Suppl. grec 263 of the Bibliothèque Nationale; see also those indicated by Björnbo, in *Abhandlungen zur Geschichte der Mathematik*, xxvi. 124, 141 f., 145. On the loss of both the Greek original and the Arabic translation, see Steinschneider, in *Zeitschrift der deutschen morgenländischen Gesellschaft*, l. 216. There is no evidence for Amari's assumption (*Storia dei Musulmani*, iii. 660) that Eugene's translation was made under Roger, nor for Steinschneider's (*E. U.*, no. 37), that it belongs to 1154.

edge of languages, as well as no mean attainment in applied mathematics, and fully justifies the characterization of our preface, *virum tam grece quam arabice lingue peritissimum, latine quoque non ignarum*.[71] His native tongue was evidently Greek, and he had sufficient mastery of it to produce fourteen hundred lines of verse which entitle him to an important place among the west-Greek writers of the Middle Ages.[72] Of the twenty-four short poems which make up this collection, the greater number are epigrams on various virtues and vices. A few deal with religious subjects, such as the Crucifixion or the ascetic life. Three are addressed to a poet-priest of Brindisi; one celebrates the seclusion of a monastic cemetery, probably that of S. Salvatore of Messina; another describes a plant in the poet's garden at Palermo. Another writer of the time appears in Roger of Otranto, who addresses certain lines to him. One of Eugene's poems is an extravagant eulogy of King William (πρὸς τὸν ἐνδοξότατον τροπαιοῦχον ῥῆγα Γουλιέλμον); another, written in prison, seems to mark the close of his public career, from which he turns to solitude and books. We are tempted to seek here some connection with the imprisonment of Aristippus, in which case the King William of the poem would be William I, to whom for other reasons it seems better suited than to William II.[73] Indeed, while our prologue

---

[71] This is also borne out by Eugene's own statement (*Optica*, ed. Govi, p. 3): 'Arabicam in grecam aut latinam transferre volenti tanto difficilius est quanto maior diversitas inter illas tam in verbis et nominibus quam in litterali compositione reperitur.'

[72] These poems are contained in a MS. of the Laurentian described by Bandini, *Catalogus Codicum MSS. Bibliothecae Mediceae Laurentianae*, i. 23–30; cf. Krumbacher, pp. 768 ff. They have been published by Sternbach, *B. Z.*, xi. 406–451 (emendations to the text, *ibid.*, xiv. 468–478, xvi. 454–459, xvii. 430–431). That the poet and the translator were the same person, which Sternbach considers uncertain, is rendered highly probable by our text, which shows that the mathematician was a Greek and lived in the period to which the poems belong. Cf. *B. Z.*, xix. 569, xx. 373–383.

[73] Krumbacher leaves the question open as among the three Williams but says, "Manches spricht für Wilhelm II." Sternbach (p. 409) decides for William II. Chronological considerations, as well as the weakness of the royal power, would seem to rule out William III, but it is not easy in the case of a eulogy of this kind to distinguish with much certainty between the other two kings of this name. On the whole, however, it does not seem that such verses, if, as seems likely, they were written at the beginning of a reign, could with much propriety or purpose have been addressed to the thirteen-year old William II, who remained under the tutelage of

places Eugene's mathematical studies in the time of William I, we cannot be certain that he was alive or, if alive, engaged in secular pursuits under William II.[74]

Eugene the admiral is likewise associated with the transmission to the West of two curious bits of Oriental literature. One is the prophecy which became widely current in the later Middle Ages under the name of the Erythraean Sibyl, an oracular forecast of the doings of kings and emperors [75] which purports to have been

his mother for five years after his accession, while there is nothing which is inapplicable to William I. Sternbach indeed argues that lines 29-35 could not relate to William I as the successor of the first king of the Norman dynasty; but one king is enough to start a royal line (βασιλικὴν τὴν ῥίζαν), and the reference to the achievements of his fathers (τὰ πατέρων βέλτιστα) does not necessarily imply that they were all kings, for Roger I was glorious enough as duke to deserve inclusion in any such comparison. Indeed the passage has more point in the case of William I, as the son of the first Sicilian king: he will enlarge his authority even more than did his father who began as duke and ended as king (μέγα τι λαβὼν κρεῖττον ἀντιπαρέχεις). On resemblances between this poem and one of George of Gallipoli, addressed to Frederick II, see Horna, in *B. Z.*, xvi. 458; and cf. Sola, *ibid.*, xvii. 430.

[74] One of his poems, it is true (no. xiv, ed. Sternbach, p. 434), mentions an abbot Onofrius, who is probably to be identified with the archimandrite of San Salvatore di Messina who appears in documents of 1175-78 (Pirro, *Sicilia sacra*, edition of 1733, ii. 979, 980; Cusa, *I diplomi greci ed arabi*, p. 371; Garufi, *I documenti inediti dell' epoca normanna in Sicilia*, p. 168). We do not, however, know in what year he became archimandrite, for the current statement (e. g., Batiffol, in *Revue des questions historiques*, xlii. 555) that he entered upon this office in 1175 has no support beyond an erroneous assertion of Pirro (p. 979) that his predecessor Lucas died in that year. Pirro says that this date is proved from the records of the monastery, but his handling of the matter does not create confidence in his citation. He quotes an obituary notice in Latin which places the death of Lucas on Saturday the third of the kalends of March in the year 6688 of the Byzantine era (= A.D. 1180), and plausibly explains the obvious impossibility of this date by a misunderstanding of the Greek computation; but he does not notice that in both 1175, the date he proposes, and in 1180 the third of the kalends of March fell, not on Saturday, but on Thursday. In order to find this coincidence before the bull of October, 1175, which mentions Onofrius, we must go back to 1171 or 1165. Now an extract from a charter of William II refers to the grant of certain lands 'in Agro' made by him and his mother (her regency ended in 1171) to Onofrius, meaning doubtless a charter of 1168 for San Salvatore (Pirro, p. 979; on the date see Chalandon, *Domination normande*, ii. 336) in which the abbot is not named. If, accordingly, Onofrius was in office in 1168 and if we can trust the obituary for the day, his predecessor, who is not mentioned in the documents subsequent to 1149, must have died at least as early as 1165, so that a poem might have been addressed to Onofrius in the reign of William I.

[75] Published by Alexandre, *Oracula Sibyllina*, ii. 291-294 (Paris, 1856); and more fully by Holder-Egger, "Italienische Prophetieen des 13. Jahrhunderts," *Neues Archiv*, xv. 155-173, xxx. 323-335 (cf. xxxiii. 97, 101, 102).

translated from the Chaldean by Doxopater and kept in the treasury of the Emperor Manuel, whence it passed westward and was translated by 'Eugene, admiral of the kingdom of Sicily.'[76]   By Doxopater is probably meant a contemporary of Eugene, Nilus Doxopatres, a Greek ecclesiastic who sojourned at Palermo and afterward appears as imperial *nomophylax* at Constantinople, and who wrote in 1143, at the instigation of Roger II, a history of the five patriarchates.[77]   In its present form, however, the Sibylline text plainly belongs to the middle of the thirteenth century and shows the influence of the Joachite friars and the movements of Frederick II's reign,[78] so that it has been usual to dismiss the attribution to Doxopater and Eugene as an attempt to support the prophetic character of the oracle by a further bit of mystification.[79]   The matter cannot, however, be so lightly set aside. While it is plain that the current version of this text belongs to Italy and the thirteenth century, it is equally clear that these oracles are of eastern origin.   Both Greeks and Saracens had such Sibylline books,[80] and we find mention of their preservation in the imperial library under Leo the Armenian and again toward the close of the eleventh century.[81]   The connection with the West

[76] Neither of the editors gives a good text of this title. The MS. of St. Mark's, Cl. X, 158, reads as follows (Valentinelli, *Bibliotheca*, iv. 108): 'Extractum de libro vasilographia in imperiali scriptura quem Sybilla erythrea babilonica ad peticionem Graecorum regis Priami edidit, quem caldaeo sermone Doxopater peritissimus transtulit, tandem de aerario Manuelis imperatoris eductum Eugenius regni Siciliae admiratus de graeco transtulit in latinum.'

[77] See Krumbacher, p. 415; Caspar, *Roger II*, pp. 346-354; Harris, *Further Researches into the History of the Ferrar-Group* (London, 1900), pp. 52 ff.

[78] Holder-Egger, *o. c.*, xv. 150, dates it 1251-54, but Kampers, *Kaiserprophetieen und Kaisersagen im Mittelalter* (Heigel and Grauert's *Historische Abhandlungen*, viii), p. 252, has shown reason for placing it a few years earlier.

[79] See the doubts expressed by Amari, *Storia dei Musulmani*, iii. 460, 660-662; Hartwig, in *Centralblatt für Bibliothekswesen*, iii. 174-176; Harris, *Further Researches*, p. 70; Steinschneider, *E. U.*, no. 37; Caspar, *Roger II*, p. 462, n. 4. The difficulty is not discussed by Holder-Egger or Kampers.

[80] Liutprand, *Legatio*, ed. Dümmler (Hanover, 1877), pp. 152-153: 'Habent Graeci et Saraceni libros quos ὁράσεις, sive visiones, Danielis vocant, ego autem Sibyllanos; in quibus scriptum reperitur, quot annis imperator quisque vivat; quae sint futura, eo imperitante, tempora; pax, an simultas; secundae Saracenorum res, an adversae.'

[81] Cont. Theophanis, i. 22, ed. Bonn, p. 36; Georgius Cedrenus, ed. Bonn, ii. 63. Cf. Alexandre, *o. c.*, ii. 287-311; Krumbacher, pp. 627 ff.

must be made at some point, and the statement that the text was brought from Manuel's treasury and was translated by Eugene is in entire accord with what we have already seen of the transmission of manuscripts and of the activity of the admiral as a translator. Even in its present form the text shows traces of Sicilian origin and of earlier elements,[82] and a comparison of all the manuscripts and a genetic study of the whole may succeed in restoring the nucleus and explaining its development.[83]

The other oriental work to which the name of the Sicilian admiral has become attached is the Sanskrit fable of Kalila and Dimna, first turned into Greek by Simeon Seth toward the close of the eleventh century under the title of Στεφανίτης καὶ Ἰχνη- λάτης and widely popular in various western versions as a treatment of the relations of princes to their subjects.[84] In one group of manuscripts of the Greek version the translator is described in the following lines:[85]

> μυθικὴ βίβλος ἐξ Ἰνδικῆς σοφίας,
> προσενεχθεῖσα πρὸς Περσικὴν παιδείαν,
> αἰνιγματωδῶς συντείνουσα τὰς πράξεις,
> πρὸς βιωτικὴν συντείνουσα τὰς πράξεις·
> ἡ μεταβληθεῖσα πρὸς γλῶσσαν τῶν Ἑλλήνων
> ἐξ Ἀραβικοῦ καὶ βαρβαρώδους ὕθλου
> παρὰ τοῦ σοφοῦ, ἐνδόξου καὶ μεγάλου
> τοῦ καὶ Ἀμηρᾶ, καὶ ῥιγὸς Σικελίας
> Καλαβρίας τε πρίνκιπος Ἰταλίας·
> οὕσπερ εὑρικὼς, ὡς γνωστικοὺς τοῖς πᾶσιν
> τοῦτο δέδωκε πρὸς ἡμᾶς τὸ βιβλίον,
> ὥσπερ δώρημα, διδασκαλίας πλέον,
> Εὐγενὴς Εὐγένιος, ὁ τῆς Πανόρμου.

[82] See *Neues Archiv*, xv. 163, 167, 168, 171, 172, 173.

[83] So Kampers arrives at the same view from a study of the thirteenth-century version: 'Mutmasslich gab es eine eryth. Sibylle, die kein Ereignis über das Jahr 1200 hinaus behandelte.' *O. c.*, p. 253.

[84] See in general Krumbacher, pp. 895–897. The Greek text is edited by Puntoni, Στεφανίτης καὶ Ἰχνηλάτης (Florence, 1889), as the second volume of the *Pubblicazioni della Società Asiatica Italiana*.

[85] Bodleian, Cod. misc. gr. 272. See Coxe, *Catalogus Codicum MSS. Bibliothecae Bodleianae*, i, c. 814; Puntoni, *o. c.*, p. vi. Puntoni entirely ignores the problem raised by these lines.

Here, while Eugene is mentioned by name only as the donor of the book, there can be no doubt that he is the 'wise and glorious admiral' to whom the translation is attributed; but, although the attribution is thus seen to be contemporary, it can hardly be correct. The divergences from the other groups of manuscripts do not appear sufficient to establish an independent translation, and when the preface goes on to explain that the Greek version was made with the assistance of 'certain men well acquainted with the Arabic tongue,'[86] we may feel reasonably sure that these are the words of Simeon Seth rather than of the learned admiral, whose familiarity with Arabic is attested by his rendering of the *Optica* as well as by the preface printed below. It would seem probable that what we have is a revision of Seth's translation at Eugene's hands, no great achievement in itself, but interesting to us as a further illustration of the range of the admiral's labors and interests.

The popularity of the Στεφανίτης καὶ Ἰχνηλάτης in Byzantine circles in the twelfth century is also seen from the following verses, which are found at the close of the copy of the fable in MS. Gr. 2231 of the Bibliothèque Nationale:[87]

Ceramei Georgii versus iambici super precedenti libro

Τοῦ κεραμέου γεωργίου στίχοι ἐπὶ τῆδε[88] τῇ βίβλῳ

Εἴπης *λιλίν[89] ἂν τὴν παροῦσαν πυκτίδα,
*ὃνυννία[89] παίζουσιν ἐκ θυμηδίας·
περσωνυμικὴν[90] ἀπιδὼν κλῆσιν φίλος,
καὶ τὴν ἐν αὐτῇ τῶν λόγων κοινὴν φράσιν·
5 δι'[91] ἧς πίθηκες καὶ λεόντων τὰ κράτη·
τῶν ἐλεφάντων καὶ κοράκων τὰ γένη

---

[86] Puntoni, *o. c.*, p. vii: ἐπὶ τούτων καί τισιν ἀνδράσι χρησάμενοι, ἀντιλαμβανομένοις τῇ ἡμῶν προθυμίᾳ, εὖ εἰδότας τῆς τῶν ἀράβων γλώσσης.

[87] On the MS. see *Catalogus Codicum MSS. Bibliothecae Regiae*, ii. 466; and Omont, *Inventaire sommaire des MSS grecs*, ii. 218. Rystenko's edition, published at Odessa in 1909 (cf. *B. Z.*, xviii. 621, xix. 569), I have not seen.

[88] Iota subscript omitted throughout in MS.

[89] In these unintelligible words there may lurk a corruption of Kalila and Dimna.

[90] Marginal gloss τὴν ἀῤῥαβικήν.

[91] Marginal gloss γρά[φεται] ἐν ῇ.

ταύρων χελωνῶν βατράχων καὶ δορκάδων
νηττῶν μυῶν τε καὶ περιστερῶν ἅμα
κιττῶν τε κύκνων ἰχθύων καὶ καρκίνων
10 καὶ τῶν σκολιῶν ἑρπετῶν ἡ κακία,
συντυγχάνουσιν οἷσπερ οὐκέστι λόγος·
εἰ δ' οὖν λογικὰ τὰ πρόσωπά μοι κρίνῃς,
καὶ τὴν ἐν αὐτοῖς σύνεσιν καταμάθῃς,
εὕρῃς ἁπάντων σωφρονέστατον βίον·
15 φεύγων ἀφορμὰς τῶν κακίστων κολάκων·
νοῶν πονηροὺς ἐκτρέπων σκαιοτρόπους·
φίλους ἀφίλους συγκρίνων διακρίνων·
καὶ πάντα πράττων εὐμαρῶς καὶ κοσμίως·
ὡς γοῦν κάλυκα περιφρουροῦσαν ῥόδον,
20 ὡς ὄστρεον μάργαρον ἐμφέρον μέγαν.
βαλάντιον σκύτινον ὡς χρυσοῦ γέμον,
κιβώτιον ξύλινον ὡς πλῆρες λίθων,
ἰάσπεών τε λυχνιτῶν ἐξανθράκων,
ἔχων τὸ παρὸν κλεινὲ Παλαιολόγε,
25 ἀγλαοφανὲς παγκλεέστατε κλάδε,
τοῦ τρισμεγίστου καὶ βριαροῦ δεσπότου,
'Ανδρόνικε κάλλιστε φυτὸν χαρίτων,
βιβλίον εὖ ἔγκυπτε τοῖς ἐγκειμένοις·
καὶ συνετίζου καὶ φρονήσει σεμνύνου
30 καὶ πάντα πράττε καθαπερεὶ συμφέρον,
ὡς ὑποδρηστὴρ τῶν μεγάλων ἀνάκτων,
δόξης ταχινώτατος ἐν τοῖς πρακτέοις·
ὡς τοῖς προσεγγίζουσί σοι κατὰ [92] γένος,
φανεὶς ἀξιάγαστος ἐν πᾶσι λόγοις·
35 ἡμῖν δ' ἀλιτροῖς οἰκέταις σοῖς ἀθλίοις,
μέγα παρηγόρημα καὶ θυμηδία.

The Andronicus to whom these lines are addressed cannot be the fifteenth-century humanist Andronicus Callistus,[93] for the MS. is of the thirteenth century. He is, moreover, a man of royal descent who holds a high place in the service of the emperor, and should doubtless be identified with the Andronicus Palaeologus who led a division of the imperial army in the war with the Nor-

[92] MS. προσεγγίζουσι σοί κατα.
[93] Besides, the humanist was not a Palaeologus. See Legrand, *Bibliographie hellénique*, i, pp. l–lvii. κάλλιστε in our text is thus an adjective, not a proper name.

mans in 1185 [94] and is addressed in one of the letters of Glycas.[95] Georgius Cerameus has a couple of lines given him in Fabricius on the basis of the mention of these verses in the Paris catalogue,[96] but nothing further is known of him unless he is the same as the distinguished preacher of the middle of the twelfth century, whom recent investigation makes archbishop of Rossano.[97] His sermons bear the name of Cerameus and most commonly of Theophanes Cerameus, but five or six other Christian names, among them George, are given in different manuscripts. Nothing can be definitely affirmed until the problem of the authorship of the sermons is straightened out, but if it should appear that Georgius Cerameus was a Calabrian archbishop, or a western Greek of any sort, another connection will thereby be established between Constantinople and the West in the twelfth century.

The mention of Euclid's *Data, Optica*, and *Catoptrica* helps to connect the Latin translations of these works likewise with the Sicilian school, if not with the translator of the *Almagest* himself. These treatises formed part of a group of texts, corresponding roughly to the 'intermediate books' of the Saracens, which formed the basis of mathematical studies in the stage between the *Elements* of Euclid and the *Almagest*.[98] Besides an unidentified version of the *Data* made from the Arabic by Gerard of Cremona,[99]

[94] Nicetas Acominatus [Choniata], ed. Bonn, p. 412; Eustathius, ed. Bonn, p. 430.

[95] Migne, *Patrologia Graeca*, clviii, coll. xxxv, 933; Krumbacher, in Munich *Sitzungsberichte*, 1894, pp. 422, 425. On the claim of the Palaeologus family to imperial descent, see Otto of Freising, *Gesta Frederici*, ed. Waitz, p. 116; Hase, in *Notices et extraits des MSS.*, ix, 2, pp. 153 ff.

[96] *Bibliotheca Graeca* (1790–1809), xi. 327, xii. 43. He is overlooked by Krumbacher.

[97] Lancia di Brolo, *Storia della chiesa in Sicilia* (Palermo, 1884), ii. 459–492; Krumbacher, pp. 172–174; Caspar, *Roger II*, pp. 459 ff.

[98] See Steinschneider, in *Z. M. Ph.*, x. 456–498, xxxi. 100–102; Menge, *Euclidis Data* (Teubner, 1896), p. liv; Heiberg, *Euclidis Optica* (Teubner, 1895), pp. xxxii, l; Cantor, *Vorlesungen*, i. 447, 705. In the fourteenth century Theodore Metochita tells us, in a passage cited by Menge and by Heiberg, that he found he could not understand the *Almagest* without the same preliminary course in the *Data, Optica*, and *Catoptrica* which was taken by our Sicilian translator.

[99] Wüstenfeld, p. 62; Steinschneider, *H. U.*, p. 510; Hultsch, in Pauly-Wissowa, xi. 1043.

the extant translations of the *Data, Optica,* and *Catoptrica* can be traced back to the beginning of the thirteenth century, and were probably made in the twelfth.[100] They were evidently made directly from the Greek, indeed the *Catoptrica* does not seem to have been known to the Arabs,[101] and the discovery that Greek texts of the three works existed in Sicily in the twelfth century points clearly to this region as the source of the Latin *interpretatio.*[102] The translator of the *Almagest* does not make quite clear the nature of his preliminary labors in the works of Euclid, but the more natural interpretation would seem to be that he not only studied them but tried his hand (*prelusi*) at turning them into Latin.

The same argument applies to the other treatise mentioned with the works of Euclid, the *De motu* of Proclus, Στοιχείωσις φυσική ἢ περὶ κινήσεως, generally known in Latin as the *Elementatio philosophica* or *Elementatio physica.* An incomplete Latin version is extant in MS. F. iv. 31 at Basel,[103] MS. Q. 290 of the Stadtbibliothek at Erfurt, [104] and MS. Lat. 6287 of the Bibliothèque Nationale; [105] the Basel manuscript is clearly of the fourteenth century, while the Erfurt manuscript is of northern origin and not later than ca. 1400, so that the translation which

---

[100] Heiberg, *Optica,* pp. xxxii, li; Steinschneider, *H. U.,* p. 512; Björnbo, in *Archiv für die Geschichte der Naturwissenschaften,* i. 390.

[101] Heiberg, *Studien über Euklid* (Leipzig, 1882), p. 152.

[102] The existence of the Greek text of the *Optica* in Sicily was already known from the prologue of Aristippus published by Rose (*Hermes,* i. 388, cf. p. 381), and the conclusion that the Latin version was of Sicilian origin was drawn therefrom by Heiberg, *Optica,* p. xxxii; *Hermes,* xlvi. 209. John Dee described one of the MSS. in his library as containing 'Euclidis Elementa Geometrica, Optica et Catoptrica, ex Arabico translata per Adellardum' (*Diary,* ed. Halliwell, Camden Society, p. 67; M. R. James, *List of MSS. formerly owned by Dr. John Dee,* Oxford, 1921, p. 16, no. 13); but there is no other reason for attributing the translation of the *Optica* and *Catoptrica* to Adelard of Bath, and the translator's name is not found with the versions of these treatises in MS. 251 of Corpus Christi College, which belonged to Dee (James, p. 30, no. 151). See ante, Chapter II, n. 66.

[103] The Basel MS. (ff. 82 v–84) which I found in 1922, has been collated by the kindness of the Oberbibliothekar, Professor G. Binz.

[104] Ff. 83 v–86. Cf. Schum, *Verzeichniss der Amplonianischen Handschriften-Sammlung,* p. 530.

[105] Ff. 21–22 v, of the fifteenth century. The three MSS. are based on the same Greek text, which is defective at the close of book i, and breaks off with ii, 4.

they contain must be anterior to the Renaissance. That this was made directly from the Greek is evident from the transfer of such words as *omogenes* and from the lettering of the demonstrations, where *abgdez* represent αβγδεζ, as well as from the closeness with which the Greek text is followed. The verbal literalness characteristic of mediaeval renderings from the Greek may be seen from the following specimen:

Incipit Elementacio Philosophica[106] Procli

Continua sunt quorum termini unum. Contingentia sunt quorum termini simul. Deinceps sunt quorum nihil medium omogenes, id est congnatum. Primum est tempus mocionis, quod nec plus nec minus mocione. Primus est locus, qui nec maior contento corpore nec minor. Quiescens est prius sicut posterius in eodem loco existens et totum et partes.

(1) *Duo individua non contingunt se invicem.* Si enim possibile, sint duo individua a͞b, contingant[107] se invicem. Contingentia vero erant quorum termini in eodem; duo ergo partium[108] termini erunt, hoc autem impossibile. Non ergo erant a͞ et b͞.

(2) *Duo individua continuum nihil faciunt.* Si enim possibile, sint duo individua a͞ et b͞ et faciant[109] continuum quod est ex ambobus. Sed omnia continua contingunt se prius adinvicem, ergo se contingunt a͞b individua existentia, quod est impossibile. Aliter: si est continuum ex a͞b individuis, vel totum totum contingit, vel totum partem, vel partes partem. Sed si totum partem vel partes partem, non erunt individua a͞b. Si vero totum totum contingit, non erunt individua sed supponetur tantum. Si ergo non erit a͞ continuum, nec vero b͞ et a͞ erunt continuum totum totum contingens.[110]

[106] The MSS. have 'ph'ica' here, but the Erfurt MS. has 'philosophica' in the *explicit*.

[107] Erfurt: *contingunt*.

[108] Based doubtless upon a text which had μέρων instead of ἀμερῶν.

[109] Paris: *faciunt*.

[110] As the printed text of the *De motu* (Paris, 1542) is not well known, I give for convenience of comparison the opening portion of the treatise from the text of Harleian MS. 5685 of the British Museum (saec. xii): (f. 133) Συνεχῆ ἐστιν ὧν τὰ πέρατα ἕν· ἁπτόμενά ἐστιν ὧν τὰ πέρατα ἅμα· ἐφεξῆς ἐστιν ὧν μηδὲν μεταξὺ ὁμογενές. πρῶτός ἐστι χρόνος κινήσεως, ὁ μήτε πλείων μήτε ἐλάττων τῆς κινήσεως. πρῶτός ἐστι τόπος, ὁ μήτε μείζων τοῦ περιεχομένου σώματος μήτε ἐλάττων. ἠρεμοῦν ἐστι τὸ πρότερον καὶ ὕστερον ἐν τῷ αὐτῷ τόπῳ ὂν καὶ αὐτὸ καὶ τὰ μέρη.

(1) Δύο ἀμερῆ οὐχ ἅψεται ἀλλήλων. εἰ γὰρ δυνατόν, δύο ἀμερῆ τὰ α͞υ ἁπτέσθωσαν ἀλλήλων· ἁπτόμενα δὲ ἦν ὧν τὰ πέρατα ἐν τῷ αὐτῷ, τῶν δύο ἄρα ἀμερῶν πέρατα ἔσται. οὐκ ἄρα ἦν ἀμερῆ τὰ α͞υ.

(2) Δύο ἀμερῆ συνεχὲς οὐδὲν ποιήσει. εἰ γὰρ δυνατόν, ἔστω δύο ἀμερῆ τὰ α͞υ καὶ ποιείτω συνεχὲς τὸ ἐξ ἀμφοῖν. ἀλλὰ πάντα τὰ συνεχῆ ἅπτεται πρότερον, τὰ ἄρα α͞υ ἅπτεται ἀλλήλων ἀμερῆ ὄντα, ὅπερ ἀδύνατον. [ἄλλως — marginal] εἰ ἐστι συνεχὲς ἐκ

If it be objected that a work of this sort could scarcely be translated otherwise, the freer style of the Renaissance may be seen in the version of Spiritus Martinus Cuneas, printed at Paris in 1542:[111]

Continua sunt quorum termini sunt unum. Contigua sunt quorum termini sunt simul. Deinceps sunt inter que nihil est eiusdem generis. Primum motus tempus est quod neque longius est eo neque brevius. Primus locus est qui neque maior neque minor est contento corpore. Quiescens est quod primo et postremo tam ipsum quam partes in eodem loco est.

THEOREMATA

I. Duo indivisibilia non tangunt se invicem. Nam (si fieri potest) duo indivisibilia *ab* tangant se invicem, at cum contigua sunt quorum termini sunt in eodem, duo indivisibilia terminos habebunt. Non igitur indivisibilia *ab*.

Not only is the mediaeval rendering closely literal, but it shows the turns of expression characteristic of the translator of the *Almagest*, such as *quoniam* for ὅτι,[112] *utique* for ἄν, *quidem* . . . *vero* for μέν . . . δέ,[113] and notably the use of *id quod* to represent the article before an attributive phrase.[114] These resemblances, when taken in connection with the mention of the *De motu* in the preface to the *Almagest*, make it probable that both translations are the work of the same scholar.

Another work of Greek mathematics which is known to have been in Sicily in the time of William I is the *Pneumatica* of Hero of Alexandria, which is mentioned by Aristippus in the introduction to his translation of the *Phaedo*.[115] All existing manuscripts are

τῶν αὖ ἀμερῶν, ἢ ὅλον ἅπτεται (f. 133v) τὸ ᾱ τοῦ ῡ, ἢ ὅλον μέρους ἢ μέρη μέρους. ἀλλ' εἰ μὲν ὅλον μέρους ἢ μέρη μέρους, οὐκ ἔσται ἀμερῆ τὰ αὖ. εἰ δὲ ὅλον ὅλου ἅπτοιτο, οὐκ ἔσται συνεχὲς ἀλλ' ἐφαρμόσει μόνον. εἰ οὖν οὐκ ἦν τὸ ᾱ συνεχὲς, οὐδὲ τὸ ῡ μετὰ τοῦ ᾱ ἔσται συνεχὲς ὅλον ὅλου ἀπτόμενον.

[111] *Procli* . . . *De Motu Libelli Duo* . . . *Spiritu Martino Cuneate interprete.* I have used the copy in the British Museum.

[112] Heiberg, in *Hermes*, xlv. 59.

[113] Supra, p. 163. These are also the regular equivalents in Boethius, and may have been taken from him by subsequent translators. See McKinlay, *Harvard Studies*, xviii. 124–128.

[114] E. g. (ii, 4), τῶν ἐπ' εὐθείας τίς κινήσεων = earum que in directo mocionum.

[115] 'Habes Eronis philosophi mechanica pre manibus, qui tam subtiliter de inani disputat quanta eius virtus quantaque per ipsum delationis celeritas': *Hermes*, i. 388. This work is not the lost *Mechanica*, preserved only in an Arabic translation

of later date, and the known Latin versions, three in number, are of the fifteenth and sixteenth centuries, so that it has been supposed that the Latin translation which is inferred from the language of Aristippus disappeared with the manuscript on which it was based.[116]   There exists, however, in the Bibliothèque Nationale [117] a translation of the abbreviated text of the *Pneumatica* [118] which not only differs from the Renaissance versions described by Schmidt,[119] but has the close literalness of a mediaeval rendering.   Its identity with the lost Sicilian translation can only be conjectured, but there would be nothing strange in the survival of the mediaeval version in the period of the humanists, who did not disdain such helps in making their own translations. The Paris text begins as follows:

Spiritalium Heronis Alexandrini Liber Primus

Cum spiritale negocium studio dignatum sit a veteribus tum philosophis tum mechanicis, illis quidem per rationes vim eius explicantibus, hiis vero et per ipsos sensibiles effectus, necessarium esse ducimus et ipsi quae ab antiquis tradita sunt in ordinem redigere et quae nos quoque adinvenimus addere; sic enim eos qui post haec in mathematicis versari volunt iuvari continget. Consequens [120] autem esse rati aqueorum horoscopiorum habitudini, quae nobis in quatuor libris descripta est, hanc tractationem esse continuam, scribemus et de ea, ut praedictum est. Per complicationem enim aeris et ignis et aquae et terrae ac dum tria elementa aut etiam quatuor complicantur, variae affectiones committuntur, quarum aliae usus vitae huic necessarios praestant, aliae stupendum aliquod miraculum ostendunt.

and containing nothing concerning the vacuum, but the *Pneumatica*, which begins with a discussion of this subject. See Rose, *Hermes*, i. 380; Schmidt, *Heronis Opera* (Teubner, 1899), i, suppl., p. 53.

[116] Schmidt, *o. c.*, pp. 52, 53.

[117] MS. Lat. 7226 B, ff. 1–43; written on paper in a French hand of the early sixteenth century, with occasional corrections in a contemporary hand and free interlinear and marginal corrections in a somewhat later humanistic hand which seeks to improve the rendering and often cites the Greek words of the original. This MS. was overlooked by Schmidt, doubtless because it is omitted from the body of the catalogue.

[118] On which see Schmidt, *o. c.*, pp. 14–23.   [119] *Ibid.*, pp. 42, 43, 49–53.

[120] The corrections, which appear with many erasures and alternative renderings, are not of sufficient importance to be reproduced in detail, but the translation of this sentence may serve as a specimen: 'Itaque cum veris certisque consecutionibus colligi (*or* confici) posse arbitremur, hanc commentationem cum horoscopiorum quae ex aqua comparantur ratione, quae iam a nobis in quatuor libris descripta est, coniunctam esse atque continuam, scribimus, etc.'

Caeterum ante ea quae dicenda sunt primum de vacuo tractandum est. Alii enim aiunt universaliter (f. 1 v) nullum esse vacuum, alii confertum quidem secundum naturam nullum esse vacuum sed sparsum per parvas particulas in aere et humore et igni et caeteris corporibus, quos potissimum sequi convenit; ex iis enim quae apparent ac sub sensum cadunt in sequentibus ostenditur id contingere. Quamquam vascula quae vulgus putat esse inania non sunt ut existimant inania sed plena aere, qui, ut iis placet qui in commentariis de natura versati sunt, pusillis ac levibus corpusculis constat quae nobis ut plurimum immanifesta sunt. Si igitur in vasculum quod videtur esse vacuum infundat quis aquam, quantum aquae in vasculum inciderit, tantundem aeris excedet. Poterit autem quis mente complecti id quod dicitur experientia tali.

There was, it is true, a mediaeval version of Hero made by William of Moerbeke which has not yet been identified, but the existence of an earlier rendering has been shown by Birkenmajer,[121] since a set of Latin extracts is cited by Richard of Fournival ca. 1250. It is not clear whether this is the version here printed or the extracts based on the longer text of Hero which Birkenmajer has found at Cracow.

One of the most obscure and one of the most important questions connected with the Greek scholars of southern Italy and Sicily is the extent of their acquaintance with Aristotle and their relation to the Latin translations of his works. It tempts our curiosity to know that the *Posterior Analytics* was in Sicily in the time of Aristippus and that the first northern author to cite it was John of Salisbury, who was a frequent visitor to the Norman kingdom; that Aristippus himself translated the fourth book of the *Meteorologica*; and that the Sicilian translator of the *Almagest* was acquainted, at least indirectly, with the Greek text of the *De caelo*. Some of these problems we shall examine more specially in another connection.[122]

Another subject which might reward further inquiry is the Biblical manuscripts of Sicilian origin. An important group of New Testament codices, the Ferrar group, has been traced to the scribes of King Roger's court,[123] but the manuscripts of the Septuagint and the Arabic translations have still to be examined with

---

[121] *Vermischte Untersuchungen* (*Beiträge*, xx, no. 5), pp. 22–30.

[122] Infra, Chapters XI, XVIII.

[123] See especially Harris, *Further Researches into the History of the Ferrar-group* (London, 1900).

reference to possible Sicilian connections. Many-tongued Sicily would be a natural centre for polyglot copies, and it is hard to conceive of any other country as the source of such a manuscript as Harleian 5786 of the British Museum,[124] written before 1153 [125] and containing the Psalter in the Vulgate and Septuagint texts and an Arabic version.

Further investigation may very likely reveal still other points of contact between Sicily and the East and other lines of influence on the intellectual life of northern Europe. Thus while Adelard of Bath doubtless got his familiarity with Saracen learning in the course of the extensive travels which took him to Syria and perhaps to Spain, it should be noted that he studied at Salerno and in Magna Graecia, and dedicated his *De eodem et diverso* to William, bishop of Syracuse, whom he credits with much mathematical knowledge.[126] John of Salisbury, who made more than one journey into southern Italy, studied with a *grecus interpres*, a native of Santa Severina, who occupied himself with Aristotle; [127] and it was probably in this region that the English humanist gained his acquaintance with the *Posterior Analytics*. John's pupil, Peter of Blois, who, like his master, advocated the cause of the classics against the rising tide of logical studies, had likewise been in Sicily.[128] Another friend of John of Salisbury, Burgundio the Pisan, the leading north-Italian Greek scholar, also made a visit to Sicily.[129] Returning in 1171 from the last of his three missions to Constantinople as an envoy of Pisa, he tells us that he stopped at Messina, Naples, and Gaeta, working all the time assiduously

---

[124] A facsimile of one page is published by the *Palaeographical Society*, i, 2, plate 132. I am indebted to Professor E. K. Rand for calling my attention to this MS.

[125] The date appears from the following entry on the last folio: '[a]nn[o] incarnat[ionis] dominice. M. C. Liii. Ind[ ] m[ensis] ianuarii die octavo die mercurii.' There is some error here, as 8 January 1153 fell on Thursday.

[126] *Supra*, Chapter II.

[127] Schaarschmidt, *Johannes Saresberiensis* (Leipzig, 1862), pp. 120–122; Rose, in *Hermes*, i. 379–381; Poole, in *Dictionary of National Biography*, xxix. 444; Webb, *Ioannis Saresberiensis Policraticus*, i, pp. xxv–xxvii, ii. 259, note.

[128] *Epistolae*, nos. 10, 46, 66, 72, 90, 93, 116, 131, in Migne, ccvii. 27, 133, 195, 221, 281, 291, 345, 386, 397.

[129] On whom see Chapter X. John of Salisbury mentions him in the *Metalogicus*, 4, 7 (*Opera*, ed. Giles, v. 163).

at his translation of Chrysostom's *Homilies* on the gospel of John.[130]

As an illustration of the amount of communication which went on in the twelfth century between Sicily and the North, and thus of the possibilities of intellectual intercourse, let us examine further the relations between Sicily and the Anglo-Norman lands.[131] The southern branches of Norman families did not lose all connection with the parent stem when conquest and colonization ceased: readers of Ordericus Vitalis will recall the interminable comings and goings of the members of the house of Grentemaisnil in the eleventh century, and as late as 1130 one of this family gave up his fiefs in the south in order to return to his relatives in Normandy.[132] The northern Normans showed pride in the achievements of their Italian kinsmen,[133] and it is characteristic that the splendor of Rouen and the glory of King Roger form the joint theme of a Latin poem.[134] No list can be attempted of the Norman and English students at Salerno [135] or of the pilgrims

[130] 'Negociis vero vice civitatis pactis, licenciam redeundi ab imperatore accipiens, Messanam veniens ibique moram faciens, manibus meis scribens librum inibi trasferre incepi. Et sic per tantam viam Neapoli et Gaete et ubicumque moram faciebam vacationem michi extorquens, iugiter transferebam et contra spem per duos continuos annos, Deo actore, totum librum de verbo ad verbum de greco in latinum transferens integre consummavi': Vatican, MS. Ottoboni Lat. 227, f. 1. Also at Merton College, MS. 30 (dated 1174); Bibliothèque Nationale, MS. Lat. 1778, ff. 74–111; Arras, MS. 229; Berlin, Cod. Elect. 332 (cf. Rose, *Verzeichnis*, ii. 122–124). Printed in part from Mabillon's copy in Martène and Durand, *Veterum scriptorum amplissima collectio*, i. 829; other extracts from this preface supra, Chapter VIII, n. 36. On the Pisan mission, see Chalandon, *Les Comnènes*, ii. 575.

[131] These two paragraphs have been revised from my articles on "England and Sicily in the Twelfth Century," *E. H. R.*, xxvi. 435–438 (1911). The general subject of Englishmen in Italy in this period is being investigated by one of my students, Dr. Paul B. Schaeffer.

[132] Alexander Telese, i, cc. 17, 20–22, in Del Re, *Cronisti e scrittori sincroni napoletani* (Naples, 1845), i. 97, 99 f.; Ordericus, iii. 455. Note also the de Lucys in Sicily: Garufi, in *Archivio storico per la Sicilia orientale*, x. 160–180 (1913).

[133] Ailred of Rievaulx, in *Chronicles of Stephen*, iii. 186; *Miracula S. Michaelis*, in *Mémoires des Antiquaires de Normandie*, xxix. 864; Ordericus Vitalis, v. 58; Robert of Torigni, i. 242; *Actus pontificum Cenomannis*, ed. Busson and Ledru, p. 417.

[134] Printed by Richard, *Notice sur l'ancienne bibliothèque des échevins de Rouen* (Rouen, 1845), p. 37; Haskins, *Norman Institutions*, p. 144.

[135] Adelard of Bath is an early instance. There are many names under 'Anglicus' in the index to the *Necrologio di S. Matteo di Salerno* (ed. Garufi, Rome, 1922). The

and crusaders who went or returned by way of Bari [136] or Messina, nor can we hope to recover many traces of the commercial intercourse which must have existed. It is, however, significant that we hear of a merchant of London at Salerno and a merchant of Brindisi at the tomb of Becket,[137] and that the money of Rouen was in common use in the region of Aversa at least as late as 1135.[138] The great monasteries of St. Eufemia, Venosa, and Mileto had been founded by monks from St. Evroult, and the *cantus Uticensis* was still sung in them in the days of Ordericus,[139] who doubtless derived his full knowledge of south-Italian affairs from the intercourse maintained with these daughter-abbeys.[140] The chroniclers of Mont-St.-Michel and Bec were likewise well informed concerning events in the South,[141] as were English historians of the close of the century; [142] and if St. Michael and St. Nicholas [143] were popular in Normandy and England, St. Thomas of Canterbury was promptly added to the Norman saints who had kept a place for themselves in the south-Italian calendars.[144] John, abbot of Telese, had studied at

supposed dedication of the 'Schola Salerni' to Robert Curthose in 1101 must, however, be given up as a result of the investigations of Sudhoff, ending in *Archiv für die Geschichte der Medizin*, xii. 149–180 (1920). On Robert's sojourn in the South, see C. W. David, *Robert Curthose* (Cambridge, 1920).

[136] *Catalogus codicum hagiog. Paris.*, ii. 422 f.

[137] Wright, *Anglo-Latin Satirical Poets*, i. 37 ff.; *Materials for the History of Thomas Becket*, i. 452. There was a William Apulus at Norwich, *Life of St. William of Norwich*, p. 31.

[138] Alexander Telese, iii, c. 8, iv, c. 1 (Del Re, pp. 133, 146); and, for a specimen coin, Sambon, in *Gazette numismatique française*, iii. 138.

[139] ii. 89–91.

[140] See Delisle's introduction, v, p. xxxvii; cf. ii. 110. Ordericus (ii. 88) also used Geoffrey Malaterra.

[141] Robert of Torigny, *passim*; *S. Nicolai in Normannia et in Apulia miracula*, by a monk of Bec, in *Cat. codd. hagiog. Paris.*, ii. 405–432.

[142] Roger of Hoveden, i. 223; William of Newburgh, ii. 428–431; Ralph de Diceto, ii. 37 f.

[143] See note 141.

[144] For St. Thomas see *Materials*, i. 165, and the mention of churches dedicated to him in 1179 in De Grossis, *Catania sacra*, p. 98 f.; and Ughelli-Coleti, *Italia sacra*, vii. 501. For the older Norman saints, see the calendar of La Trinità di Venosa, now MS. 334 of the library of Monte Cassino, printed in Gattola, *Accessiones ad historiam Cassinensem*, ii. 843 ff.; and the so-called *Missale Gallicum* of the cathedral of Palermo (MS. 544), where the entry, in a later hand, of 'Jorlandi episcopi' opposite the

Bec;[145] while Albold of St. Edmund's, Robert of St. Frideswide's, and Warin of St. Albans were heads of English religious houses who had spent more or less time in southern Italy.[146] Many men of Norman birth received ecclesiastical preferment in Sicily, not only in the period of reorganization following the conquest, but in the time of Roger II and his immediate successors. William, bishop of Syracuse, the friend of Adelard of Bath, would seem to have been a Norman,[147] as likewise, a generation later, the archbishop of Palermo, Roger Fescan.[148] We find John of Lincoln and Herbert of Braose among the canons of Girgenti in 1127,[149] and Richard of Hereford and William of Caen (?) among those of Palermo in 1158.[150] Under William the Good four prelates of English origin are known: Richard Palmer, bishop of Syracuse and archbishop of Messina, Walter, archbishop of Palermo, and his brother Bartholomew, bishop of Girgenti,[11] and Herbert of Middlesex, archbishop of Compsa.[152] Doubtless, if our sources of information were fuller, other names could be added to this list, for the presence of Englishmen and Normans in the South was due quite as much to royal policy as to other causes.

'Sancti Laudi Episcopi' of the twelfth century original (f. 251 v) shows how the St. Lô of the Norman calendar has given way to St. Gerland of Girgenti. The use of Rouen was employed in Sicily down to the sixteenth century. See La Mantia, *Ordines judiciorum Dei nel missale gallicano del XII secolo nella cattedrale di Palermo* (Palermo-Turin, 1892), p. 4; and the MSS. of Norman origin in Madrid described by Delisle in *Journal des savants*, 1908, pp. 43–49; and by Karl Young, in *Publications of the Modern Language Association of America*, xxiv. 325. On the importation of canonistic material from the North, see E. Besta, "Di una collezione canonistica palermitana," in *Circolo giuridico*, xl (1909); and H. Niese, *Die Gesetzgebung der normannischen Dynastie im Regnum Siciliae* (Halle, 1910), pp. 46–49, 73–76, 80, 93 f., 113 f., 185 f. [145] Eadmer, p. 96.

[146] *Cat. codd. hagiog. Paris.*, ii. 422; *Gesta abbatum S. Albani*, i. 194 f.; supra, notes 50–67.

[147] Pirro, *Sicilia sacra* (ed. 1733), i. 620; supra, Chapter II, n. 8.

[148] This seems to me likely, not so much because of Pirro's statement (i. 86–88) but because the name Fescan (Cusa, *I diplomi greci ed arabi di Sicilia*, i. 17, 27) is the contemporary form for Fécamp.

[149] Palermo, Biblioteca comunale, MS. Qq. H. 6, f. 7, printed incorrectly by Gregorio, *Considerazioni sopra la storia di Sicilia*, bk. i. ch. 3, n. 16. Their bishop Walter was also 'francigena' (*Archivio storico siciliano*, xxviii. 148).

[150] *Documenti per la storia di Sicilia*, first series, i. 20.

[151] On these three consult the index to Chalandon.

[152] Ralph of Diceto, ii. 37; Ughelli, vi. 811.

While King Roger's court was cosmopolitan, he showed a preference for the French and did not forget the ties which bound him to those of Norman blood.[153] Robert of Selby, chancellor during the greater part of the reign and in war as in peace a trusted agent of the king, was an Englishman by birth and dispensed a lavish hospitality to his fellow countrymen. St. William of York, possibly a kinsman of the king, visited Robert at Palermo when exiled from his see, and John of Salisbury drank the chancellor's heavy wines to his undoing.[154] Among the Sicilian prelates whose assiduity at the court scandalized the archbishop of Canterbury,[155] those of English origin were preëminent. Richard Palmer, *vir litteratissimus et eloquens*,[156] occupied a leading position in the *curia* in the later years of William the Bad, and he, with the two other English prelates, Walter Offamil and Bartholomew, were members of the small junto which managed the government during the succeeding reign.[157]   In the north the archbishop of Rouen and the bishop of Bayeux were relatives of William II,[158] while Becket corresponded with Queen Margaret and the principal officers of the court.[159]   Like John of Salisbury, John Belmeis, treasurer of York and bishop of Poitiers, doubtless owed much of his eminence as a linguist to his sojourn in Apulia; [160] Simon of Apulia, later dean of York and bishop of Exeter, was *valde carus et familiaris* to Henry II; [161] and if Gervase of Tilbury passed from the English court to service under William II,[162] Peter of Blois, 'the intimate friend' of both kings,[163] began his career as tutor of William and *sigillarius* in his chan-

---

[153] Hugo Falcandus, p. 6; Romualdus, in *M. G. H.*, *Scriptores*, xix. 426; ibn-al Atir, in Amari, *Biblioteca arabo-sicula*, i. 450; supra, n. 5.

[154] Kehr, *Urkunden*, pp. 49, 75–77, 511.

[155] Migne, *Patrologia*, cc. 1461.

[156] Hugo Falcandus, p. 63.

[157] Chalandon, ii, passim.

[158] Hugo Falcandus, p. 109; Stubbs, *Lectures on Mediaeval and Modern History*, p. 149.

[159] *Materials*, v. 247; vi. 396; vii. 142 f.

[160] Webb, *Joannis Saresberiensis Policraticus*, ii. 271.

[161] Giraldus Cambrensis, iv. 383; cf. *Epp. Cantuarienses*, p. 276.

[162] *M. G. H.*, *Scriptores*, xxvii. 385; Pauli in *Nachrichten* of the Göttingen Academy, 1882, pp. 313–315.

[163] Stubbs, introduction to Roger of Hoveden, ii, p. xcii.

cery.[164] The relations of these sovereigns, always friendly, were firmly cemented by the marriage of William to the Princess Joanna in 1177, an event which served as the occasion of still closer contact between the courts. Florius de Camerata, a justiciar under three kings, acted as one of the envoys who were sent to fetch the princess, while of the bishops and courtiers who preceded and accompanied her to Palermo John of Norwich and Osbert, clerk of the king's *camera*, are especially noteworthy as officers of the royal administration.[165] It is plain that both William the Good and Henry II had ample opportunity to keep informed regarding current conditions in each other's kingdom, while with respect to the administrative system of King Roger's time, Henry had an ever-ready source of information in a Sicilian official whom he had called to his side, his almoner and confidential adviser Master Thomas Brown, who as 'Kaid Brun' and μάστρο θωμᾶ τοῦ βρούνου appears as a royal chaplain in Sicily from 1137 to 1149, and has an important place at the English court from 1158 to 1180.[166]

How far such connections affected the world of learning, we can only guess. It is, of course, essential not to exaggerate the importance of the Sicilian movement. In spite of its more immediate contact with Greek sources, it shows less vitality than the contemporary humanism of the North, and its translations were less important, both in quantity and in influence, than the great body of material which came through the Saracens of Spain. Still, these Sicilian scholars have an honorable place in the history of European learning. At a time when Latin Europe was just advancing from the Boethian and pseudo-Boethian manuals to Euclid's *Elements*, they were familiar with geometrical analysis and applied mathematics as presented in the most advanced works of Euclid and in Ptolemy's *Optics*, Proclus, and Hero; and

---

[164] *Epp.* 10, 46, 66, 72, 90, 93, 116, 131, in Migne, ccvii.

[165] Chalandon, ii. 367 f., 376–378; Ramsay, *Angevin Empire*, p. 193; and the sources there cited. Careful copies of the marriage settlement are given by Roger of Hoveden, ii. 95; Benedict of Peterborough, i. 169; Gervase of Canterbury, i. 263; see also Robert of Torigny, ed. Delisle, ii. 75; and Martène and Durand, *Veterum scriptorum collectio*, i. 902, from MS. Vat. Reg. 980, f. 171.

[166] For his biography, see my discussion in *E. H. R.*, xxvi. 438–443.

they had come into possession of the chief work of ancient astronomy, the *Almagest*.   In philosophy they appear to have acquired the *New Logic* of Aristotle somewhat earlier than their northern contemporaries, and they had likewise an acquaintance with certain dialogues of Plato and with Diogenes Laertius.   Theology and ecclesiastical history were not neglected, and a group of New Testament manuscripts has been traced to Sicilian copyists. The school of Salerno was the leading medical school of Europe. Libraries existed, and the search for ancient manuscripts was carried on.   Sicilian scholars could write decent Greek, and — when they were not translating — decent Latin, and they could venture, not without success, into the field of original verse. Within its limits the intellectual movement at the court of King Roger and his son had many of the elements of a Renaissance, and like the great revival of the fourteenth century, it owed much to princely favor.   It was at the kings' request that translations were undertaken and the works of Nilus and Edrisi were written, and it was no accident that two such scholars as Aristippus and Eugene of Palermo occupied high places in the royal administration.   In their patronage of learning, as well as in the enlightened and anti-feudal character of their government, the Sicilian sovereigns, from Roger to Frederick II,[167] belong to the age of the new statecraft and the humanistic revival.

[167] For Frederick II, see Chapters XII–XIV.

Preface to the Sicilian Version of the Almagest [1]

Eam pingendi Gratias antiqui feruntur habuisse consuetudinem, ut unam quidem vultum aversam, due quibus illa manum porrigeret aspectarent. Cuius misterii non ignarus dudum memoriter teneo gratiam simplam profectam duplam reverti oportere. Tui ergo boni
5 muneris memor, quo earum quas Aristotiles [2] acrivestatas vocat arcium doctrina quasi haustu aque vive animum sicientem liberaliter imbuisti, olim quidem anxie queritabam quid tue dignum benivolentie referre possem. Nec enim eis que philosophie tardus assecla longo pauperis exercitio vene conquisieram purus ingenii torrens fons et domus arcium
10 pectus penitus indigebat. Opes quoque apud earum contemptorem minimum promereri non dubius intelligebam. Angebatur ergo in dies magis ac magis animus meus eo molestius sustinens votum quo complendi voti absolutius facultas negabatur. Verum diutini clamorem desiderii superna tandem pietas exaudivit, dignum ut arbitror plene
15 tribuens remunerationis instrumentum, quod tuum tanto, ut tua pace loquar, precedit munus, quanto finis eo quod ad finem laudabilius. Nec enim tuum latet acumen quod omni sapienti liquet, numerorum mensurarumque scientiam ad eam que astrorum quasi quandam [3] introductionis prestruere pontem. Huius vero eam partem que siderum motus
20 speculatur, veterum lima speculum modernorum Claudius Ptolomeus astrorum scientie [4] peritissimus .XIII. perscripsit libris. Qui a Grecis quidem mathematica seu megisti sintaxis, a Sarracenis vero almegesti corrupto nomine appellantur. Hos autem cum Salerni [5] medicine insudassem audiens quendam ex nunciis regis Sicilie quos ipse Constanti-
25 nopolim miserat agnomine Aristipum largitione susceptos imperatoria Panormum transvexisse, rei diu multumque desiderate spe succensus, Scilleos latratus non exhorrui, Caribdim permeavi, ignea Ethne fluenta circuivi, eum queritans a quo mei finem sperabam desiderii. Quem tandem inventum prope Pergusam fontem Ethnea miracula satis cum
30 periculo perscrutantem, cum occulte quidem alia, manifeste vero mens scientie siderum expers prefatum mihi opus transferre prohiberent, grecis ego litteris diligentissime preinstructus, primo quidem in Euclidis Dedomenis, Opticis, et Catoptricis,[6] Phisicaque Procli Elementatione prelusi. Dehinc vero prefatum Ptolomei aggressus opus, expositorem
35 propicium divina mihi gratia providente Eugenium, virum tam grece quam arabice lingue peritissimum, latine quoque non ignarum, illud contra viri discoli voluntatem latine dedi orationi. In quo nimirum mea mens infando pressa labore inceptum sepe destituisset opus, nisi superande difficultatis auctor potentissimus amor tui tuumque munus
40 animum crebra mutui repetitione[7] pulsarent. Neque enim questus spe

---

[1] For the MSS. see above, pp. 157–159. I have here followed MS. D, Wolfenbüttel Gud. 147, f. 2, which gives the best text of the preface. For the principal variants of A and C, see *Harvard Studies*, xxi. 99–102, xxiii. 156 f.
[2] *De caelo*, 3, 7, 306 A.   [4] Om. C D.   [6] D, *Catoplicis*.
[3] So all the MSS.   [5] MS. D, *Sarelni*.   [7] MS. D, *repetitioni*.

motus aut gloria istum potui laborem substinere, cum liquido constet
spei locum artifici non relinqui, ubi ars ludibrio et dedecori est. Neque
enim artificem mirari potest qui artem non miratur. Sensisti vero et tu
nonnullos hiis in temporibus cause quam ignorant iudices audacissimos,
45 qui, ne minus scientes videantur, quecunque nesciunt inutilia predicant
aut profana. Iuxta quod Arabes dicunt: Nullus maior artis inimicus
quam qui eius expers est. Eoque pertinacius criminandis artibus instant
quo ab earum laude impericie probrum certius sibi conspiciunt immi-
nere. Eos omitto qui honestatis zelo honesta quoque studia persecun-
50 tur. Quos pie peccare recte dixerim dum nocivam curarum putredinem
recidere contendentes, a sanarum altrice curarum philosophia manum
minime continent indiscretam, sed et eam ipsius partem graviori crim-
inatione persecuntur que ingeniis exquisita clarissimis et exculta quo
defecatior ac purior est, eo sapientie vocabulo dignior, eo gratiori qua-
55 dam compede speculationis iocundissime animos hominum continet alli-
gatos. Horum siquidem error sive coloratus honesto malicioso quoque
predictorum testimonio fretus, apud imperitos quorum maxima est multitudo in bonarum neglectum arcium efficacissime peroravit, ut iam numerorum quidem mensurarumque scientia omnino superflua et inutilis,
60 astrorum vero studium ydolatria estimetur. Ita nimirum sentiebat vir [8]
religiosus ac prudens cum dicebat: Hoc est igitur illud quadruvium
quo his viandum est qui a sensibus procreatis nobiscum ad certiora in-
telligentie perduci volunt. Eisdem quoque attestatur Remigius [9] dicens:
Cum omnes artes pessumdate essent, aput Egyptios Abraham eos as-
65 trologiam docuisse. Sed et sanctum Moysen sanctumque Danielem
Dominus credo ob astrorum scientiam reprobavit. Stultum quippe
creatoris opera contemplari, eorumque speculatione ineffabilem ipsius
potentiam ac sapientiam delectabilius admirari. Nefarium quoque
penitusque liquet illicitum ad Conditoris cognitionem conditorumque
70 cognitione animum sublevare, Creatorem insensibilem sensibilium speculatione sibi quodam modo sensibilem comparare. O mentes cecas
viamque philosophandi penitus ignorantes! A creatura siquidem mundi
invisibilia Dei intellecta conspiciuntur,[10] nec satis insensibilium veritatem percipere potest mens humana ni ad eam preludio sensibilium sibi
75 viam facultatemque preexcuderit. Hinc a sapientibus institutus est ac
diligenter observatus hic studiorum ordo, ut primo quidem ingenite
ruditatis nebulas diligenti creatorum disquisitione serenarent, omnibus
quidem sed eis potissimum invigilando disciplinis que ipsam sine omni
erroris devio sine omni dubitationis scrupulo veritatem contemplantes

[8] Boethius, *De institutione arithmetica*, 1, 1 (ed. Friedlein, Teubner, 1867, p. 9,
ll. 28 ff.): 'Hoc igitur illud quadruvium est, quo his viandum sit, quibus excellentior
animus a nobiscum procreatis sensibus ad intellegentiae certiora perducitur.'

[9] Probably from the unpublished commentary of Remigius on Martianus Capella. Cf. *De nuptiis Philologiae et Mercurii*, 8, 812.

[10] *Rom.* 1, 20: 'Invisibilia enim ipsius a creatura mundi per ea quae facta sunt
intellecta conspiciuntur.'

80 occulum mentis Boetio [11] teste rursus illuminant, dehinc vero robore hoc
animati in theologica exercitate mentis aciem fiducialiter intendebant.
Unde et ab ordine docendi et discendi theologiam metaphisicam nomi-
nabant. Verum nostri nimirum aquile hoc quasi quodam molimine
giganteo minime indigent sine omni creaturarum adminiculo radiis
85 summe lucis oculos infigere potentissimi atque summe secreta veritatis
efficaciter penetrare. Vix rudimentis a puerilibus celum involant terras-
que habitare dedignantur. Super nubes eorum conversatio, atque in
ipso summe sinu sapientie sese requiescere gloriantur. Mundanam
desipiunt sapientiam, eique vacantium deliramenta subsannant. Tibi
90 vero, vir mentis serenissime, longe alia mens est. Tu sacras artes et
propter se appetendas, scientibus dulces, insciis adorandas rectissime
arbitraris. Nec vero tuum fallit acumen quoniam perfectio beatitudinis
in plenitudine consistit cognitionis, quo sciendo proficimus, hoc acces-
sum ad beatitudinem fieri, presertim cum ocio quidem mens corrumpa-
95 tur, studium vero virtuti sit amicum. Preclarum quoque tibi credo
videtur, in quo prestat [12] homo ceteris animalibus hominem homini pre-
nitere. Hinc insurgendum summisque viribus iudicas incumbendum ut
omni scientie genere mens illustretur, ad beatitudinem preparetur, suo
proprio bono decoretur. Tui ergo tuique similium gratia presentem
100 hunc laborem ego suscepi, quibus si placeo intentio quoque mihi mea
perfecta est. Rideant et insultent artium inimici, ignota iudicent, astrorum studium insaniam predicent. Michi confiteor hec insania dulcis,
michi dulce clamare cum Nasone: [13]

105
    Felices anime quibus hec agnoscere primum
    Inque domos superas scandere cura fuit!

Faveas ergo summisque tibi vigiliis opus elaboratum benignus queso
suscipias. Illud tamen unum super omnia moneo ac rogo ut ea qua
et in geometricis usus es edocendis discretione collaudanda ad huius
operis lectionem dignos admittas, indignos abigas. Suam quippe rebus
110 dignis adimet dignitatem, siquis eas communicaverit indignis.

[11] Boethius, *De institutione arithmetica*, I, I (ed. Friedlein, p. 10, ll. 1–7): 'Sunt
enim quidam gradus certaeque progressionum dimensiones, quibus ascendi progredi-
que possit, ut animi illum oculum . . demersum orbatumque corporeis sensibus
hae disciplinae rursus inluminent.'

[12] Cf. Sallust, *De coniuratione Catilinae*, I, I; Cicero, *De inventione*, I, 4.

[13] Ovid, *Fasti*, i, 297–298. In the text of Merkel-Ehwald (Teubner, 1889):

    'Felices animae, quibus haec cognoscere primis
    Inque domus superas scandere cura fuit!'

CHAPTER X

NORTH-ITALIAN TRANSLATORS OF THE
TWELFTH CENTURY[1]

THE history of Greek studies in northern Italy in the twelfth century lacks the coherence and definiteness which we have found in Sicily. The north had no resident Greek population, no Greek monasteries, no university like Salerno, no royal library of Greek manuscripts. It had likewise no political unity, and the connections of its several regions with the East arose chiefly out of the trade of the commercial republics, the contacts of the Crusades, the diplomatic negotiations between the two empires, directed chiefly against the Sicilian kingdom, and the related negotiations of the eastern emperor and the Roman church. Of these the first are probably the most significant, creating as they did the Venetian and Pisan quarters at Constantinople and bringing into residence there a number of scholars who learned Greek and transmitted a certain amount of Greek learning to the West. Some of these, we know, were engaged in permanent service in the imperial household. Besides these more continuous connections, however, the various embassies must be noted, not only as giving us fleeting glimpses of the Italian colony, but as furnishing occasions for the transmission of eastern learning to the West. Especially under Manuel Comnenus do we find a steady procession of missions to Constantinople, papal, imperial, French, Pisan, and other, and a scarcely less continuous succession of Greek embassies to the West, reminding us of the Greeks in Italy in the

[1] Besides newly discovered material, this chapter utilizes my earlier studies on Moses of Bergamo, *B. Z.*, xxiii. 133–142 (1914); Leo Tuscus, *E. H. R.*, xxv. 492–496 (1918), and *B. Z.*, xxiv. 43–47 (1923); and Burgundio, *American Historical Review*, xxv. 607–610 (1920). The article on "Leo Tuscus" was sent to the *B. Z.* in July, 1914, but the cessation of this journal during the war led me to send a revised copy in 1918 to the *E. H. R.*, where it appeared in October. In 1922 the *B. Z.*, without my knowledge, and in evident ignorance of its previous publication, printed the original article.

early fifteenth century.[2] It was an opportunity for any scholars who were interested in Greek learning, and occasionally there was a man like the Pisan Burgundio who made good use of it for something besides theology.

Characteristic of these missions are the occasions they furnished for theological disputation over the differences between the two churches; indeed the reports of such discussions are sometimes our best evidence of what was going on in the world of learning.[3] As early as 1112 we find the archbishop of Milan, Peter Chrysolanus, disputing before the Emperor Alexius with Eustratius of Nicaea and others, as recorded in various Greek texts and in a fragment of the Latin *libellus*.[4] From 1136 to 1155 the chief figure was Anselm, bishop of Havelberg since 1129, and from 1155 to his death in 1158 archbishop of Ravenna.[5] Sent by the Emperor Lothair in 1136, he took occasion to thresh out theological matters with Nicetas, archbishop of Nicomedia, and others.[6] He was again in Constantinople in 1153 and 1154, and on his way back in April, 1155, he debated with Basil of Achrida at Thessalonica.[7] Henry, archbishop of Benevento, who was in Constantinople on behalf of Alexander III in 1161 and again in

[2] See in general Chalandon, *Les Comnènes*, ii. 161–173, 259–262, 343–361, 555–608; and the literature there cited.

[3] See in general Hergenröther, *Photius* (Regensburg, 1869), iii. 789 ff.

[4] For the speeches of Eustratius and John Phurnes, see Demetracopoulos, *Bibliotheca ecclesiastica* (Leipzig, 1866), i. 36 ff. (cf. Dräseke, in *B. Z.*, v. 328–331); for the Greek text of Chrysolanus, Migne, *Patrologia Graeca*, cxxvii. 911–920. The Latin fragment of Chrysolanus is at Monte Cassino, MS. 220, f. 149, printed in *Bibliotheca Cassinensis*, iv. 351–358. Cf. Chalandon, *Les Comnènes*, i. 263, n; Krumbacher, p. 85; Tiraboschi, *Storia della litteratura italiana* (1787), iii. 324–327; Hergenröther, *Photius*, iii. 799–803; Hurter, *Nomenclator*, ii. 12 f.

For similar instances in the eleventh century, see Petrus Diaconus, in Migne, *P. L.*, clxxiii. 1027, 1043; and cf. Manitius, *Lateinische Litteratur*, ii. 384 f.

[5] E. Dombrowski, *Anselm von Havelberg* (Königsberg, 1880); J. Dräseke, in *Zeitschrift für Kirchengeschichte*, xxi. 160–185 (1900).

[6] For Nicholas of Methone in 1136, see Dräseke, *B. Z.*, i. 458 ff.

[7] For the discussion as to the date of these missions, see especially Kap-Herr, *Die abendländische Politik Kaiser Manuels* (Strasburg diss., 1881), pp. 148–151; Simonsfeld, *Jahrbücher Kaiser Friedrichs I*, i. 200, 231, 300; Chalandon, ii. 344 f. For the debate with Basil, see Josef Schmidt, *Des Basilius aus Achrida bisher unedierte Dialoge* (Munich, 1901); and for Basil in general, Vasilievskiï, in *Vizantiiskii Vremmenik*, i. 55–132 (1894).

1165 and 1166, seems also to have opposed Basil.[8]   In 1169–1170 the patriarch of Constantinople, Michael Anchialou, addressed to the Emperor Manuel a dialogue against the Latins, apparently directed at the two cardinals then on a mission in the East,[9] and Andronicus Kamateros puts a similar dialogue in the mouth of the Emperor, who interested himself actively in such debates.[10]   At some time between 1130 and 1182 Henry, patriarch of Grado, had a friendly discussion with Theorianus.[11]   Down to about 1166 Nicholas of Methone [12] was an outstanding figure in these controversies with the Latins, first with Anselm and later with a resident Pisan, Hugo Eterianus, to whom we shall come below; while in the period just preceding 1179 we shall find Hugo collecting materials from the Greek Fathers for the benefit of an emissary of Frederick Barbarossa.   As late as the Fourth Crusade an anonymous Greek records his earlier contentions with Hugo.[13] Many other undated polemics of this period might be listed,[14] Greek polemists even appearing at Rome in 1150 and at the Lateran council of 1179.[15]   On the Latin side it is obvious that northern Italy had a noteworthy share in these theological controversies.

Certain of these discussions seem to have been stenographically reported,[16] and in any case they are set forth at length in Greek manuscripts, many of which have now been published.   The discourses of the Latins are less well known, being sometimes recorded only in the Greek reports.   Still we have the fragment of

---

[8] Schmidt, pp. 27 f.; Chalandon, ii. 559, 563 f.

[9] Loparev, in *Vizantiskii Vremmenik*, xiv. 334–357 (1907).

[10] Migne, *Patrologia Graeca*, cxli. 395; Hergenröther, *Photius*, iii. 811–814.

[11] Migne, *P. G.*, xciv. 404–409.

[12] Dräseke, in *Zeitschrift für Kirchengeschichte*, ix. 405–431, 565–590 (1888); idem, *B. Z.*, i. 438–478 (1892); cf. vi. 412.

[13] Arsenii, as noted in *B. Z.*, iv. 370, n.

[14] Hergenröther, iii. 803 f.; Krumbacher, pp. 87–91; Chalandon, ii. 653; dialogues of Nicetas of Maronea in *Bessarione*, xvi–xix (1912–15).

[15] Migne, *P. L.*, clxxxviii. 1139; Nectarius of Casule at the Lateran Council, Baronius, *an.* 1179, nos. x–xii. Cf. Nicholas of Casule at Constantinople ca. 1205: Engdahl, *Beiträge zur Kenntnis der byzantinischen Liturgie* (Berlin, 1908), pp. 85 f., and references.

[16] *B. Z.*, xv. 358.

Chrysolanus,[17] the *Dialogi* of Anselm of Havelberg,[18] written out
at the Pope's request fourteen years after the disputation of 1136,
and the controversial writings of Hugo Eterianus.[19]   Theology of
a less contentious sort found its way westward in the translations
of Burgundio and in an anonymous treatise *De diversitate nature
et persone*.[20] Interpreters were needed for such debates, as well as
for the diplomatic negotiations of the missions;[21] sometimes they
accompany the emissaries, and again they are chosen from the
resident Latins, and it is among these men who knew Greek that
we can most profitably seek evidence of intellectual connections
with the West.   The best example is found in the account which
Anselm of Havelberg gives of his public debate with Nicetas, held
in the Pisan quarter at Constantinople in April, 1136.   Among
the multitude present

> Aderant quoque non pauci Latini, inter quos fuerunt tres viri sapientes
> in utraque lingua periti et litterarum doctissimi, Iacobus nomine Veneticus
> natione, Burgundio nomine Pisanus natione, tertius inter alios precipuus
> grecarum et latinarum litterarum doctrina apud utramque gentem clarissi-
> mus Moyses nomine Italus natione ex civitate Pergamo; iste ab universis
> electus est, ut utrimque fidus esset interpres.

Each of these interpreters is otherwise known as a translator.
Let us begin with the one whom Anselm considered the most
eminent. Moses of Bergamo, though he has long held a place in
Italian historiography, is as yet unknown as a grammarian and
translator, and his position as intermediary between Greek and
western learning requires further study. The principal source of
information respecting him is his letter, written probably in 1130,
to his brother Peter de Brolo,[22] provost of the church of S. Alessandro at Bergamo.[23] Moses is then resident at Constantinople

---

[17] Supra, n. 4. See also Masnovo, in *Archivio storico lombardo*, xlix. 1 (1922).

[18] D'Achery, *Spicilegium* (Paris, 1723), i. 161–207; Migne, clxxxviii. 1139–1248.

[19] Infra, n. 121.

[20] Infra, n. 108.

[21] E. g., 'Gibertus interpres imperii' in 1170: *M. G. H., Scriptores*, xviii. 86.

[22] Also known as Peter di San Matteo; cf. Capasso, in *Archivio storico lombardo*,
fourth series, vi. 301.

[23] Lupi and Ronchetti, *Codex diplomaticus civitatis et ecclesiae Bergomatis* (Bergamo, 1790–99), ii. 949–962, where the date is discussed. Cf. the analysis given by
Capasso and Pesenti in the articles cited below.

and engaged in the emperor's service,[24] which has recently taken him to Thessalonica. He has various relatives and friends in and about Bergamo whom he hopes soon to visit; he has not forgotten the churches of his native city in distributing funds at his disposal, and the cathedral receives four pallia by his gift.[25] In Venice he is in relations with the abbot of S. Niccolò and with Domenico Bassedelli, *iudex et maximus terre vir*, master of the ship which had brought the relics of St. Stephen from Constantinople in 1110,[26] either of whom will forward the young relative whom he asks his brother to send in place of their deceased nephew. Peter's last letters had come at the hands of a certain John the Roman, who had been sent on a mission by Milan and whose shabby appearance and undignified conduct were particularly offensive. At Constantinople Moses is a man of some wealth with a position to sustain, but in the burning of the Venetian quarter he has recently lost the greater part of his fortune, to the value of more than 500 bezants, including his whole collection of Greek manuscripts, brought together by long effort at the price of three pounds of gold.[27]

This remarkable zeal for collecting manuscripts entirely accords with Anselm's account of Moses' learning and leads the way to an inquiry concerning his literary labors. His most important work is the so-called *Pergaminus*, a poem in three hundred and seventy-two rhyming hexameters descriptive of the city of Ber-

---

[24] 'Me principis violentia percinctum laborem subire coegit.' We can only conjecture the nature of his employment, unless we attach some weight to the note in the MS. of the *Pergaminus* which calls him 'valens et probus homo in scriptura in curia imperatoris Cplani.' Moses mentions his influence at the imperial *vestiarium*.

[25] See the "Indiculus de codicibus et ecclesiasticis supellectilibus a Petro preposito comparatis" in Lupi, ii. 921.

[26] *Translatio S. Stephani*, in Cornelius, *Ecclesiae Venetae* (Venice, 1749), viii. 106. Monticolo's forthcoming edition will doubtless identify more fully the numerous Venetians mentioned in this narrative. Bassedelli witnesses a Venetian document of 1124 in Gloria, *Codice diplomatico padovano*, no. 162.

[27] 'Combusti sunt igitur omnes libri greci quos multo dudum labore conquisiveram precii trium librarum auri et reliqua universa nisi siquid in auri puri moneta fuit, que mihi iactura damni plus D. byzantiis intulit.' The fire is not mentioned by the Venetian chroniclers: Heyd, *Histoire du commerce du Levant* (1885), i. 196 n. On the Venetian quarter in this period, see now Horatio F. Brown, in *Journal of Hellenic Studies*, xl. 68–88 (1920).

gamo and constituting a source of prime importance for the early history of the commune.[28] First published under the name of a Moses Muzio or Mozzi with the date of 707 and a dedication to Justinian II, it was shorn of these fictitious adornments by the criticism of Muratori and can now be placed with reasonable certainty in the early years of the twelfth century.[29] In the unique manuscript of the fifteenth century the treatise is anonymous, but it is cited two hundred years earlier as the work of 'Magister Moyses,' and a contemporary gloss in the manuscript calls h'm 'Magister Moyses Pergamensis valens et probus homo in scriptura . . . in curia imperatoris Constantinopolitani.' The identity with the author of the letter to the provost Peter has been further established by the stylistic resemblances between the two works and particularly by Grecisms in the text of the *Pergaminus*. For its age the poem gives evidence of some literary skill and a respectable Latin culture.

The editors and critics of the *Pergaminus* have been acquainted with no other literary work of Moses. Tiraboschi, however, long ago attributed to him an *Expositio* of the Greek words in the biblical prefaces of St. Jerome which four manuscripts mentioned as the work of a certain Moses,[30] and this treatise, first described by Hauréau,[31] was edited by Pitra in 1888[32] and, more critically, by Gustafsson in 1897.[33] Oddly enough, none of these scholars

[28] Muratori, *Rerum Italicarum Scriptores*, v. 521–536. See especially the studies of Capasso, "Il 'Pergaminus' e la prima età communale a Bergamo," in *Archivio storico lombardo*, 4th series, vi. 269–350 (1906); and Pesenti, "Il Pergaminus," in *Bollettino della civica biblioteca di Bergamo*, vi. 121–151, vii. 1–22 (1912–13). The suggestion of Giesebrecht (Munich *Sitzungsberichte*, 1879, ii. 279) that Moses was also the author of the poem of 1162–66 now edited by Monaci as *Gesta di Federigo I in Italia* (Rome, 1887) has been refuted by Monaci on chronological and other grounds.

[29] Pesenti argues from the mention of Bishop Ambrose without his title that the poem is anterior to his consecration in 1111, but the argument does not seem to me decisive.

[30] *Storia della letteratura italiana* (1787), iii. 351, citing a MS. of the Marciana and three from Leipzig and Paris catalogues. Pesenti had a vain search made for the lost MS. of the Marciana, but went no further in his efforts and knows nothing of the editions.

[31] *Notices et extraits des MSS.*, xxxiii, 1, p. 244; *Notices et extraits de quelques MSS.*, i. 122. [32] *Analecta sacra*, v. 125–134.

[33] *Moysi Expositio*, in *Acta Societatis Fennicae*, xxii, no. 3; cf. *B. Z.*, vi. 461.

thought of identifying the Moses of the title with the Bergamask writer of this name, and as 'Magister Moyses de Grecia' he has secured a separate entry in bibliographical literature.[34]   This Moses was otherwise unknown to Hauréau; Pitra attached him conjecturally to the school of Scotus Eriugena and the Irish masters of the ninth century; Gustafsson inclined to the twelfth century because of the *copiae litterarum vere largae* manifest in the work, but could neither identify or place him.   No known manuscript describes the author more definitely, yet Tiraboschi's identification is highly probable.   This Moses cannot be later than the twelfth century, the date of the earliest manuscripts, and his learning and style could not well have been found earlier.   He has lived in the East long enough to be called *grecus* and to get an acquaintance with Byzantine writers and a very considerable knowledge of the language, yet he handles Latin easily and correctly and cites Caesar, Lucan, Terence, Horace, and Virgil.   All this agrees entirely with Moses of Bergamo and, so far as we now know, with no one else of the name.[35]   Moreover, as we shall see, there are parallelisms with another work specifically ascribed to the Bergamask author.   The treatise does not discuss all the Greek words in Jerome's prefaces, but it covers the most obvious difficulties and adds various illustrations and amplifications, of which the longest, the chapter devoted to *Homerocentones*, is a definite contribution to our information on this subject.   The author's knowledge of Greek and Latin grammar is rarely at fault and amply confirms Anselm's opinion of his attainments in the two languages.[36]   Probably the *Expositio* is not the earliest of his

Gustafsson has made a wider but not a complete use of the MSS. and bases his edition upon two at Munich and two at Leipzig. To his list should be added the lost Venetian codex and the extracts in Add. MS. 35091, ff. 115-116, of the British Museum.

[34] E. g., Chevalier, *Bio-bibliographie*[2], col. 3271.

[35] The German gloss *antfriston* in c. 13, in spite of Traube's opinion that it was probably in the archetype (Gustafsson, p. 9), does not seem to me sufficiently established as part of the original to serve as a basis of inference concerning the author.

[36] Gustafsson says (p. 9): 'Aliquantulum . . . inter vulgares magistellorum greges eminet et rerum copia et praeceptorum prudentia et sinceritate quadam sermonis.'

literary labors; in any case it had its origin in an inquiry concerning *Homerocentones* made many years before in a letter from a British clerk named Paganus,[37] at a time accordingly when Moses had already acquired a certain reputation for Greek learning.

Another evidence of the literary activity of Moses of Bergamo is found in a treatise hitherto unknown which is definitely ascribed to him in the only manuscript which I have been able to discover, MS. 52 of the Bibliothèque de Nîmes:[38]

Moyses Pergameni prologus in presens opusculum quod ipse de greco transtulit

Cum sapientis cuiusdam grece lingue librum necessaria quedam querendo percurrerem, contigit hunc quoque me circa finem repperire libellum. Cuius titulo mox percurso tanto protinus eum quoque legendi sum desiderio tactus ut, iis intermissis quorum mihi fuerat occasione repertus, ad ipsum me tota mentis aviditate converterem. Cum vero diligenter eum finetenus perlegissem, quamvis et frigus ingens velut circa mensis decembris initium foret et occupationes alie me plurime circumstarent, nocturnis me vigiliis et translationis laboribus tradidi, ne pretiosum repertum thesaurum solus possidens invidie vel inertie merito ceu piger et nequam servus arguer, cum presertim grecas litteras propter id potissimum didicisse me sim sepe testatus, ut ex eis in nostras siquid utile reperirem quod nobis minus ante fuisset debita devotione transverterem. Gratias igitur ago Deo quia, sicut ait apostolus,[39] qui dedit velle dedit et perficere pro bona voluntate. Te vero, lector amice, devote rogo, quisquis hunc labosculum nostrum transcribere forte dignaberis, ne transscriptum cum suo prototypo [40] conferre graveris,[41] nec turbere queso si cum titulum materie legeris auctoris nomen suppositum non inveneris. Quamvis enim conditoris nomen in fronte de more non gerat, nichil in eo tamen [42] videri debet apocryphum, cum totum quicquid id est de sacra sit pariter canonicaque scriptura collectum. De me quoque qui transtuli proemio supplicando subiungo quatinus ego Moyses videlicet pergamenus cum per me tibi tradita legeris orationibus tuis seu vivus seu luci subtractus interserar.

Exceptio compendiosa de divinitus inspirata scriptura sive argumentum orthodoxe fidei de Sancta Trinitate quod in tribus est personis deitas et quod ante secula Filius et Spiritus et quam divina scriptura quanque quidem de essentia natura scilicet nos doceat deitas quanque vero de diversis personis ipsius. De Sancta Trinitate Moyses in Genesi: [43] Et dixit Deus, Faciamus hominem ad imaginem et similitudinem nostram . . . .

---

[37] 'Quidam clericus nomine Paganus Britannus genere.' To an Italian this can hardly mean Breton, as Hauréau interprets it.

[38] Saec. xiii ineuntis, ff. 96–126. See *Catalogue général des MSS. des départements*, old series, vii. 557.

[39] Philippians, ii. 13. [41] MS. *gravis*. [43] Genesis, i. 26.

[40] MS. *prototylo*. [42] MS. *tam*.

What treatise is here translated is a question which I must leave to specialists in Greek theology. In itself the work is of slight interest, being little more than a catena of passages, largely from the Old Testament, dealing with the life of Christ. Gregory and Chrysostom are cited, and on two occasions the author comments on Greek words after the manner of the *Expositio*. In one of these (Sirach, li. 9) he finds it necessary to distinguish between ἱκεσίαν and οἴκησιν, in the other (Habbakuk, iii. 2) he explains the difference between ζωή and ζῷον.[44] In the preface the writer's fondness for *ceu* and for locutions like *me sim sepe testatus* can also be paralleled in the *Expositio*,[45] while the request for the reader's prayers is noteworthy in both.[46] One new fact is here brought out besides the explicit mention of the writer's name, the fact that Moses was a translator as well as a grammarian, and learned Greek for the special purpose of turning into Latin works not previously known in the West. Further search may perhaps disclose more significant examples of his work in this field.

Meanwhile we may with high probability identify another specimen of his grammatical exegesis. In MS. 22 of the Bibliothèque Nationale at Luxembourg[47] the *Expositio* is followed in the same hand by a brief treatise written in answer to an inquiry respecting the accentuation of the oblique cases of *character*. The reply first gives the principal parts of χαράσσω and a list of its derivatives with their Latin equivalents, and then accompanies the declension of χαρακτήρ with a discussion of the inflexion of nouns in -ηρ which is based directly on the Κανόνες of Theodosius of Alexandria. The whole treatment is in the manner of the *Expositio* and the glosses in the Nîmes manuscript, and there are

---

[44] Ff. 97 v, 115. Cf. in the *Expositio*, cc. 10, 20, 30, 39, the accentuation of ἀσπίς and the distinction between ἔτυμος and ἑτοῖμος, συκή and σῦκον, σῖτος and σῖτον.

[45] *Ceu nescius*, p. 16, l. 19; *ceu puto*, p. 17, l. 22; *sim sepe rogatus* in the Luxembourg version of the epilogue (van Werveke, *Catalogue des MSS.*, p. 42).

[46] See the prologue of the *Expositio* and the more developed conclusion of the Luxembourg MS.: 'vovens et petens pariter per orationes eorum iuvari quibus hec per me nota profuerunt.'

[47] Saec. xiii, ff. 179–180; extracts in van Werveke, *Catalogue*, p. 42. For kind assistance in securing photographs I am greatly indebted to the librarian, M. d'Huart.

parallelisms in phraseology.[48]   Unfortunately this folio of the manuscript has been injured just where we should expect to find the name of the author and a further description of the addressee, so that no writer is named.   The *Item*, however, which connects this tract with the *Expositio* creates a strong presumption in favor of Moses, which is confirmed by the style and mode of treatment. The brother Alexander of the dedication is otherwise unknown, but the text is corrupt, and we are justified in suspecting a scribe's confusion with the name of the church, S. Alessandro, of which Peter de Brolo was provost; we may conjecture that the treatise was addressed to Peter, whose literary and theological interests are known from the library which he collected.[49]   The mention of Dacia would seem to point to the Danubian campaigns of John Comnenus in 1128,[50] on which Moses may have accompanied him in some secretarial position such as he seems to have held at the court.   The text reads:

Item ad Alexandrum prepositum ex Datia

Quesivit a me nuper prudentia tua, Alexander domine mi frater atque dig [51]   (f. 179 v)
nomen per oblicos casus proferri deberet in penultima silliba.   Libenter ergo tibi Deo donante declarabitur protinus quod quesisti cum prius tamen patuerit quale sit hoc nomen vel unde sit tractum.   *Charasso* sive *caracto*, nam per .s. geminum solent apud Grecos huius modi verba scribi sive per .t., per .s. secundum linguam communem per .t. secundum atticam, ut *thalassa* sive *thalatta*, hoc est mare, *philasso* sive *philato*, hoc est custodio vel servo. Sunt huius verbi duo preterita perfecta, primum quidem *parakeimenon* id est adiacens, quod est *kecharacha*, secundum vero quod dicitur *aoriston* id est infinitum, id est *ekaraxa*, sicque [52] duo quoque sunt eius infinita, a *parakeimeno* quidem [53] *kekarakene* ab infinito vero *caraxe* vel *caraxein* communiter vel *caracten* attice.   Significat autem hoc verbum fodere vel cavare sive signare.   Derivatur ex hoc *carax*, id est corona amminiculorum infixorum circa vitem per que possit ipsa sustentari ne propria debilitate corruat vel canabis vel cuiuslibet talis in campo vel crista fosse circa locum quemvis muniminis causa quam nos vallum vocare solemus.   Hinc aliud verbale

---

[48] E. g., the use of *protinus* (cf. the Nîmes prologue), *sufficientissime respondisse* (cf. *Expositio*, p. 29, l. 22), and the discussion of πλάσσω (*Expositio*, p. 18, l. 7).

[49] *Codex diplomaticus Bergomatis*, ii. 919–924.   Peter's name is connected with S. Alessandro in both the *Indiculus* and the letter of Moses, so that the basis of the scribe's confusion could easily have existed in the address of the treatise.

[50] See Chalandon, *Les Comnènes*, ii. 58–62.   Cf. the *percinctum laborem* and the journey to Thessalonica in the letter to his brother.

[51] One and a half lines gone.   [52] MS. *sed que*.   [53] MS. *quod*.

nomen *characoma* quod nos recte vallationem seu vallamen possumus dicere. Hinc etiam *charagma* quod simpliciter quodlibet signum significat vel insignitionem [54] monete que de hoc equidem verbo femmino quoque genere *charage* grece dicuntur. Hinc *parachasimon* [55] *nomisma* dicitur adulterine monete. Hinc corrupte latinum verbum dicitur tractum charaxare, quod est minutim fleobotomo id est ferro vene cesorio plagas infligere quibus ventose superponuntur ad eliciendum sanguinem. Ex hoc ergo verbo grece quod est *charatein* sive *charassein* [56] derivatur verbale nomen *charactes*, id est signator, sicut *apoen* [57] *plasso* quod est fingo *plistes*, id est fictor,[58] *character* quoque pariter, id est effigies vel effigiatio sive statua, unde est:

Cuius ad effigiem non tantum meiere [59] fas est.[60]

signum vel in ovibus vel in ceteris animalibus cuius impressione dominis suis cercius cognoscantur. Est autem nomen hoc apud grecos *ixitonon*, id est quod acuto circa finem profertur accentu, per oblicos vero casus universos circumflexo tono profertur in penultima sillaba, id est *tu caracteros to caracteri ton caractera* genitivum dativum accusativum singulares, *hoi caracteres tus caracteras o caracteres* nominativum et accusativum et vocativum plurales. Nam genitivus pluralis *ton caracteron* acuitur in penultima sillaba eo quod ultima sillaba producitur per ꝏ. mega ratione regule que communis est et nobis et Grecis quia in polisillabis dictionibus si penultima natura longa est ultima vero brevis, penultima circumflectitur, ut in superioribus obliquis id est *caracteros caracteri* monstratum est. Si vero ambe longe sint, acutus accentus est in penultima, ut huius et e et o Muse. Hoc autem nomen id est *karacter* in notatione [61] grece per ηeta scribitur que semper longa est quamque nos semper in e longam vertimus, ut Criητη Crete Mytylηnη, et econtrario Greci nostrum .e. longum sepe vertunt in eta suum longum similiter, ut rex rηx reges rηges. Ut autem nomen hoc in fine nominativi casus et in penultimis obliquorum circumflectitur, talis apud Grecos de ipso vel ceteris similibus regula est. Eorum in ηr oxitonorum sunt quotquot quidem habent .t. per etam declinantur, ut *luter luteros*, id est vas in quo lavantur, ut in Moysi lege sepe legitur quod nos latine labium vel labellum dicimus seu vas significet seu partem eris, ex verbo lavo vel luo sicut grece *lutηr* ex verbo *luein* seu *luse* quod nos similiter dicimus luere; *capter* [62] *capteros*, id est flexus sive flexura (f. 180) vel meta circensium ludorum circa quam [reg]imen currus flectitur que *captos* quoque dicitur, sicut nos quoque flexuram omnem vel angulum vocare solemus, de verbo *capto* id est flecto quod nos cambire vel campsare dicere con suevimus; *elater elatéros* id est agitator, de verbo *elan* id est agitare. Excipi

---

[54] MS. *insigninoɱ*.   [55] I. e., παραχαράξιμον.   [56] MS. *charasseim*.

[57] *Sic.* Perhaps some reference to ποιεῖν ποιητής has dropped out.

[58] Cf. *Expositio*, p. 18, l. 7: 'πλάσσω grecum verbum est quod latine proprie fingo dicitur. Hinc nomen verbale πλάστης vel πλαστήρ id est fictor.'

[59] MS. *melere*.

[60] Juvenal, *Sat.*, i, 131. A space of about ten letters is gone before *signum*.

[61] MS. *notō*. Hereafter the MS. regularly has η written above the *eta* of the Latin text.

[62] I. e., καμπτήρ, καμπτός, κάμπτω.

tur *pathr patéros* quod per .e. breve nostrum simul ac ipsorum commune scribitur in penultima per casus omnes obliquos *pateros pateri patera pateron patéres* acuta penultima sed [63] correpta. Quotquot vero non habent .t. per .e. breve scribuntur, ut *ether ethéros daer daeros*, id est frater mariti, *aer aeros*, apud eos acuta penultima cum sit ultima brevis. Excipiuntur *spinter spinteros*, id est scintilla, *eleuter*, id est ventor *eleuteros*. Et hoc est canon tricesimus secundus masculinorum nominum apud Grecos indeclinabilibus nominum de oxitonis in *ηr*.[64] Tricesimus vero tertius de varotonis in *ηr* similiter: [65] *o piηr tu pieros o iber tu iberos*.[66] In *ηr* per eta varitonorum quecunque quidem longa penultima sunt per .e. tenue breve scilicet declinantur, ut *frater frateros*, id est frater fratris, *piηr pieros* acuta antepenultima per obliquos. Quecumque vero brevem habent penultimam per eta id est .e. longum declinantur, ut *ibηr iberos* [67] similiter acuta antepenultima sed producta scilicet propter .η. perpetuo longum, apud nos vero circumflexa cum sit longa ante breve secundum superiorem regulam polisemarum dictionum. Lucanus: [68]

Interea Caesar victis remeabat Iberis.

Eadem est .η. penultima sive Iber Iberis declines sive Iberus Iberi. Et Virgilius: [69]

Haut impacatos a tergo horrebis Iberos.

Panther [70]   *ke τu theros*,[71]
id est fera, quam [72]   per mutationem in .e.
nostrum longum acuitur apud eos in penultima nominativi in obliquis vero in antepenultima nisi in genitivo plurali ubi penultima acuitur *ton pantheron*, apud nos vero circumflectitur penultima per omnes obliquos cum ipsa longa sit et ultima brevis ratione polisemarum dictionum per omnes casus singulares et plurales, ut genitivus *panteris* dativus *panteri* accusativus *pantera* nominativus pluralis *panteres* genitivus *pa[n]terum* accusativus *panteras* vocativus *panteres*, quorum omnium brevis est, dativus et ablativus pluralis acuitur in antepenultima cum ipsa sit longa due vero sequentes breves. Sciendum preterea quod pater et mater et frater ex eta greco sicut in latinum .e. versa rectius producuntur quam brevientur, quamvis ea frequens consuetudo breviet, ut Inde toro pater Eneas,[73] et Frater ad alloquium,[74] et Mater et Enee.[75]

Quare [76] hoc quod tibi, dilecte frater, de multis nominibus devote sit oblatum munusculum, quamvis tu de uno solo quesiveris. Ego munus meum non

---

[63] MS. *si.*

[64] Theodosii Alexandrini *Canones*, ed. Hilgard (Leipzig, 1889), p. 23, c. 32: Ὁ λουτήρ τοῦ λουτῆρος, ὁ αἰθήρ τοῦ αἰθέρος: τῶν εἰς ῆρ ὀξυτόνων ὅσα μὲν ἔχει τὸ τ̄ διὰ τοῦ η̄ κλίνεται, καμπτῆρος ἐλατῆρος, σεσημειωμένου τοῦ πατέρος ἀστέρος· ὅσα δὲ μὴ ἔχει τὸ τ̄ διὰ τοῦ ε̄ κλίνεται, αἰθέρος δαέρος, σεσημειωμένου τοῦ σπινθῆρος Ἐλευθῆρος.

[65] Idem, c. 33: Ὁ Πίηρ τοῦ Πίερος, ὁ Ἴβηρ τοῦ Ἴβηρος: τῶν εἰς ῆρ βαρυτόνων ὅσα μὲν μακρᾷ παραλήγεται διὰ τοῦ ε̄ κλίνεται, φράτερος Πίερος, ὅσα δὲ βραχείᾳ παραλήγεται διὰ τοῦ η κλίνεται, Ἴβηρος ἐρίηρος· τὸ δὲ πάνθηρ τοῦ ἀπλοῦ τὴν κλίσιν ἐδέξατο.

[66] MS. *inηr ineros.*   [69] *Georgica*, 3, 408.   [72] Half a line gone.

[67] MS. *uiηr uieros.*   [70] One line gone.   [73] Virgil, *Aeneid*, 2, 2.

[68] *Pharsalia*, 5, 237.   [71] MS. *thuros.*

[74] I have not identified this quotation.

[75] Ovid, *Ars amatoria*, 1, 60: 'mater in Aeneae.'   [76] MS. *Q;*

soleo verbis ornare velut quidam cum de *in* prepositione regulam quesitus dixit, "Egregiam vobis scribo regulam," volens ut credo munus suum maius his qui quesiverant facere quam ipsi forsitan essent facturi. Fecisti mihi nuper et alteram questionem prolixam satis et acute compositam de duobus nostre salutis muneribus, sed [cum] per multos magnos sepe et claros [77] viros sit diserte soluta tuque circa finem sis tibi visus sufficientissime respondisse, satius mihi videtur penitus inde tacere quam que per eos habunde dicta sunt vel nulla potius redarare. Sit ergo opusculum sicut petisti si nichil melius per me forte possit [78] tuo nomini dedicatum. Explicit.

The literary reputation of Moses and the nature of his writings indicate that the works which have thus far come to light are only fragmentary remains of a many-sided activity. A Latin poet, a translator from the Greek, a grammarian, and a collector of Greek manuscripts, he might almost hold his own three hundred years later. We cannot call him a humanist, for his culture reflects rather the theological preoccupations of his age, but he was at least a Hellenist and is entitled to an honorable place in conjunction with the renaissance of the twelfth century.

Of the two other Latins mentioned by Anselm of Havelberg, James of Venice is known only as the translator of Aristotle's *New Logic*, and we shall have occasion to examine his work in that connection.[79] Burgundio the Pisan is more celebrated, by reason of his public career as well as of his indefatigable zeal as a translator.[80] Appearing first at the debate of 1136 in Constantinople, he is found in legal documents at Pisa from 1147 to 1180, first as an advocate and later as a judge; he is sent on diplomatic missions to Ragusa in 1169 and to Constantinople in 1172,[81] and is present at the Lateran Council of 1179; and he died at a ripe old age in 1193. The sonorous inscription on his tomb is still preserved, celebrating this *doctor doctorum, gemma magistrorum,* eminent alike in law, in medicine, and in Greek and Latin letters;

---

[77] MS. *cloros.*   [78] MS. *pᵗ.*   [79] Infra, Chapter XI.

[80] See particularly G. M. Mazzuchelli, *Gli scrittori d'Italia* (Brescia, 1753), ii, 3, pp. 1768–1770; [Fabroni], *Memorie istoriche di più uomini illustri pisani* (Pisa, 1790), i. 71–104; Savigny, *Geschichte des römischen Rechts im Mittelalter* (1850), iv. 394–410; F. Buonamici, "Burgundio Pisano," in *Annali delle università toscane*, xxviii (1908); P. H. Dausend, "Zur Uebersetzungsweise Burgundios von Pisa," in *Wiener Studien*, xxxv. 353–369 (1913).

[81] Besides the documents cited by Savigny, see G. Müller, *Documenti sulle relazioni delle città toscane coll' Oriente* (Florence, 1879), pp. 18, 416 ff.

and this reputation is confirmed by the surviving manuscripts of his work.[82] Translation was evidently not the principal occupation of this distinguished career, indeed Burgundio tells us that one of his versions required the spare time of two years, but his long life made possible a very considerable literary output. Theology held the first place: John of Damascus, *De orthodoxa fide* (1148–50), which had been "preached for four centuries as the theological code of the Greek church";[83] the *Homilies* of John Chrysostom on Matthew (1151)[84] and John (1173)[85] and perhaps on Genesis (incomplete in 1179);[86] St. Basil on Isaiah (before 1154);[87] Nemesius, *De natura hominis*, dedicated to Frederick Barbarossa on his Italian expedition of 1155;[88] perhaps others.[89] Two of these versions were dedicated to Pope Eugene III, who secured a manuscript of Chrysostom from the patriarch of Antioch and persuaded Burgundio to undertake the task of turning it into Latin.[90] His results were used by the great theologians of the Western Church, such as Peter Lombard and Thomas Aquinas;[91] indeed he "made accessible to the West works which exercised

[82] Cf. his survey of previous translations, ancient and mediaeval, from the Greek, supra, Chapter VIII, note 36. For the epitaph see Buonamici.

[83] J. Ghellinck, "Les Oeuvres de Jean de Damas en Occident au XIIe Siècle," in *Revue des questions historiques*, lxxxviii. 149–160, reprinted in his *Le mouvement théologique du XIIe Siècle* (Paris, 1914), pp. 245–275, where further studies of Burgundio are promised. Cf. M. Grabmann, *Geschichte der scholastischen Methode*, ii. 93; Duhem, iii. 37; Minges, in *Theologische Quartalschrift*, 1914, pp. 234 ff.

[84] Preface in Martène and Durand, *Veterum scriptorum amplissima collectio* (Paris, 1724), i. 817. On the date, cf. Dausend, in *Wiener Studien*, xxxv. 357.

[85] Preface, incomplete, Martène and Durand, p. 828; see Chapter VIII, n. 36, Chapter IX, n. 130.

[86] Robert of Torigni, ed. Delisle, ii. 109. Cf. C. Baur, *S. Jean Chrysostome*, p. 62.

[87] Savigny, iv. 401; supra, Chapter VIII, n. 36, where a version of the Psalter is also mentioned.

[88] Preface in Martène and Durand, i. 827; preface and text, ed. C. Burkhard, Vienna programmes, 1891–1902; on the MSS. see Diels, Berlin *Abhandlungen*, 1906, pp. 67 f.

[89] Commentary of St. Paul, inferred from the sepulchral inscription; Athanasius, *De Fide*, conjectured by Bandini, *Catalogus*, iv. 455; St. Basil on Genesis (*ibid.*, iv. 437; *Codices Urbinates Latini*, i. 78); Chrysostom on Acts, R. Sabbadini, *Le scoperte dei codici: nuove ricerche* (Florence, 1914), p. 264; work on meteorology announced in preface to Nemesius.

[90] Martène and Durand, i. 817.

[91] Ghellinck, *loc. cit.*; G. Mercati, *Note di letteratura biblica* (Rome, 1901), pp. 141–144.

great influence on the scholastics, the exegetes, the mystics, and the orators of the Middle Ages." [92]   In medicine, Burgundio's name is attached to the Latin versions of ten works of Galen: [93] *De sectis medicorum*, dedicated in 1185 to 'King Henry,' doubtless the newly knighted son of the emperor, the future Henry VI, [94] *De temperamentis*, [95] *De virtutibus naturalibus*, [96] *De sanitate tuenda*, [97] *De differentiis febrium*, [98] *De locis affectis*, [99] *De compendiositate pulsus*, [100] *De differentiis pulsuum*, [101] *De crisibus*, [102] and *Therapeutica* (*Methodi medendi*); [103] while his translation of the *Aphorisms* of Hippocrates is cited in the thirteenth century as preferable to

[92] Mercati, p. 142. His Chrysostom is cited as late as Poggio; *Sitzungsberichte* of the Vienna Academy, lxi. 409.

[93] The elaborate catalogue of Greek MSS. and translations of Galen published by H. Diels, "Die Handschriften der antiken Aertzte," in *Abhandlungen* of the Berlin Academy (1905), pt. i, pp. 58–150, does not ordinarily indicate the authorship of the Latin versions, which in many cases still remains to be investigated. Evidently some of Burgundio's work was revised in the fourteenth century by Nicholas of Reggio and Peter of Abano. For Nicholas see F. Lo Parco, "Niccolò da Reggio," in *Atti della R. Accademia di Archeologia di Napoli*, n. s., ii, pt. 2, pp. 241–317; for Peter, Thorndike, ii, ch. 70. There may be some confusion with Johannes de Burgundia, better known as Sir John Mandeville, to whom is ascribed a treatise *De morbo epidemie* (e. g., Trinity College, Cambridge, MS. 1102, f. 53, MS. 1144, f. 110 v; Caius College, MS. 336, f. 114 v); see Mrs. Singer in *Proceedings of the Royal Society of Medicine*, ix. 162–173 (1916); and Mrs. Singer and Levy, in *Annals of Medical History*, i. 395–411 (1918).

[94] 'Translatio greca est Burgundionis.' Bibliothèque Nationale, MS. Lat. 6865, f. 81; Diels, p. 60. 'De greco in latinum domino Henrico regi a Burgundione iudice Pisano anno incarnationis M.C.LXXXV. fideliter translatus': MS. Montpellier 18, f. 95, where the *Archiv für die Geschichte der Medizin* (ii. 16) has incorrectly 1184.

[95] 'Explicit liber Galieni de complexionibus translatus a Burgundione cive Pisano secundum novam translationem.' Vatican, MS. Barberini Lat. 179, f. 14 v; MS. unknown to Diels, p. 64.

[96] Prague, Public Library, MS. 1404; not in Diels, p. 66.

[97] Diels, p. 75; Lo Parco, "Niccolò da Reggio," pp. 282 ff.   [98] Diels, p. 80.

[99] 'Explicit liber Galieni de interioribus secundum novam translationem Burgundii.' Vatican, MS. Barb. Lat. 179, f. 36 v; MS. not in Diels, p. 85.

[100] 'Finis libri qui est de compendio pulsus a Burgundione iudice cive Pisano de greco in latinum translati.' Bibliothèque Nationale, MS. Lat. 15460, f. 111 v; MS. not in Diels, p. 86.

[101] Diels, p. 87.   [102] Munich, Cod. Lat. 35; Diels, p. 90.

[103] 'Expletus est liber tarapeutice cum additionibus magistri Petri de Ebano que deficiunt ex translatione Burgundionis civis Pisani.' Vatican, MS. Barb. Lat. 178, f. 44 v; not in Diels, p. 92. Cf. G. Valentinelli, *Bibliotheca manuscripta ad S. Marci Venetiarum*, v. 79, and MS. Madrid 1978 (L. 60), f. 45 v.

that from the Arabic.[104] In a quite different field, he turned into Latin a treatise on the culture of the vine,[105] doubtless for the practical benefit of his native Tuscany, just as a Strasburg scholar of the sixteenth century sought to help the vineyards of the Rhine by translating extracts from the same *Geoponica*.[106] Still another scientific work is promised in the preface to Nemesius, an account namely of the heavens, winds, storms, earthquakes, and waters, and why the sea is salt — the content of Aristotle's *Meteorology* and more, though hardly this work itself, a promise which he may not have carried out. As a lawyer, too, he had opportunity to apply his knowledge of Greek to translating the Greek quotations in the *Digest*,[107] for which he appears to have used the text of the famous Pisan manuscript. He is freely credited with the Latin version by the glossators of the thirteenth century, and, as in the case of his theological and medical translations, the results of his work passed into the general tradition of the later Middle Ages.

With Burgundio we have passed far into the second half of the twelfth century and well beyond the times of Anselm of Havelberg. In approaching the Constantinople of this period we may well begin with another emissary of Frederick Barbarossa, apparently also a German, who visited the Greek capital in 1179 and shortly before. Let us start with his preface, as preserved in a contemporary codex of the University of Cambridge: [108]

[104] Puccinotti, *Storia della medicina* (Leghorn, 1850), ii, 2, p. 290; Neuburger, *Geschichte der Medizin* (Stuttgart, 1906), ii, 1, p. 375. As cited by Diels, pp. 14–16, the Latin MSS. do not mention Burgundio.

[105] Edited by Buonamici, in *Annali delle università toscane*, xxviii (1908). Incomplete MS. also in the Ambrosian, MS. C. 10. sup., f. 118 v; also formerly at Erfurt (W. Schum, *Beschreibendes Verzeichniss der Amplonianischen Handschriften-Sammlung*, p. 802) and at Peterhouse, Cambridge (James, *Catalogue*, p. 11).

[106] *Serapeum*, xvii. 287 ff.

[107] Savigny, iv. 403–410; Mommsen, *Digesta*, editio maior (1876), i. 35*; H. Fitting, "Bernardus Cremonensis und die lateinische Uebersetzung des Griechischen in den Digesten," in Berlin *Sitzungsberichte*, 1894, ii. 813–820; N. Tamassia, "Per la storia dell' Autentico," in *Atti del R. Istituto Veneto*, lvi. 607–610 (1898). I agree with Savigny that there is no evidence that Burgundio translated the *Novels*, and that the reference to them in the preface to his translation of Chrysostom's St. John (see Chapter VIII, n. 36) shows that Burgundio accepted the extant version as a literal translation made at Justinian's order.

[108] MS. Ii. iv. 27, ff. 129–130 v.

Incipit liber de diversitate nature et persone proprietatumque personalium non tam Latinorum quam ex Grecorum auctoritatibus extractus.

Circumspicienti mihi quanta sit in humanis studiis varietas, in varietate dissensio, in dissensione contradictio, in contradictione obstinatio, inutile duxi causis horum investigandis operam dare, cum manifestum sit ex variis animorum affectionibus studiorum evenire varietatem, ex errore ignorantię dissensionem, ex tumore iactantię contradictionem, ex conatu inprudentię obstinationem. Quorum et primum et secundum est humanum, tercium ceca temeritate, quarta pertinaci contumacia plenum. Ideoque duobus in prioribus facilis est recursio, in tercio difficilis revocatio, in quarto irrevocabilis exorbitatio. In illis lapsus ex simplicitate miserabilis venia meretur, in istis ex perversitate dampnabilis in perniciem precipitatur. Considerans igitur a nostris studiis multos dissentire scolis plerosque contradicere et inpetulantię sue obsequium aliquos arroganter illudere, obstupui vehementer admirans unde vel illi vel nos in tantam impericię coruissemus insaniam quod, ut taceam de philosophicis opinionibus, circa theologię secreta tam inextricabilem non modo pateremur sed et excitaremus discordiam. Quam ob rem beatissime divinitati, in qua omnes thesauri sapientię consistunt et in mortalia pectora pro sua bonitate dividuntur, supplicari cepi ut viam veritatis mihi panderet et, si labi ex simplicitate contingeret, ex perversitate maligni sensus precipitari in perniciem non permitteret. Et quoniam ex Grecorum fontibus omnes Latinorum discipline profluxerunt, precibus meis adieci ut eius opitulante gratia, si quo modo fieri posset, per auctoritates irrefragabiles sapientiam Grecię nostrarum dissensionum decisionem aliquando consequi mererer.

Hoc ineffabiliter estuans desiderio forte legatione Frederici gloriosissimi Romanorum imperatoris functus ad Manuelem Constantinopolitanum basileon regum orientalium potentissimum, hilariter in Illiricum et avide viam nullis laboribus et periculis meis inviam arripui. Consistens ergo in urbe regia priori legatione mense uno et diebus .vii. tempore scismatis, posteriori vero mensibus duobus tempore pacis anno quo Lateranense concilium in vere celebratum fuit, priscorum sanctorum Grecię doctorum interpretante [109] Ugone Etheriano litteris grecis et latinis peritissimo, diu desideratam propositi mei letus accepi consummationem. Libellum secundum questiones in priori legatione a me propositas de diversitate nature et persone in posteriori dedit magni Basilii et Gregorii Nazanzeni aliorumque sanctorum auctoritatibus fulcitum, quem non modo ad meas preces sed et viri eloquentissimi Petri scolastici in florentissimo Austrię oppido de voluminibus Grecorum cum multa diligentia et cautela collegit. Preterea librum de immortali Deo addidit quem contra modernorum Grecorum opinionem de Spiritus Sancti processione de Patre et Filio compositum et antiquorum Grecię doctorum scriptis communitum Alexandro pape transmisit, in quo personalium proprietatum et personarum essentięque diversitatem aptissimis beatorum episcoporum olim in Grecia theologizantium documentis declaravit. Qui cum et ipso confitente audivi Alberici cuiusdam in dialecticis fuisset auditor in Francia aliorumque a studiis nostris in theologia dissidentium viam pub-

---

[109] *interprete?*

licam trivisset, prefatorum virorum et aliorum clarissimorum Grecię doctorum sanctitate coactus est in latinum transferre sermonem, unde suam propriam quam de Gallia et Italia in Achaiam detulerat convinceret opinionem.

Accepta hec ab illo munera super aurum et topazion preciosa velut opes Cresi amplexatus sum. Cumque reversus in Germaniam ad Fredericum victoriosissimum Romani imperii principem Petro venerabili Tusculano episcopo tunc ibidem legatione sedis apostolicę fungenti apportatum libellorum meorum thesaurum demonstrassem ipseque sanctissimas illorum sententias diligenter ruminasset, admiratus plane fuit tantam in Gisilberto Pictaviensi episcopo sapientiam quod cum Grecorum volumina tanquam lingue eorum ignarus nunquam legisset, in illorum tamen intellectu tam scriptis quam dictis totus fuisset, statimque illos transcribi iussit. Latebat tamen eum quod beati Theoderiti et Sophronii scripta in latinum translata sepe revolvisset cum aliorum libris sive Grecorum sive Latinorum et maxime Athanasii et Hylarii, quorum suffragiis in concilio Remensi coram papa Eugenio contra suorum emulorum oblocutiones usus fuit cum gloria. Gratias ergo quantas potero pietati divine agere non cessabo quę longis suspiriis et sollicitudini meę finem hunc facere dignata est, ut iam cum Cirillo Alexandrino sentire debeam et Iohanne Damasceno non idem esse personam et naturam, cumque magno Basilio et Gregorio theologo non idem esse personales proprietates personas et essentiam. Quod quidem supranominatus Pictaviensis episcopus ab antecessore suo Hylario non discordans in expositione Boetii de Trinitate evidenter asseruit, quibus tamen auctoribus uteretur non declaravit, exercitatis divinarum scripturarum lectoribus laudem horum inveniendorum relinquens. Quos ad investigandorum illorum studium et amorem invitat dum in operis sui prologo testatur diligentibus ipsarum rimatoribus posse videri ea que dixit sua furta potius esse quam inventa.

Denique quia Latinos latet quanta evidentia de his rebus Grecorum loquatur sapientia, opere precium duxi in publicas aures proferre quod ab orthodoxis doctoribus eorum divina opitulatione percepi, quatinus per illos pateat et a veritatis tramite tam non exorbitasse et emulos suos in ignorantię nebulis aberrasse frustraque in eius declinatione laborasse quem summis et inconcussis Grecię columpnis constat suffultum fuisse. Sed sicut sanctus Hylarius precatur, postulare presumo ut quisquis hec legenda et cognoscenda susceperit modum sibi atque mihi patientie fidelis indulgeat et usque ad absolutionem universa percenseat. Iniquum enim est non comperta usque ad finem ratione dictorum preiudicatam sentenciam ex unciis eorum quorum adhuc causa ignoretur afferre, cum non de inchoatis ad cognoscendum sed de absolutis ad cognitionem sit iudicandum. Est etenim michi non de piis lectoribus metus ac benignis auditoribus sed de quibusdam nimium apud se cautis et prudentibus non intelligentibus per beatum apostolum sibi ne superbe saperet preceptum, quos vereor nolle omnia ea quorum absolutio a me in consummatione erit prestanda cognoscere dum verum intelligere ex his que absolventur evitant tanquam inclementes et iniqui alienorum dictorum iudices atque consueti servare sola ea dogmata non que rationabiliter didicerunt sed que ex consuetudine tenuerunt. Quorum plurima turba est non considerantium quid vere vel convenienter sed quid ad aurium suorum pruritum sibiletur. Quam ob rem antequam attingam propositum, quoniam expedit quid ad officium spectet de rebus divinis disserentium diffinire atque distinguere non sit lectori tediosum.

The purpose of this treatise is thus fairly clear, not the usual controversy with the Greeks, but to find in orthodox Greek theologians support for the doctrines of the author's own school in relation to the Trinity. He begins with a discussion of the type of men who should write on theology, and the manuscript breaks off in the midst of a discussion of substance and essence.[110] Then comes a treatise *De ignorantia* of a different sort.[111] The *De diversitate nature* is, however, preceded in this codex by a *Liber de homoysyon et homoeysion* [112] which is in the same style and may well be by the same author.

The date of the *Liber de diversitate* can be fixed in the first instance by its references to the schism which ended in 1177 and to the Lateran council of March, 1179. No mission from Barbarossa to Constantinople in the latter year is mentioned by the modern students of their relations, yet George of Corfu at this time represented Manuel in Italy,[113] so that diplomatic negotiations were still going on. The meeting with Peter of Pavia, cardinal bishop of Tusculum from 1179 to 1182, took place in 1180, when this cardinal is known to have been with the emperor 18 March at Constance,[114] having apparently passed through Carinthia on his way.[115] The identity of the author does not appear, nor does that of the Austrian *scolasticus* Peter who accompanied him. Anselm

---

[110] F. 130 v: '*Ad officium theologi spectat rerum veritatem et verba congrua observare*. Archana theologię investigare volenti . . . maneat quicquid eternaliter existit' (f. 176 v).

[111] Ff. 177–187: 'Quid ignorantia sit multi ignorant . . . delinquere venaliter dicetur.'

[112] Ff. 1–128 v: 'Sanctus Hylarius Pictavorum episcopus in libro de synodis . . .'

[113] Baronius, *Annales*, ad an. 1178, nos. xiii–xvi; 1179, nos. x–xii; A. Mustoxidi, *Illustrazioni Corciresi* (Milan, 1811–14), ii. 181–184, and app.; cf. W. Norden, *Papstthum und Byzanz* (Berlin, 1903), pp. 112 f. The two bishops George of Corfu have not been wholly disentangled: Krumbacher, p. 770.

[114] Stumpf, *Reichskänzler*, nos. 4314–16; Giesebrecht, *Deutsche Kaiserzeit* (1895), vi. 576.

[115] *Archiv für Kunde oesterreichischer Geschichtsquellen*, xi. 320. Peter stayed in Germany until 1181: *Chronica regia Coloniensis*, ed. Waitz, p. 323; Delehaye, in *Revue des questions historiques*, xlix. 49–56 (1891).

of Havelberg, conjectured by the Cambridge *Catalogue*,[116] is, of course, chronologically impossible, as he died in 1158.

Respecting western matters, the preface shows the author as an opponent of Albericus, perhaps the Albericus of Rheims who died in 1141.[117] He appears also as a staunch supporter of Gilbert de la Porrée, recalling the favorable judgments of John of Salisbury and Otto of Freising. Our author is not the only Gilbertine who dabbled in Greek theology, for Paul Fournier has made known the anonymous author of a *Liber de vera philosophia*, written ca. 1180–90, apparently in southern France, who had visited Jerusalem and cites freely the Greek Fathers; he also cites the treatise of a Master A., canon of Valence, who had explored the libraries of Greece, as well as the West, for material in support of his thesis.[118] Though ignorant of Greek, Gilbert had used Greek authorities in presenting his argument at the council of Rheims (1148). Further interest in the results of Greek studies is seen in the dedication to a Gilbert, apparently Gilbert de la Porrée, of the *Differentie* of a certain Guillelmus Corborensis, a series of etymologies *de pelago greci ydiomatis* which in alphabetical order explains to the Latin world the difference between similar roots like *alchos* and *archos*.[119]

As regards the East, our preface introduces us to Hugo Eterianus, the principal Latin at this time engaged in theological controversy with the Greeks.[120] A Pisan by birth, Hugo, as we here

---

[116] iii. 464.   [117] Grabmann, *Geschichte der scholastischen Methode*, ii. 138–140.

[118] *Études sur Joachim de Flore* (Paris, 1909), pp. 51–78; cf. Grabmann, ii. 434–437. The Gilbertine *Sententiae* edited by Geyer (*Beiträge*, vii, no. 2–3) lack this Greek element. On Gilbert's use of Greek, see, however, Hofmeister in *Neues Archiv*, xxxvii. 693 (1912).

[119] 'Quamquam non dubitem te, incordialis [sic] et intime Gilleberte, per incitamentum subtilis ingenii et de blandimento capacis memorie dictionum latinarum differentias vigilantissime cognovisse. . . . Alchos et archos differunt . . .' Wolfenbüttel, MS. Gud. lat. 326, f. 1; B. N., MS. lat. 7100, f. 32 v. I hope to notice more specially this and one or two other mediaeval glossaries overlooked by Loewe and Götz.

[120] Gradenigo, *Lettera intorno agli Italiani che seppero di greco*, ed. Calogierà, pp. 50–55; [Fabroni], *Memorie di più uomini illustri pisani* (Pisa, 1790), ii. 59–68, iv. 151–153; Fabricius-Harles, *Bibliotheca Graeca*, viii. 563, xi. 483; Fabricius, *Bibliotheca mediae Latinitatis*, iii. 292 (ed. 1754); G. Müller, *Documenti sulle relazioni delle città toscane coll' Oriente*, pp. 384 f.; Hergenröther, *Photius*, iii. 175–177, 814 f., 833–837.

learn, had studied dialectic in France with Albericus and others before going to Constantinople, where his theological activity has long been known. His *De sancto et immortali Deo*, here mentioned, was finished in 1177, when Alexander III acknowledged its receipt.[121] He had also written, before 1173, a *Liber de anima corpore iam exuta*, or *De regressu animarum ab inferis*, at the request of the Pisan clergy.[122] Other evidence of his activity is found in a lost treatise *De Filii hominis minoritate ad Patrem Deum* mentioned by his brother Leo; [123] in a set of extracts from his works containing accusations of all kinds against the Greeks; [124] and in an unpublished dispute with Nicholas of Methone preserved in Greek at Brescia.[125] He was obviously fitted to collect and interpret material for our author's purpose; indeed his mastery of Greek theology has been recognized.[126] From his first dated appearance in 1166 [127] down to his death in 1182, Hugo kept up his controversies, and his vigorous advocacy of Latin doctrine against the Greeks won him commendation from Alexander III [128] and,

---

[121] Jaffé-Löwenfeld, *Regesta*, no. 12957; Baronius, *Annales*, xix. 512. The treatise, also known as *De heresibus Grecorum* and *De processione Spiritus Sancti*. will be found in Migne, ccii. 227–396. MSS. are common, e. g., Vatican, Codd, Vat. lat. 820, 821, Urb. lat. 106; Laurentian, MS. xxiii. dext. 3 (Bandini, iv. 631); Assisi, MS. 90, f. 53 (Mazzatinti, *Inventari*, iv. 38); Subiaco, MS. 265 (Mazzatinti, i. 210); B. N., MS. Lat. 2948; Troyes, MS. 844; Cambridge, Corpus Christi College, MS. 207. The *De heresibus* was also issued in Greek; for a reply cf. *B. Z.*, iv. 370.

[122] Migne, ccii. 167–226. There is a copy of ca. 1200 in the Archives of the Crown of Aragon at Barcelona, MS. Ripoll 204, ff. 106–192. The date is fixed by the mention of Albert as consul.

[123] See his preface printed below, p. 217.

[124] *Maxima bibliotheca patrum* (Lyons, 1677), xxvii. 608 ff. Cf. Hergenröther, iii. 175 ff., 833 ff.

[125] Martini, *Catalogo dei MSS. greci*, i. 251; cf. *B. Z.*, vi. 412.

[126] Hergenröther, iii. 814 f.

[127] See his letter to the consuls of Pisa in Müller, *Documenti*, no. 10, dated 1166 by the editor, although the text of the epitaph there cited clearly gives 1176. That Hugo was at Constantinople by 1166 is otherwise known: see below, p. 216, the preface of Leo here printed, and Hugo's reference to his relations with the cardinals who came from Rome in that year (Migne, *Patrologia Latina*, ccii. 233). In the letter to the Pisans Hugo says that his theological opinions had already made him unpopular, and the disputes with Nicholas of Methone doubtless fall before this year.

[128] Jaffé-Löwenfeld, no. 12957.

just before his death, a cardinal's hat from Lucius III.[129] Though he does not appear with any official title, he was in relations with the Greek emperor and on one occasion accompanied him into Cappadocia and the Turkish territory.[130]

Closely associated with Hugo, though in a different field of translation, was his brother Leo, generally known as Leo Tuscus, who was assisted further by their nephew Fabricius. Leo, already *invicti principis egregius interpres* in 1166,[131] is in 1182 still *imperialium epistolarum interpres*,[132] and can in the meantime be traced in Manuel's service during the Asiatic campaigns, as we learn in general terms from Hugo's *De heresibus* [133] and more definitely from the preface printed below. Besides assisting Hugo in his literary labors,[134] Leo executed two translations from the Greek. One, a version of the mass of St. Chrysostom,[135] was made at the request of a recent visitor to Constantinople, the noble Rainaldus de Monte Catano, to whom it is dedicated, subject to the criticism of

frater et preceptor meus Vgo Eterianus sua gravitate gravior, nam is Grecorum loquela perplexa internodia olorum evincentia melos verborumque murmura, que pene Maronis pectus fatigarent ac Ciceronis, intrepida excussione [136] inspectis narrationum radicibus mirifice discriminat.

[129] *Ibid.*, no. 14712.
[130] 'Quod propriis oculis imperatorem sequendo per Cappadociam Persarumque regiones intuitus sum': *Bibliotheca patrum*, xxvii. 609.
[131] Müller, no. 10. On the date see n. 127. Cf. Migne, ccii. 167 'imperialis aule interpretis egregii.'
[132] Müller, no. 21.
[133] Migne, ccii. 274.
[134] 'Qui est ingenii mei acumen huiusque suscepti laboris incentivum,' says Hugh: Migne, ccii. 274.
[135] It is printed, with the preface, in Claudius de Sainctes, *Liturgiae sive Missae sanctorum patrum* (Antwerp, 1562), f. 49; cf. Swainson, *The Greek Liturgies*, pp. 100, 144. There is a copy in the Bibliothèque Nationale, MS. Lat. 1002, f. 1: 'Magistri Leonis Tusci prologus ad factam Grecorum missam ab eo verbis Latinis divulgatam ad quendam Raynaldum. Cum venisses Constantinopolim . . .' Engdahl, *Beiträge zur Kenntnis der byzantinischen Liturgie*, in Bonwetsch and Seeberg's *Neue Studien*, v. 35, 84 (1908), has used only an incomplete Karlsruhe MS. of the translation which does not contain the preface. Leo's translation is mentioned by Nicholas of Otranto in the preface to his translation of the mass of St. Basil: Engdahl, p. 43; MS. Lat. 1002, f. 22 v.
[136] So Allatius, who cites this passage, *De ecclesiae consensione*, p. 654. MS. Lat. 1002 has *exursione*, the printed text *excursione*.

The other of Leo's translations is a version of the *Oneirocriticon* of Ahmed ben Sirin, important both for the vernacular renderings which were based upon it in the sixteenth century and for the establishment of the Greek text, of which it represents a tradition older than the extant manuscripts.[137] The preface, which is addressed to Hugo, and exhibits, like the preface to the version of the mass, marked resemblance of style to his writings, sheds further light on Hugo's activity, since it shows him engaged in the controversy over the subordination of the Son to the Father which was started by Demetrius of Lampe, and, if we are to believe Leo, exerting an influence upon the emperor's decision. The mention of Manuel's campaign against the Turks in Bithynia and Lycaonia offers a means of dating the work.[138] The campaign of 1146 being obviously too early, opinion seems to have decided for that of 1160–61; at least all scholars who mention the version, from Rigault and Casiri to Steinschneider, Krumbacher, and Drexl, though without discussing the question, give 1160 as the date. This seems to me untenable, partly because the expedition of this year can scarcely be said to have reached Lycaonia, but chiefly because the Demetrian controversy began only in 1160, and the imperial decree which put an end to it (*augustalis clementie decretum*) is of the year 1166.[139] All of this is already well in the past (*ex eo igitur tempore*), and the emperor engaged in no further Turkish campaigns except the unsuccessful enterprise of 1176. Now we know from Hugo's *De heresibus*, completed in 1177, [140] that its composition was interrupted by Leo's absence in Asia Minor with the emperor,[141] and it is accordingly to 1176 that the

---

[137] See Steinschneider, "Ibn Shahin und Ibn Sirin," in *Zeitschrift der deutschen morgenländischen Gesellschaft*, xvii. 227–244; and *E. U.*, nos. 77, 130; Krumbacher, p. 630; Drexl, *Achmets Traumbuch* (*Einleitung und Probe eines kritischen Textes*), Munich dissertation, 1909, who gives an account of the manuscripts preliminary to the preparation of a critical edition. None of these writers appears to have examined the preface. See now Thorndike, ch. 50.

[138] On these campaigns see Chalandon, *Les Comnènes*, ii. 247–257, 456–459, 503–513.

[139] Chalandon, ii. 644–651.

[140] As seen from the date of Alexander III's letter acknowledging it: Migne, ccii. 227; Jaffé-Löwenfeld, no. 12957.

[141] Migne, ccii. 274.

translation of Ahmed should be assigned. The following text of the preface is from MS. 2917 of Wolfenbüttel: [142]

Ad Hugonem Eterianum doctorem suum et utraque origine fratrem Leo
Tuscus imperatoriarum epistolarum interpres de sompniis et oraculis

Quamquam, optime preceptor, invictum imperatorem Manuel per fines sequar Bithinie Licaonieque fugantem Persas flexipedum hederarum [143] complectentes vestigia, tamen memorandi non sum oblitus sompnii a te visi quod dictum inexpugnabilem virum eneo in equo supra columpnam [144] quam Traces dicunt Augustiana Bizancii sito nobiliter sedere conspicabaris, eodem autem in loco doctissimis quibusdam astantibus Latinis Romana oratione cum in quodam legeret libello interpellanti tibi soli favorem prestitisse visus est. Latuit tunc utrumque nostrum ea quidem quid portenderet visio, at vero eiusmodi oraculum editus per te de Filii hominis minoritate ad Patrem Deum libellus tempore post revelavit sub tegumentis. Profecto eneus ille sonipes anima carens altissime sonantissimeque questionis erat que inter Grecos versabatur ventilatio, verbum scilicet Dei secundum quod incarnatum Patri equale prestans rationis veritatisque radicitus expers ut quadrupes nominatus. Solvit autem illam controversiam clamitante illo libello augustalis clemencie decretum pauco scandali fomento contra voluntatem illius relicto. Ex eo igitur tempore pectus sollicitudine percussi, sub corde ignitos versavi carbones, cogitando utile esse si onirocriti Grecorum philosophis ariolanti loqui latine persuaderem enucleatim atque inoffensam perspicuitatem figmenti sompnialis tuo favore nostrorum Tuscorum desiderio breviter reserarem. Quos quidem fluctu percupio aspergi undiosiore ut irrigentur affatim efficianturque fecundiores, nam Seres, ut fertur, arbores suas undis aspergunt quando uberiorem lanuginem quam sericum creat admittere nituntur. Ceterum haut facile est in huiusmodi versari pelago cuius latitudo ad aures usque dehiscit non sponte remigem asciscens invalidum. Non solum enim subtilibus expositum investigationibus et illos repellit qui debilitate pedum serpunt, ut antipodes, et eos qui non movent linguas, ut pleraque aquatilium, set neque monoxilo se navigari lintre patitur. Quamobrem loquelam imperatoriorum interpretationibus apicum obsequentem per excubias interdum huic translationi non irrita [145] spe addixi, totum opus sapiencie tue dicaturus iudicio, mei quidem auctoris, tui vero probatoris equilibre pensans meritum. Nam tuum examen cognoscere [146] non sum ambi-

---

[142] Ff. 1–20 (saec. xiii). Also in the Bodleian, Digby MS. 103, ff. 59–127 v, saec. xiii; modern copy in Ashmolean MS. 179. There is a copy of the fourteenth century in the British Museum, Harleian MS. 4025, ff. 8–78; another in the Biblioteca Casanatense, MS. C. vi, 5 (1178); without the preface the translation is found in Vat. MS. Lat. 4094, ff. 1–32 v. Thorndike (ii. 292) also notes B. N. MS. Lat. 7337, p. 141; and Vienna, MS. 5221.

[143] Ovid, *Metamorphoses*, 10, 99.

[144] The statue of Justinian called Augusteion, in the place of the same name. See Du Cange, *Constantinopolis Christiana*, bk. i, c. 24; Unger, *Quellen der byzantinischen Kunstgeschichte* (Vienna, 1878), pp. 137 ff.

[145] So the Digby MS. Wolfenbüttel: *unita*.   [146] MS. Digby: *discernere*.

guus quicquid arida exsanguisque poscit ratiocinatio. Desiderantissimus enim nepos Fabricius,[147] grecarum sciolus et ipse litterarum sompnialium figmentorum odoratus rosaria, scribendi assiduitate me a confluentibus elevat prestatque non mediocre adiumentum, atque idcirco neque nomen sine subiecto neque sine viribus erit edicio, Sidoneis Tirrenisque sagittis parum penetrabilis apparitura ut arbitror. Ergo quisquis nodosorum sompniorum fatigatur involucris, si per aliquod hic scriptorum absolvi postulet, caveat pretemptare plus nosse quam sat est, ne titulos depravet[148] Apollinee urbis ambagum rimis herbidisque sentibus. Ego autem tui solius utrarumque linguarum peritissimo examini volumen hoc subpono, ut in eo que arescunt ac caligant per te illustrata orbi demum succincta perfectione[149] vulgentur.

Another Italian writer appears at Constantinople in this period in the person of a certain Pascalis Romanus, who also shared the interest in signs and wonders which prevailed at Manuel's court. His *Liber thesauri occulti*, with an introduction citing Aristotle's *De naturis animalium*, Hippocrates, and 'Cato noster,' is a dreambook compiled at Constantinople in 1165, if we may believe the author, from Latin, Greek, and Oriental sources:[150]

Incipit liber thesauri occulti a Pascale Romano editus Constantinopolis anno mundi .vi. dc. lxxiiii. anno Christi .m. c. lxv.

Tesaurus occultus requiescit in corde sapientis et ideo desiderabilis, set in thesauro occulto et in sapiencia abscondita nulla pene utilitas, ergo revelanda sunt abscondita et patefacienda que sunt occulta. Quare de plurimis ignotis et occultis unius tantummodo elegi tegumentum aptamque revelacionem describere, videlicet sompnii secundum genus et species eius quo res profunda et fere inscrutabilis ad summum patenti ordine distinguatur. Eius namque doctrina philosophis et doctis viris valde necessaria est, ne forte cum exquisiti fuerint muti vel fallaces inveniantur . . . (f. 43) Collectus autem est liber iste ex divina et humana scriptura tam ex usu experimenti quam ex ratione rei de Latinis et Grecis et Caldaicis et Persis et Pharaonis et Nabugodonosor annalibus in quibus multipharie sompnia eorum sunt exposita. . . . Non itaque longitudo prohemii nos amplius protrahat nec responsio aliqua impediat, set omni cura seposita succincte ad thesaurum desiderabilem aperiendum properemus.

Sompnium itaque est figura quam ymaginatur dormiens . . .

---

[147] Fabricius was a member of the papal household in 1182, when he was sent to Constantinople by Lucius III: Müller, no. 21. Another learned friend, Caciareda, is mentioned in the *De heresibus* (Migne, ccii. 333 f.).

[148] Wolfenbüttel: *degravet.*

[149] Digby: *profussione.*

[150] Digby MS. 103, ff. 41–58 v, preceding Leo's *Oneirocriticon.* The first of the two books of the treatise is also in the British Museum, Harleian MS. 4025, f. 1. See also B. N., MS. lat. 16610, f. 2 v (Thorndike, ii. 297).

Paschal the Roman can also be almost certainly identified with
the translator from the Greek, in 1169, of the curious book known
as Kiranides. This strange compend of ancient lore respecting
the virtues of animals, stones, and plants is well known in the
Greek, from which it has been edited and translated by Mély and
Ruelle,[151] but the Latin version has not been specially studied. At
least five Latin manuscripts are known,[152] all with the following
preface, showing that the translation was made by request of some
Latin:[153]

In Christi nomine amen incipit liber Kirannis Ypocrationis filie.[154]
Eruditissimo domino magistro[155] Ka. Pa. infimus clericus. Admiror et
commendo sagacitatem tue prudentie[156] que cum docta et experta sit in
hiis que super naturam nostri circuli sunt et que iam quasi ultra .vii. celos
contemplando penetravit, modo etiam infima experimenta terrena conspic-
ere non dedignatur. Rogasti enim me ut hunc librum medicinalem de
greco eloquio in latinum sermonem[157] transferrem. Res quidem facilis fuit
ad dicendum sed difficilis ad perficiendum, verum caritativo amore tuoque
beneficio permotus obedire non renui. Et quoniam diverse sunt translationes de agarenica lingua in grecam,[158] ut nosti, librum grecum quem mihi
dedisti studiose et fideliter per omnia emulatus sum, ipsos etiam duos pro-
logos quamvis asperos velud de antiquissimis titulis abstractos preterire
nolui, non verba, que de sterilitate barbarica sunt, sed sensum utilitatis re-
colligendo. Si quid ergo reperieris alienatum,[159] non infidelitati vel malicie
sed communi errori deputetur.[160] Nullus enim tam sapiens qui absque titulo

[151] F. de Mély, *Les lapidaires de l'antiquité et du moyen âge*, ii, iii (Paris, 1898–
1902). For discussions of these confused texts, see P. Tannery, in *Revue des études
grecques*, xvii. 335–349; Cumont, in *Bulletin de la Société des antiquaires de France*,
1919, pp. 175–181; R. Ganszyniec, in *Byzantinisch-Neugriechische Jahrbücher*, i.
353–367, ii. 56–65, 445–452 (1920–21); Thorndike, ii, ch. 46.

[152] Vatican, MS. Reg. lat. 773, f. 21 (ca. 1300); MS. Vat. lat. 4864, f. 18 (in
a humanist hand of ca. 1400); MS. Pal. lat. 1273, f. 121, in a northern hand of
the fifteenth century ('translatus a magistro Gerardo Cremonensi de arabico in
latinum'); Montpellier, Ecole de Médecine, MS. 277, f. 41 (saec. xv); Bodleian,
MS. Ashmole 1471, f. 143 v (saec. xiv). There are two early editions (Leipzig,
1638; Frankfort, 1681) and a French translation (Arsenal, MS. 2872, ff. 38–57).
There is a fragment at Wolfenbüttel, MS. 1014, f. 102. The fragment 'De virtute
aquile' at Merton College (MS. 324, f. 142), also in Bodleian, E Musaeo MS. 219,
f. 138 v, translated by Willelmus Anglicus, is, as Thorndike (ii. 93, 487) conjectured, from Kiranides (3, 1).

[153] The text is here based on the best two of the foregoing manuscripts, Reg. lat.
773 (A) and Montpellier 277 (B).

[154] Title not in A.   [157] Om. B.   [159] *ab communi sensu*, A.
[155] Om. B.   [158] Om. A.   [160] *deputantur*, A.
[156] *evidentie*, A.

inscientie reperiatur.[161] Volo tamen te scire [162] quod est apud Grecos quidam liber Alexandri magni de .vii. herbis .vii. planetarum, et alter qui dicitur Thessali misterium ad Hermem, id est Mercurium, de [163] .xii. herbis .xii. signis attributis et de .vii. aliis herbis per .vii. alias stellas, qui si forte pervenerint ad manus meas vel tuas, quia celestem dignitatem imitantur, recte [164] huic operi preponentur. Transfertur itaque liber iste Constantinopoli Manuele imperante [165] anno mundi vi̅m̅ dclxxvii, anno Christi m. c. lxix. indictione secunda.[166]

Liber phisicalium virtutum, compassionum, et curationum collectus ex libris duobus, ex primo videlicet Kirannidarum Kiranni regis Persarum et ex libro Apocrationis Alexandri ad propriam filiam. Habebat autem primus liber Kiranni sic sicut et supponemus: Dei donum magnum angelorum accipiens fuit Hermes trimegistus deus hominibus omnibus notus. . . .

Everything turns on the interpretation of 'Ka. Pa.' The author of the Montpellier catalogue [167] read 'Ha. pa.,' which Pansier made into 'Ha[driano] Pa[pe],' though Pope Hadrian had died ten years before. The scribe of the Ashmolean manuscript extended the second word to 'Parrissen.,' which led Thorndike [168] to make 'cancellario Parisiensi' out of the whole. MS. Pal. lat. 1273 has 'Ra. Pa.', which Vat. 4864 makes into 'Raynaldo Parissino.' There can, however, be no doubt that 'Ka. Pa.' stood in the original text, and one would expect, as usual, the first to denote the addressee and the second the writer. Whoever may have been the 'Ka.' for whom the translator labored, no other 'Pa.' is known in Constantinople at this time, whereas Paschal the Roman we have found there four years earlier engaged on a similar task and using an exactly parallel form of date.[169] Moreover

---

[161] *Nullus enim tam sapiens reperitur qui absque titullo inscientie sit*, B.

[162] *volo te transsire*, A.

[163] From this point A is injured for the first half of eight lines.

[164] *certe*, B.   [165] *imperatore*, B.

[166] The year A.D. is faint in A. MS. Pal. lat. 1273 has the same date as A and B. Vat. lat. 4864 has 'anno Domini Ihesu Cristi milesimo c.lx. indictione ii^a, anno vero mundi dclxxvii.' Ashmole 1471 has: 'anno mundi   anno Christi m° cc°.lxxx°. alias m°. c°. lxix°. indictione secunda.' B adds, 'Explicit epistola, incipit prologus.'

[167] *Catalogue des MSS. des départements*, old series, i. 395; Pansier, in *Archiv für die Geschichte der Medizin*, ii. 25.

[168] ii. 230. E. Meyer, *Geschichte der Botanik* (Königsberg, 1855), ii. 349 ff., followed by Cumont in *Revue de philologie*, 1918, p. 88, conjectured that the translator was Raymond Lull or one of his disciples.

[169] There may be some connection with the mission of two cardinals to Constantinople in 1169: Chalandon, *Les Comnènes*, ii. 566. Can 'Ka.' be the Caciareda of note 147 ?

the monogram PASGALIS stands at the head of the Palatine MS. 1273.

The translator of Kiranides knows of other works in Greek on the magical virtues of herbs and planets, which he even places before Kiranides itself. Latin versions of these appear in several manuscripts,[170] sometimes along with Kiranides,[171] but with no indication of the translator, who was perhaps also Paschal the Roman.

Another Roman in the East appears in the Master Philip, friend and physician of Alexander III, who is sent with the letter of that Pope to Prester John 27 September 1177.[172] Moreover, the well known letter of Prester John to Manuel purports to have been transmitted by Manuel to Frederick Barbarossa and to have been done into Latin by Christian, archbishop of Mainz,[173] Frederick's lieutenant in Italy, which would bring us around once more to the intellectual contacts between the two empires. But as this letter of Prester John is clearly a western fabrication, we here pass beyond the realm of historical fact into that outer penumbra of Greco-Latin literary relations which still awaits the explorer.

The interest in divination and astrology at the Byzantine court [174] was reflected in the contents of the imperial library, from which a brief catalogue has reached us of a score of occult works of restricted circulation.[175] How many such found their way westward through Greek manuscripts or Latin versions from the Greek, we do not know. One at least we have in the two books of the *De revolutionibus nativitatum* of abu Ma'ashar (Apomasar), of which a Latin version from the Greek, not later than the

---

[170] Thorndike, ii. 233 f., who does not mention the edition of the seven herbs and seven planets in Sathas, *Documents inédits relatifs à l'histoire de la Grèce au moyen âge*, vii, pp. lxiii–lxvii (from St. Mark's, Cod. gr. iv. 57, suppl.). See H. Haupt, in *Philologus*, xlviii. 371–374; Cumont, in *Revue de philologie*, 1918, pp. 85–108.

[171] E. g., MS. Montpellier 277.

[172] Jaffe-Löwenfeld, no. 12942.

[173] F. Zarncke, *Der Priester Johannes* (Leipzig, 1879) ; cf. Thorndike, ii, ch. 47.

[174] Cf. Krumbacher, p. 627; *Catalogus codicum astrologicorum Graecorum*, v, 1, pp. 106 ff.; Oeconomos, *La vie religieuse dans l'empire byzantin* (Paris, 1918), pp. 70–93.

[175] Edited from the Angelica MS. 29 in *Catalogus codd. astr.*, i. 83 f. Note also the *Almagest*: supra, Chapter IX.

thirteenth century, is preserved in several manuscripts.[176]   The
prophecy of the Erythraean Sibyl, we have seen, also purports to
have been derived from a book in Manuel's library.[177]   We touch
this shadowy realm again in certain treatises on alchemy, where
we find the name of the Emperor Manuel, joined in one instance
to that of Frederick.[178]

Surer ground is reached with the Latin treatise on ophthal-
mology compiled from Greek sources by a certain Zacharias who
studied and practised at Constantinople in Manuel's reign, gain-
ing there from a court physician, Theophilus, "for the love of
God and money, knowledge which he could acquire from none
of the Latins." [179]

Other discoveries doubtless remain to be made in relation to
the north-Italian translators.   So far as their work has been recovered, it is largely concerned with theological material, both in
the form of controversy between the two churches and in versions
of earlier Greek writers, who, like John of Damascus, might thus
come to exercise an important influence on the West.   Logic and
grammar also appear in the case of James of Venice and Moses of
Bergamo, while medicine treads close on theology in the versions
of Burgundio and reappears in Zacharias.   Leo the Pisan and
Paschal the Roman are important chiefly in relation to the
occult.   The mathematical and astronomical interests of the
Sicilian school are strikingly absent.

---

[176] 'De revolutionibus nativitatum liber primus translatus de greco in latinum.
Sole nativitatis tempore . . .': B. N., MS. lat. 7320[2] (saec. xiii); Vatican, MS. Vat.
lat. 5713, f. 61; MS. Pal. lat. 1406, f. 45.   For the Greek original see C. Ruelle, in
Comptes-rendus de l'Académie des Inscriptions, 1910, pp. 34–39; F. Boll, in Heidelberg Sitzungsberichte, 1912, no. 18.

A MS. of the Laurentian, MS. Strozzi 61 (saec. xii) contains an 'Ars astrologie
translata de greco secundum Phtolomeum.   Doctrinales scripturi libros . . .'

[177] Supra, Chapter IX, n. 76.

[178] J. Wood Brown, Michael Scot, pp. 83 f.

[179] "Magistri Zachariae tractatus de passionibus oculorum qui vocatur Sisi-
lacera, id est Secreta secretorum," in P. Pansier, Collectio opthalmicorum veterum
auctorum (Paris, 1907), v. 59–94; cf. Neuburger, Geschichte der Medizin, ii, 1, pp.
314 f.

CHAPTER XI

VERSIONS OF ARISTOTLE'S POSTERIOR ANALYTICS[1]

In the intellectual history of the Middle Ages one of the most fundamental facts is the persistent and pervasive influence of the writings of Aristotle. Always considerable, this influence grew and spread as new groups of the master's works became available to the scholars of western Europe, and it can be measured and defined only as we can ascertain accurately the date, the character, and the diffusion of the different Latin versions of each portion of the Aristotelian *corpus*. In a general way it is well understood that the *Categories* and the *De interpretatione* were accessible throughout the Middle Ages in the translations of Boethius; that the other logical works were quite unknown to the earlier period and came to be used only in the second quarter of the twelfth century, whence they were called the *New Logic*; that the *Physics*, *Metaphysics*, and *Parva naturalia* reached the West about 1200; and that the *Rhetoric*, *Ethics*, and *Politics* make their appearance in the course of the next two generations.[2] There are, however, many obscure and doubtful points in this process, and the doubt and obscurity are greatest with reference to the period of the twelfth century. Thus we know nothing definite of the channels by which the *Metaphysics* suddenly reached Paris at the begin-

---

[1] Revised from *Harvard Studies in Classical Philology*, xxv. 87-105 (1914). For the resulting discussion see A. Hofmeister, in *Neues Archiv*, xl. 454-456 (1915); Baeumker, in *Philosophisches Jahrbuch*, xxviii. 320-326 (1915); Geyer, *ibid.*, xxx. 25-43 (1917).

[2] See in general Jourdain, *Recherches*; Baeumker, "Zur Reception des Aristoteles im lateinischen Mittelalter," in *Philosophisches Jahrbuch*, xxvii. 478-487 (1914); Grabmann, *Forschungen über die lateinischen Aristotelesübersetzungen des XIII. Jahrhunderts (Beiträge*, xvii, 1916); supplemented for the *Ethics* by Pelzer, *Revue néo-scolastique*, 1921, pp. 316-341, 378-400; and for the *Metaphysics* by F. Pelster in *Festgabe Baeumker* (Münster, 1923), pp. 89-118. Brief accounts in Sandys, *History of Classical Scholarship*[3], i, especially pp. 527, 567-569, 587 f.; and P. Mandonnet, *Siger de Brabant*[2] (Louvain, 1911), pp. 9-15. "La storia dell' aristotelismo è ancora da farsi," says Marchesi, *L'Etica Nicomachea nella tradizione latina medievale* (Messina, 1904), p. 1.

ning of the thirteenth century, and we are ignorant of the date
and authorship of the two versions, one from the Greek and one
from the Arabic, through which it was thereafter known. With
regard to the *Physics*, it is still necessary, not only to determine
the exact time when the version from the Arabic reached Latin
Europe,[3] but also to investigate the problem of possible earlier
translations from the Greek. An incomplete copy in the Vatican
which cannot be later than the very beginning of the thirteenth
century establishes the existence of a version of the *De physico
auditu* made from the Greek but differing from the Greco-Latin
version later current,[4] and there are traces of some acquaintance

[3] In the translation of Gerard of Cremona; cf. the text in MS. Lat. VI, 37 of
St. Mark's (Valentinelli, *Bibliotheca manuscripta*, v. 9): 'secundum translationem
Gerardi.' On the dates when these treatises reached Paris, see Chapter XVIII;
Mandonnet, *op. cit.*, pp. 13–15; Minges, in *Archivum Franciscanum historicum*, vi.
17. It is dangerous to use catalogues of manuscripts as evidence of such dates. Thus
MS. 221 of Avranches, containing the *Physics*, which is ascribed by Delisle to the
twelfth century, is more probably of the thirteenth, as is clearly MS. 428 of
the Biblioteca Antoniana at Padua. So MS. 421 of the Antoniana, containing the
Metaphysics and likewise placed in the twelfth century by the printed catalogues, is
not earlier than the fourteenth; cf. now Minges, *loc. cit.*, p. 16, who puts the MS.
earlier than I should. A copy of the *Meteorologica* in the Laurentian (MS. Strozzi
22), also attributed to the twelfth century, is plainly of the thirteenth. For similar
mistakes with respect to manuscripts of the *New Logic*, see below, n. 36.

[4] MS. Regina, 1855, ff. 88–94 v; cf. *Harvard Studies in Classical Philology*, xxiii.
164 (1912). Although my former attribution of this MS. to the twelfth century was
confirmed by excellent palaeographical authority, further examination shows that
it cannot with certainty be placed earlier than the opening years of the thirteenth
century. I have found no other copy of this version, which begins as follows: '*Aristotilis physice acroaseos. A.* Quoniam agnoscere et scire circa methodos omnes accidit quarum sunt principia vel causę vel elementa, ex eorum cognitione tunc enim
unumquodque cognoscere putabimur cum causas agnoverimus primas et principia
prima et usque ad ęlementa; palam quia et de natura scientię temptandum est
diffinire primum quę circa principia sunt. Apta vero a notioribus nobis via et manifestioribus ad manifestiora natura et notiora. Non enim eadem nobis nota et simpliciter. Ideoque hoc modo procedere et necesse de inmanifestioribus quidem natura nobis vero manifestioribus ad manifestiora natura et notiora. Sunt autem nobis
primum aperta et manifesta confusa magis, posterius autem ex his fiunt nota elementa et principia dividunt ea. Quapropter ab universalibus ad singularia oportet
progredi. . . . Ergo quia sunt principia et quę et quot numero determinatum sit
nobis ita. Rursum aliud incoantes principium dicimus. *Aristotilis phisicę acroaseos
.A. explicit.*' Book ii begins as follows on f. 94, but breaks off abruptly on the verso:
'Entium alia quidem sunt natura alia causas propter alias. Natura vero dicimus
esse animalia et eorum partes atque plantas ac alia corporum ut terram ignem et

with its contents in the twelfth century.[5] Certainly the current rendering of the fourth book of the *Meteorologica* was made from the Greek by Henricus Aristippus in Sicily before 1162; [6] there is evidence that the Greek text of the *De caelo* was known there in the same period; [7] and further research may quite possibly carry back other works of which versions from the Greek are known in manuscripts of the thirteenth century.[8]

The place of the *New Logic* in the thought of the twelfth century is better known, but there are intricate and perplexing problems connected with it, and fresh evidence is much needed. The history of the *Posterior Analytics* offers the greatest difficulty, yet it cannot be considered apart from the other members of this group of treatises, and any new light which may be shed upon it will make correspondingly clear some points connected with the *Prior Analytics*, the *Topics*, and the *Elenchi*. Moreover, since it was considered the most advanced and the most difficult of these works, its diffusion and assimilation serve to measure the range and depth of Aristotelian studies throughout the period.

The reception of the *New Logic* was the privilege of the genera-

---

aerem atque aquam; hęc enim et similia natura dicimus esse. . . .' Cf. Grabmann, *Forschungen*, pp. 173 f. For specimens of the current translations from the Greek and the Arabic, see Jourdain, pp. 405–407; for traces of the *Physics* in the twelfth century, Chapter V, nn. 58 ff. The version of MS. Reg. 1855 is probably of south-Italian or Sicilian origin, and should perhaps be connected with the occurrence of a Greek MS. of the first book of the *Physics* in the oldest catalogues of the papal library, the Greek part of which collection was probably derived from the library of the Sicilian kings. For the MS. see the catalogue of 1295 in *Archiv für Litteratur- und Kirchengeschichte des Mittelalters*, i. 41, no. 442; and the catalogue of 1311 in Ehrle, *Historia Bibliothecae Romanorum Pontificum*, i. 97, no. 610. For the origin of the Greek MSS. of the papal library see Chapter IX, n. 35.

⁵ Supra, Chapter V.

⁶ Rose, in *Hermes*, i. 385. The explicit statement concerning the authors of the translation of the *Meteorologica* will also be found in MS. 1428, f. 171, and MS. 9726, f. 58 v, of the Biblioteca Nacional at Madrid.

⁷ Supra, Chapter IX, pp. 183, 191. Cf. Heiberg, in *Hermes*, xlvi. 210; Mortet, in *B. E. C.*, lxxiv. 364.

⁸ See particularly Baeumker, *Die Stellung des Alfred von Sareshel*, in Munich *Sitzungsberichte*, 1913, no. 9, especially pp. 33 ff., where evidence is given of early translations of the *De anima* and the *Parva naturalia* from the Greek. Note the versions of the *Metaphysics*, *Ethics*, *De generatione*, and *De caelo* from the Greek in Bodleian MS. Selden supra 24, of the early thirteenth century.

tion living between ca. 1121 and 1158.[9] When Abelard wrote his
Dialectic, the Latin world knew none of the logical works of
Aristotle except the *Categories* and the *De interpretatione*, but he
elsewhere cites the *Sophistici Elenchi* and *Prior Analytics*.[10] His
contemporary Gilbert de la Porrée refers his readers to the *Prior
Analytics*. Otto of Freising, a student at Paris ca. 1130 and in
close touch with philosophical developments in France and Italy
until his death in 1158, became acquainted with all parts of the
New Logic, which he was the first to introduce into Germany. His
master, Thierry of Chartres, who lived until 1155, or shortly before, but taught at Paris for some years before 1141,[11] reproduces
the whole *Organum*, save only the *Posterior Analytics* and the
second book of the *Priora*; while the *Posteriora*, cited in Sicily in
the same period, comes to its own in the North in the analysis
given by Thierry's pupil John of Salisbury in his *Metalogicus* in
1159. The later emergence of the *Posterior Analytics* does not
necessarily indicate a reception distinct from the allied works,
but is rather to be explained by its difficulty, *paucis ingeniis pervia*, and the corruption of the Latin text;[12] and it is altogether

[9] On these questions see Prantl, *Geschichte der Logik im Abendlande*[2], ii. 98 ff.;
Grabmann, *Geschichte der scholastischen Methode* (Freiburg, 1909–11), i. 149–151,
ii. 66–81; Mandonnet, *Siger de Brabant*[2], pp. 9 f.; Schmidlin, "Die Philosophie
Ottos von Freising," in *Philosophisches Jahrbuch*, xviii. 160–175 (1905); Hofmeister,
"Studien zu Otto von Freising," in *Neues Archiv*, xxxvii, especially pp. 654–681
(1911); Webb, *Ioannis Saresberiensis Policraticus*, i, pp. xxiii–xxvii; A. Schneider,
in *Beiträge*, xvii, no. 4, pp. 10–18 (1915); Grabmann, *Forschungen*, pp. 1 ff.

[10] 'Aristotelis enim duos tantum, predicamentorum scilicet et Periermenias,
libros usus adhuc Latinorum cognovit': Cousin, *Ouvrages inédits d'Abélard*, p. 228.
See now Geyer, in *Philosophisches Jahrbuch*, xxx. 31–39, who is still vague on the
chronology of Abelard's writings. The history of the *Analytics* in the earlier Middle
Ages might appear in a new light if we could explain a passage in John the Scot
which cites the *Analytics*, where the *Metaphysics* is probably meant. E. K. Rand,
Johannes Scottus (Munich, 1906), pp. 6, 42.

[11] Cf. Poole, *E. H. R.*, xxxv. 338 f. (1920). I agree with Hofmeister in denying
the force of the argument of Clerval (*Les écoles de Chartres*, p. 245) for dating the
Eptatheuchon of Thierry before 1141. Geyer does not take up Thierry, though he
eliminates Adam du Petit-pont from the discussion.

[12] John of Salisbury, *Metalogicus*, 4, 6, in Migne, *Patrologia*, cxcix. 919: 'Posteriorum vero Analyticorum subtilis quidem scientia est et paucis ingeniis pervia,
quod quidem ex causis pluribus evenire perspicuum est. Continet enim artem demonstrandi, que pre ceteris rationibus disserendi ardua est. Deinde hec utentium
raritate iam fere in desuetudinem abiit, eo quod demonstrationis usus vix apud solos

likely that the arrival of the *New Logic* is to be placed in the earlier, rather than in the later, years of the period with which we are dealing. In any case its sudden appearance in the logical and philosophical literature of the second quarter of the twelfth century should be brought into relation to a much-discussed notice of the year 1128. Under that year we read in the chronicle of Robert of Torigni, abbot of Mont-Saint-Michel: [13]

Iacobus clericus de Venecia transtulit de greco in latinum quosdam libros Aristotilis et commentatus est, scilicet Topica, Analyticos Priores et Posteriores, et Elencos, quamvis antiquior translatio super eosdem libros haberetur.

This entry is not found in the earliest redaction of the chronicle, completed in 1156–57, but appears in the redactions of 1169 and 1182, for the latter of which we have the author's own copy, and there can be no doubt that it emanated from Robert himself, who was by no means ignorant of what went on in Italy and who on more than one occasion takes the opportunity of mentioning significant facts of literary history.[14] Although the entry is not strictly contemporary, it is by a well informed contemporary writer, and while the date may not be absolutely exact, it falls within a few years of the only other known reference to James of Venice, which mentions him at Constantinople in 1136.[15] In the passage of Robert two important points stand out: the existence of an earlier version of the *Topics, Analytics,* and *Elenchi,* and the new rendering, with its accompanying commentary. Nothing is

mathematicos est, et in his fere apud geometras duntaxat; sed et huius quoque discipline non est celebris usus apud nos, nisi forte in tractu Ibero vel confinio Africe. Etenim gentes iste astronomie causa geometriam exercent pre ceteris, similiter Egyptus et nonnulle gentes Arabie. Ad hec liber quo demonstrativa traditur disciplina ceteris longe turbatior est, et transpositione sermonum traiectione litterarum desuetudine exemplorum que a diversis disciplinis mutuata sunt. Et postremo, quod non attingit auctorem, adeo scriptorum depravatus est vitio ut fere quot capita tot obstacula habeat, et bene quidem ubi non sunt obstacula capitibus plura. Unde a plerisque in interpretem difficultatis culpa refunditur, asserentibus librum ad nos non recte translatum pervenisse.'

[13] Ed. Delisle, Société de l'Histoire de Normandie, i. 177; also in *M. G. H., Scriptores,* vi. 489.

[14] See the well informed notices of Gratian (i. 183), Master Vacarius (i. 250), Burgundio of Pisa (i. 270; ii. 109), and Gilbert de la Porrée (i. 288).

[15] Anselm of Havelberg, *Dialogi,* 2, 1, printed above, p. 197. Geyer, *Jahrbuch,* xxx. 38 f., rests his whole argument upon the year 1128.

said respecting the author of the earlier translation, but in the absence of any other known version it has generally been identified with that of Boethius. We have then to explain the main problem in the Aristotelian tradition of the early Middle Ages, namely why, if these works were translated by Boethius, they remained unknown from the sixth to the twelfth centuries, only to come to light at the very moment when they were also translated by James of Venice. Recently a solution has been sought, first by denying that any such translations were made by Boethius [16] or, at least, that they survived, and then by maintaining that the versions current in the later Middle Ages under his name were really the work of James of Venice, in whose time they first emerge.[17] James of Venice is himself a riddle. His learning, his knowledge of Greek, and his opportunity of access to Greek texts of Aristotle [18] are known to us from Anselm of Havelberg's account of the disputation at Constantinople in 1136,[19] but he is mentioned by no other chronicler, and no translations have been found in his name. With the field thus free for conjecture, some have cast doubt upon the statement of Robert of Torigni,[20] while others have made of James the chief intermediary in the transmission of the *New Logic* to Latin Europe. Neither of these views seems to me a sound interpretation of existing evidence, and both are invalidated by a new source of information.

In the library of the chapter of Toledo there is preserved a manuscript of the thirteenth century [21] containing three transla-

---

[16] In view of the explicit statements of Boethius on this point (*In Topica Ciceronis*, Migne, lxiv. 1051, 1052; *De differentiis topicis, ibid.*, coll. 1173, 1184, 1193, 1216), this denial of authorship (Schmidlin, p. 169; Grabmann, ii. 71) cannot be taken seriously. Cf. Brandt, "Entstehungszeit und zeitliche Folge der Werke von Boethius," in *Philologus*, lxii. 250, 261; Mandonnet, *Siger de Brabant*[2], p. 8.

[17] This attribution to James was suggested by Rose, in *Hermes*, i. 381 f. (1866). Schmidlin and Grabmann succeed in convincing themselves that it has really been proved. Hofmeister (*Neues Archiv*, xxxvii. 657, 659, 663) is more cautious on this point, while denying positively the Boethian authorship of the current version.

[18] On Aristotelian studies at Constantinople in the eleventh and twelfth centuries see Grabmann, ii. 74 f., and the literature there cited.

[19] Supra, n. 15.   [20] So Jourdain, p. 50.

[21] MS. 17–14, containing seventy-seven folios in different hands of the thirteenth century. The title of the volume at the top of f. 1 has been cut off. The MS. begins with the preface to the unknown translation discussed in this chapter, this transla-

tions of the *Posterior Analytics* and a version of the commentary of Themistius. One of the translations is the mediaeval version from the Greek commonly attributed to Boethius, another the ordinary version from the Arabic. The third [22] contains a text which I have not succeeded in finding elsewhere, accompanied by a preface of exceptional interest:

[V]allatum multis occupationibus me dilectio vestra compulit ut Posteriores Analeticos Aristotelis de greco in latinum transferrem. Quod eo affectuosius agressus sum quod cognoscebam librum illum multos in se sciencie fructus continere et certum erat noticiam eius nostris temporibus Latinis non patere. Nam translatio Boecii apud nos integra non invenitur, et id ipsum quod de ea reperitur vitio corruptionis obfuscatur. Translationem vero Iacobi obscuritatis tenebris involvi silentio suo peribent Francie magistri, qui quamvis illam translacionem et commentarios ab eodem Iacobo translatos [23] habeant, tamen noticiam illius libri non audent profiteri. Eapropter siquid utilitatis ex mea translatione sibi noverit latinitas provenire, postulationi vestre debebit imputare. Non enim spe lucri aut inanis glorie ad transferendum accessi, sed ut aliquid [24] conferens latinitati vestre morem gererem voluntati. Ceterum si in aliquo visus fuero rationis tramitem excessisse, vestra vel aliorum doctorum ammonitione non erubescam emendare.

Here at last is a new bit of evidence regarding James of Venice: his translation included both the *Posterior Analytics* and commentaries thereon; it has reached the centres of learning in France, but, apparently because they have not conquered its difficulties,

tion ending on f. 11 v. Ff. 13–28 v have 'Translatio Posteriorum Analyticorum Aristotilis s[ecundum]' with a letter effaced, i. e., the version current under the name of Boethius. F. 29, 'Translatio Posteriorum Analyticorum Aristotilis secundum Tthom [*sic*; cf. Geyer, p. 40, n.]. Omnis doctrina et omnis disciplina cogitativa non fit nisi ex cognitione. . . .' (= the ordinary version from the Arabic; see Jourdain, p. 404). F. 54, 'Explicit liber Posteriorum Analyticorum Aristotilis secundum translationem Th. Incipit commentum Themistii super eandem translationem Posteriorum Analyticorum. Scio quod si intendo . . .' (Jourdain, p. 405; see below, n. 63.) The treatise breaks off abruptly at the bottom of f. 77 v.

MS. 17–14 is not described by José Octavio de Toledo, *Catálogo de la librería del cabildo toledano*, supplement to *Revista de Archivos*, viii and ix, and separately, Madrid, 1903. This catalogue, made in the library at the time of the revolution of 1869, has been printed without verification or completion and without any indication of the important MSS. at that time transferred to the Biblioteca Nacional at Madrid, where they still are. I examined MS. 17–14 at Toledo during the hour when the library was open May 2 and 14, 1913, but repeated efforts of friends to secure collations on the spot have been met with the statement that the MS. has been misplaced and can no longer be found. It will doubtless appear in due time, when the problems left open can be determined by certain collations.

[22] F. 1.   [23] So corrected in margin from *translationem*.   [24] Or *aliud?* MS. *a'd*.

the masters make no public use of it. This disposes at once of the theory that the version of James is apocryphal, while it also makes clear that this version was not the basis of the revival of the *Analytics*, and also renders it unlikely that it passed into general use and can thus be identified with the current translation. Robert of Torigni is also confirmed at another point, namely in his assertion, which some have sought to explain away,[25] that there was an older version already in existence. This our preface ascribes to Boethius, thus adding one more to the number of those who in the twelfth century accepted this attribution.[26] An explanation is also suggested why the Boethian translation came but slowly into use: it is incomplete, and the text is corrupt. This agrees exactly with John of Salisbury, who says of the current version, *adeo scriptorum depravatus est vitio ut fere quot capita tot obstacula habeat, et bene quidem ubi non sunt obstacula capitibus plura;* [27] and the statement is amply confirmed by existing manuscripts, where to take only the instances where a Greek word was left standing in the Latin, we find in some cases merely *grecum*, while in others the word has become hopelessly corrupt.[28] Thus in 1, 2 (Bekker, p. 71, l. 18), where ἐπιστημονικόν was carried over and explained as *facientem scire*, we find in MS. R. 55 sup. of the Ambrosian (f. 194) *grecum* corrected to *apiteticon* in the first instance and in the second instance *ginitvopikoli*, while MS. H. IX, 2 of Siena (f. 130 v) has what seems intended for *epinuorikon*. In 1, 4 (Bekker, p. 73, l. 40) ἰσόπλευρον καὶ ἑτερόμηκες becomes in the Siena MS. (f. 132 v) *jjodniyipop* quod est equilaterum *kHedorinke* id est altera parte longius; in the Ambrosian (f. 195 v) *gyodtinkipo* quod est isopleros equilaterum *gkθuθcdeli*; in MS. VIII, 168 of St. Mark's (f. 94), *iodnapop* and *kaisodeorrylie*. In 1, 5 (Bekker, p. 74, l. 27) ἰσόπλευρον becomes *iodHaaqoH* and *kaiodpaapor* in the Siena MS. (ff. 133 v, 134), and *ortoniegobon* in the Ambrosian (f. 196 v), while σκαληνές is represented respectively by *kokaajyon* and *okaanor*. In 1, 7 the Greek text reads (Bekker, p. 75, l. 15): οἷον τὰ ὀπτικὰ πρὸς γεωμετρίαν καὶ τὰ

---

[25] Schaarschmidt, *Johannes Saresberiensis*, p. 122; Hofmeister, in *Neues Archiv*, xxxvii. 658 f.   [26] See below, nn. 31–33.   [27] *Metalogicus*, 4, 6, supra, n. 12.

[28] MS. Avranches 227 commonly has *grecum* in the passages cited in the text.

ἁρμονικὰ πρὸς ἀριθμητικήν. This becomes in the Siena MS. (f. 135): ut *onti kay* perspectiva ad geometriam *kaaita apiHoyka* id est consonativa ad arimeticam. The Ambrosian MS. (f. 197 v) has *kagroapinopika*; MS. 557 of the Biblioteca Antoniana at Padua has *Rait^a apruopil'ia*.

The existence of these passages does not, of course, go to prove that the translation in which they occur was the work of Boethius, but the whole trend of the available evidence seems to me to lead to that conclusion. Boethius tells us specifically that he translated both *Analytics* as well as the *Topics*.[29] These, however, pass out of use in the early Middle Ages, and as late as the time of Sigebert of Gembloux, who died in 1112, he is known as the translator of the *Categories* and the *De interpretatione* only.[30] Then comes the revival of the *New Logic* in the second quarter of the twelfth century, and at once men begin to ascribe its Latin form to Boethius. Our translator is clear on this point; Otto of Freising evidently held the same view;[31] the anonymous poet on the seven liberal arts in an Alençon manuscript is quite explicit,[32] and so is Burgundio the Pisan.[33] It is certainly significant that the generation which first possessed the *New Logic* considered Boethius to have been its translator. Moreover, when writers of this period quote passages from Aristotle they use the current version which in later manuscripts is regularly attributed to Boethius. This is notably true of Otto of Freising[34] and of John of Salisbury.[35] While in these cases the Latin text is not cited as being the work of Boethius, neither is it ascribed to any one else, and in the absence of twelfth-century manuscripts of the *New Logic*[36] fur-

---

[29] See above, n. 16.   [30] Migne, clx. 555.

[31] *Chronicon*, 5, 1 (ed. Hofmeister, p. 230).

[32] MS. 10, in Ravaisson, *Rapports sur les bibliothèques de l'Ouest* (Paris, 1841), p. 406: 'Transtulit hanc resolvendo binis analeticis.' Cf. Prantl, *Geschichte der Logik*[2], ii. 105; Hofmeister, in *Neues Archiv*, xxxvii. 672.

[33] Infra, n. 37.

[34] This is shown by Schmidlin, pp. 172–175, by means of a collation of MSS. Thierry of Chartres may use a different version of the *Prior Analytics* (Webb, *Ioannis Saresberiensis Policraticus*, i, p. xxv) but elsewhere uses the *vulgata* (Geyer, p. 30).   [35] Jourdain, pp. 254–256.

[36] Assertions of the catalogues to the contrary are without foundation in the case of Cod. Lat. Monacensis 16123 and MS. 401 of the Biblioteca Antoniana, both of

ther evidence is not at hand. While later copies frequently mentioned Boethius as the translator, none refer to James of Venice, who after the three contemporary notices which have been cited disappears — *obscuritatis tenebris involvitur.*[37] We know furthermore that the current version cannot be that of our anonymous translator, which is quite different, nor can it be the *nova translatio* cited by John of Salisbury,[38] who distinguishes the two. Until some definite evidence is produced to the contrary, we are justified in regarding the current mediaeval version as the work of Boethius.[39]

It has indeed been urged by Grabmann [40] that Boethius could not have been the author of the translation of the *New Logic* because its Latinity is unworthy of so accomplished a stylist. The defect of this argument of course lies, apart from the ignorance of Boethius which it betrays, in overlooking the difference between translation and independent composition. Boethius translated

which are of the fourteenth century. I have verified Grabmann's statement (*Methode*, ii. 78) that there are in Paris no MSS. of the *New Logic* anterior to the thirteenth century, and have searched in vain for such MSS. elsewhere. For mention of Aristotle in contemporary catalogues of the twelfth century see Manitius, *Geschichte der lateinischen Litteratur des Mittelalters*, i. 30; Grabmann, ii. 78. Except for the occasional occurrence of the translation from the Arabic, the MSS. of the thirteenth and fourteenth centuries give regularly the Boethian versions. Delisle is in error in saying that MSS. 224 and 227 of Avranches (*Catalogue des MSS. des départements*, x. 103, 106) contain a different version.

[37] Curiously enough, James is not mentioned by his acquaintance Burgundio the Pisan in his review of translations from the Greek in 1173, where we read merely: 'Sed et Boetius clarissimus philosophus Porphirium et Aristotilem in Categoriis et Periemeniis, in Topicis et Elenchis et Nichomachum arismeticis transferens de verbo ad verbum ex greca latine reddidit lingue' (MS. Ottoboni 227, f. 2; cf. Chapter VIII, note 36).

[38] See below, n. 53.

[39] The citations of Aristotle by Boethius are too few to serve as a basis for identifying the translation, but it is noteworthy that the definition quoted in the περὶ Ἑρμενείας, 2, 6 (ed. Meiser, ii. 122), from the beginning of the *Prior Analytics* ('Propositio ergo est . . .') corresponds exactly with the current version. This is overlooked by Geyer, pp. 39 ff., who regards James of Venice as the author of the current version but brings forward no new evidence on this point.

[40] ii. 71: 'Ein Schriftsteller nun, dem solche Qualitäten als Stilisten und Latinisten von berufenster Seite zugesprochen werden, kann doch unmöglich die Latinität, die uns in den Aristoteleszitaten des Otto von Freising und in den Analytiken, der Logik und der Elenchik der scholastischen Schullogik entgegentritt, hervorgebracht und sich etwa grammatische Verstösse wie *parvissimum* geleistet haben.'

like a schoolboy because to him, as to the Middle Ages after him, faithful translation must be absolutely literal (*verbum verbo expressum comparatumque*), its purpose being *non luculentae orationis lepos sed incorrupta veritas.*[41] Hence the much more frequent occurrence of Grecisms in the translations than in his other works. Statistical comparisons, it is true, show stylistic variations among the several Boethian translations, as for example between the *Prior* and the *Posterior Analytics*;[42] but these do not go so far as to indicate difference of authorship and cannot be safely used when made upon the uncertain basis of the present printed text. In any event a writer who can create a genitive of comparison to render a passage in Aristotle's *Categories*[43] cannot be deprived of the version of the *Elenchi* because he sees fit to render μικρότατον by *parvissimum.*[44] If the argument proved anything, it would prove too much, for it would compel us to give up Boethius as a translator.

There remains still the problem why, with the translation of Boethius in existence, the *New Logic* was neglected until the twelfth century, and why it was so suddenly revived.[45] For an answer we have at present only guesses. One may easily suppose that in an age which had use for only elementary logic, as it had for only "the slenderest of lawbooks," the advanced treatises fell into neglect and the manuscript tradition was correspondingly attenuated. In the revival of dialectic in the twelfth century men begin to seek additions to the store of logical writings and they discover the Boethian text. It is incomplete and corrupt, and attempts are made, at least two in number, to provide a better

---

[41] Boethius, *In Isagogen Porphyrii*, 1 (ed. Brandt, p. 135).

[42] See McKinlay's caːeful investigation in *Harvard Studies*, xviii. 123–156.

[43] Migne, lxiv. 210; cf. McKinlay, p. 125.

[44] 2, 9, as quoted by Otto of Freising, *Chronicon*, 2, 8 (ed. Hofmeister, p.76). There is, of course, classical authority (e. g., Lucretius, 1, 615, 621; 3, 199) for the *parvissimum* which shocks Grabmann. The retouching of the mediaeval version in the printed text (Migne, lxiv. 1040) is well illustrated in this whole passage.

[45] There is also the problem as to what became of the Boethian commentaries on these works; cf. Brandt in *Philologus*, lxii. 250. Schmidlin (p. 169) uses the absence of such commentaries as an argument against the Boethian authorship of the translations, but similar reasoning might be used against his attribution of the translations to James of Venice, for we are expressly told that the version of James was accompanied by a commentary. See above, p. 229.

rendering. None of these attempts, however, succeeds in passing into general use, and the old translation, completed and perhaps improved but still in spots unintelligible, becomes the received version upon which mediaeval knowledge of the higher logic depends.

The character of the version of the Toledo manuscript will be clearer when it is seen beside the text of the current version which is given below in the second column. The first book begins:

Omnis didascalia et omnis disciplina deliberativa [46] ex preexistenti fit cognitione. Manifestum autem hoc contemplantibus in cunctis. Etenim mathematice discipline per hunc modum veniunt et aliarum unaqueque artium. Similiter autem et circa orationes et que per sillogismos et que per inductionem; etenim utreque per precognita faciunt didascaliam, hee quidem accipientes ut ab intellectis, ille autem monstrantes universale per hoc quod manifestum est singulare. Similiter autem et rethorici persuadent, aut enim per exemplum,[47] quod est inductio, aut per enthimemata, quod est sillogismus. . . .

Omnis doctrina et omnis disciplina intellectiva ex preexistenti fit cognitione. Manifestum est autem hoc speculantibus in omnes. Mathematice enim scientiarum per hunc modum fiunt et aliarum unaqueque artium. Similiter autem et circa orationes que per sillogismos et que per inductionem fiunt; utreque enim per prius nota faciunt doctrinam, he autem incipientes tanquam a notis, ille vero demonstrantes universale per id quod manifestum est singulare. Similiter autem et rhetorice persuadent, aut enim per exemplum, quod est inductio, aut per entimema, quod vere est sillogismus. . . .

Book ii begins and ends:

Quesita sunt equalia numero quot scimus. Querimus autem quatuor: quod, propter quod, an est, quid est. Etenim quando prius quidem hoc aut hoc querimus in numerum ponentes, sicut utrum deficit sol aut non.

. . . .

Questiones sunt equales numero his quecumque vere scimus. Querimus autem quatuor: quod est, propter quod est, si est, quid est. Cum quidem enim utrum hoc aut hoc sit querimus in numerum ponentes, ut utrum sol deficiat aut non, ipsum quod querimus.

. . . .

Si igitur nullum aliud preter scientiam genus habemus verum, intellectus sit scientie principium, et hoc quidem principium principii sit. Hoc autem omne similiter se habet ad [omnem] rem.

Si igitur nullum aliud genus preter scientiam habemus verum, intellectus utique scientie erit et hoc quidem principium principii utique erit. Hoc autem omne similiter se habet ad omne rerum genus.

[46] MS. *delibatā*.

[47] Gloss: *vel exempla.*

Both renderings have the extreme literalness characteristic of mediaeval translations from the Greek, but the Toledo text is distinctly the closer of the two, as seen in the omission of the predicate and the carrying over of such words as *didascalia*. Other characteristics of this version are the use of *autem* instead of *vero* for δέ, the insertion of a superfluous relative to represent the article in an attributive phrase,[48] and the rendering of the optative with ἄν by the subjunctive in cases where Boethius uses *utique* with the future indicative.[49] Though he had Boethius before him, the author still shows some independence, judged by mediaeval standards; his work is not that of an unskilled hand; and the fact that the preface contains no suggestion of ignorance or inexperience, such as is frequent in such prologues, makes it probable that this was not his first labor of translation.

No clew is given to the name of the translator or the friend to whom his work is dedicated, but the preface must have been written between the appearance of the translation of James of Venice ca. 1128 and the close of the twelfth century, when a new version had been made from the Arabic by Gerard of Cremona (d. 1187), and when the *Posterior Analytics* had begun to influence the teaching of logic at the University of Paris.[50] Moreover, in all probability it is anterior to 1159, when the *Metalogicus* of John of Salisbury shows that the knowledge of the *Posteriora* was already "open to the Latin world," and can thus be placed in the generation which first received the *New Logic*. The author is in touch with the teaching of the French schools, yet he speaks of their masters (*Francie magistri*) in a way which implies that he was not a Frenchman; and his knowledge of Greek and access to the Greek text would imply that, if not an Italian, he was at least for the time being resident in Italy. We know that two of the Italian translators of this period were acquainted with the *Posteriora*, the Pisan Burgundio, whom John of Salisbury cites in the *Metalogicus* [51] as an authority for a statement concerning Aris-

---

[48] Thus τόδε τὸ ἐν τῷ ἡμικυκλίῳ τρίγονον (Bekker, p. 71, l. 20) becomes 'hic qui in semicirculo triangulus.'

[49] Cf. also the translation of the Almagest: supra, Chapter IX. [50] See below, n.64.

[51] 4, 7 (Migne, cxcix. 920): 'Fuit autem apud Peripateticos tante auctoritatis scientia demonstrandi ut Aristoteles, qui alios fere omnes et fere in omnibus philo-

totle, and the Sicilian Henricus Aristippus, who in the preface to his version of the *Phaedo*, written in 1156, singles out the *Apodiptica* as one of the notable works to which scholars have access in Sicily; [52] but both of these are excluded from the authorship of the Toledo preface by its style and by the familiarity it betrays with French learning. Aristippus, it is true, has, on the basis of the passage just cited, been set down as a translator of the *Posteriora*, and further conjecture has made him the source of John of Salisbury's acquaintance with this treatise and the author of the *nova translatio* which John cites in a passage of the *Metalogicus*.[53] There is, however, no reason for believing that Aristippus translated all the Greek writings which he cites in his prefaces, nor is there the least basis for identifying him with the *grecus interpres* with whom John of Salisbury studied in Apulia and from whom he is, without any warrant, supposed to have obtained the *nova translatio*. John's familiarity with the *Posteriora*, which he is one of the first northern authors to cite,[54] may well have been the result of his frequent journeys to Italy, perhaps

sophos superabat, hinc commune nomen sibi quodam proprietatis iure vindicaret quod demonstrativam tradiderat disciplinam. Ideo enim, ut aiunt, in ipso nomen philosophi sedit. Si mihi non creditur, audiatur vel Burgundio Pisanus, a quo istud accepi.' The passage does not show personal familiarity with the *Posteriora* on the part of Burgundio but merely knowledge of the Byzantine tradition, such as he doubtless acquired in the course of his visits to Constantinople. On Burgundio see Chapter X.

    [52] *Hermes*, i. 388: 'Habes de scientiarum principiis Aristotelis Apodicticen, in qua supra naturam et sensum de axiomatis a natura et sensu sumptis disceptat.' On Aristippus see Chapter IX.

    [53] 2, 20 (Migne, col. 885): 'Gaudeant, inquit Aristoteles [*Anal. Post.*, 1, 22, Bekker, p. 83, l. 33], species; monstra enim sunt, vel secundum novam translationem cicadationes enim sunt; aut si sunt, nihil ad rationem.' Cf. Rose, in *Hermes*, i. 381 ff. The identification of Aristippus with the *grecus interpres* and the author of the *nova translatio* was first advanced by Rose on the basis of an ingenious combination of conjectures. It has been accepted without indicating its conjectural character by Grabmann and Schmidlin, and by Baeumker, in *Allgemeine Geschichte der Philosophie (Die Kultur der Gegenwart*[2], i, 5), p. 363; Hofmeister and Mandonnet are more cautious. Webb gives a sober résumé of this *quaestio difficillima*. What is most needed is more facts. Geyer, p. 42, suggests that John may refer to a translation of this single term only.

    [54] He is usually considered the first, but the *Posteriora* seems to have been used, in a translation which requires investigation, by the author of the *De intellectibus*, which belongs to the school of Abelard. Prantl, *Geschichte der Logik*[2], ii. 104, n. 19; Geyer, p. 37.

even of his sojourns in Apulia, but he quotes the "new translation" only once, and his steady reliance is on the current version.
When the Toledo manuscript again becomes accessible to scholars, it will be easy to determine whether it contains the rendering
of τερετίσματα by *cicadationes* which earmarks the *nova translatio* of the *Metalogicus*. Meanwhile, since in this period we hear
of a text of the *Posteriora* in Sicily only, it would seem that the
home of the Toledo version should be sought there, while its
author's acquaintance with the French schools points to one of
the scholars from beyond the Alps who are found not infrequently
as visitors to the southern kingdom.

The collation of another passage may very likely determine the
relation of the Toledo version to still another translation from the
Greek, cited as the work of a certain John by Albertus Magnus,
who in one instance prefers it to the Boethian rendering.[55] The
conjecture that the name is an error for James [56] is not supported
by the manuscripts, and the identification with John of Basingstoke [57] has to explain the silence of Grosseteste, who, if a translation by his friend Basingstoke had been in existence, would
certainly have made use of it in his commentary on the *Logic*. Another John who was concerned with the *Posterior Analytics* is John
of Cornwall, under whose name a series of *Questiones* is preserved
in a manuscript of Magdalen College, Oxford.[58] Inasmuch, however, as this work constantly cites Lincolniensis, it cannot be the
work of John of Salisbury's contemporary of that name,[59] whose
writings moreover betray no familiarity with Greek; and even if
we crowd the chronology sufficiently to admit the citation of

[55] *In Analytica posteriora*, 1, 4, 9; 2, 2, 5; *Opera* (Lyons, 1651), i. 579, 624. See
Jourdain, p. 310.

[56] Jourdain, p. 59. I have collated MS. Vat. Lat. 2118, f. 140; and MS. Lat.
16080, f. 101 v, of the Bibliothèque Nationale.

[57] Prantl, *Geschichte der Logik*, iii. 5.

[58] MS. 162, ff. 183–245 v; cf. Coxe, *Catalogus*, ii. 75. The treatise begins and
ends: 'Scire autem opinamur unumquodque cum causam recognoscamus . . . licet
alia non cognoscatur nisi tantum in universali.' Then follow 'Tituli questionum
Cornubiensis' to the number of forty-seven, with this explicit: 'Expliciunt questiones et tituli tam primi libri quam secundi Posteriorum Analeticorum dati a
domino Johanne de Sancto Germano de Cornubia. Amen.' There was a copy at
Canterbury ca. 1500: Historical MSS. Commission, *Various Collections*, i. 225.

[59] On whom see Kingsford, in *Dictionary of National Biography*, xxix. 438.

Grosseteste on the one hand and the use of the *Questiones* by Albert on the other, there is, in such portions of the text as I have been able to examine by means of photographs, no indication that any save the ordinary translation was used in the *Questiones*. For the present we must leave the problem of John's version unsolved.

Likewise of the twelfth century is the first translation of the *Posteriora* from the Arabic, which appears in the long list of works turned into Latin by that indefatigable translator Gerard of Cremona, who died in 1187.[60] No copy of this translation has been found under Gerard's name,[61] but if it acquired anything of the popularity enjoyed by his other versions, we are justified in identifying it with a version which occurs not infrequently in manuscripts of the thirteenth century and is plainly derived from the Arabic.[62] The list of Gerard's translations also includes the commentary of Themistius on the *Posteriora*, of which we have copies which are clearly based upon an Arabic original.[63]

By the close of the twelfth century, accordingly, there had been produced at least four Latin versions of the *Posterior Analytics*, the work respectively of Boethius, James of Venice, the anonymous translator of the Toledo manuscript, and Gerard of Cremona; while further investigation is required to determine whether the *nova translatio* cited by John of Salisbury and the version of the unknown John should be added as a fifth and a sixth or are to be identified in one or both cases with those of James of Venice and of the Toledo text.

[60] Boncompagni, in *Atti dell' Accademia dei Lincei*, iv. 388 (1851); Wüstenfeld, p. 58; Steinschneider, *E. U.*, no. 46(8, 38).

[61] It is, however, cited by Richard of Furnival, ca. 1250: Delisle, *Cabinet des MSS.*, ii. 525; Birkenmajer, *Ryszarda di Fournival*, p. 44, no. 14.

[62] Jourdain, p. 404, gives a specimen.

[63] See the specimen in Jourdain, p. 405; and cf. MS. Lat. 14700 of the Bibliothèque Nationale; MS. 17–14 of Toledo, f. 54; Cod. Lat. Monacensis 317 (*Catalogus codicum MSS. Latinorum*, edition of 1892, i. 80). Probably this is the commentary mentioned in the mediaeval catalogue of the Sorbonne: Delisle, *Cabinet des MSS.*, iii. 57.

It may be observed in this connection that the MSS. themselves give no support to Valentinelli's statement (*Bibliotheca manuscripta*, iv. 13–15) that the translation of the *Topica* and *Elenchi* in two codices of St. Mark's is the work of Abraham de Balmes, the physician of Cardinal Grimani. The MSS. are anterior to Abraham's time, and the text has the *incipits* of the current mediaeval version.

As a subject of academic study the *Posterior Analytics* found its way slowly into the mediaeval universities. Alexander Neckam, who can hardly have begun his studies at Paris before 1175, describes the change in the teaching of logic there produced by its introduction,[64] and Roger Bacon speaks of the first lectures on it at Oxford as given in his time by a certain Master Hugh.[65] Elaborate commentaries were, however, prepared by the great schoolmen of the thirteenth century, some of whom took pains to collate the different versions. Grosseteste, though relying mainly upon the current Boethian translation, also cites *alie translationes* and the commentary of Themistius.[66] The *Questiones* of John of Cornwall, whoever he may have been, seems to follow Grosseteste and the current version.[67] Albertus Magnus is careful to compare this version, which he ascribes to Boethius, with that from the Arabic and with that of the unknown John, and cites the works of Themistius and John the Grammarian, as well as the Arab commentators.[68] The commentary of Thomas Aquinas on the *Posteriora* [69] is, like his other commentaries, less discursive and follows with some closeness the current text, corrected in at least one instance by reference to the Greek.[70] The ordinary version is also followed by the later schoolmen, Egidio Colonna, Albert of Saxony, and Walter of Burley.[71]

[64] *De naturis rerum*, ed. Wright, p. 293: 'Antequam legeretur liber ille asserebant doctores Parisienses nullam negativam esse immediatam. Sed hic error sublatus est de medio per beneficium Apodixeos.' Cf. Chapter XVIII.

[65] Rashdall, *Universities of the Middle Ages*, ii. 754; Sandys, *History of Classical Scholarship* [3], i. 570.

[66] Baur, *Die philosophischen Werke des Robert Grosseteste (Beiträge*, ix), p. 18 *. I have examined MS. Borghese 306 of the Vatican.

[67] Supra, n. 58.

[68] See his commentary in *Opera* (Lyons, 1651), i. 513–658; ed. Borgnet (Paris, 1890), ii. 1–232; and cf. Jourdain, pp. 308–310.

[69] *Opera* (Rome, 1882), i. 129–403.

[70] Bk. i, lect. 6, according to the text of Jourdain, p. 396. I can find no evidence that, as Mandonnet says (pp. 11, 40–42), William of Moerbeke translated the logical works for the benefit of St. Thomas. The passages cited from contemporary writers do not mention these among William's Aristotelian translations, nor is any copy of them known. Cf. Grabmann, ii. 70.

[71] These commentaries exist in various early editions. That of Albertus de Saxonia is in MS. 227 of Avranches (*Catalogue des MSS. des départements*, x. 106); see further *Beiträge*, xxii, no. 3–4, p. 48.

It is characteristic of the place which Aristotle still held in European thought that he should have been one of the earliest authors at whom the humanists tried their hand. Roberto de' Rossi, the first pupil of Chrysoloras, busied himself with the works of the Stagyrite, seeking to soften the bare harshness of the literal version of Boethius,[72] and we have from his pen a rendering of the *Posterior Analytics* which can be definitely assigned to the close of the year 1406. Voigt, it is true, knows of Robert's translations only through their mention by Guarino of Verona and says they do not occur in the manuscript catalogues;[73] but MS. 231 of the *Fondo antico* of St. Mark's[74] contains *Aristotelis Posteriorum Analeticorum nova Roberti translatio*, accompanied by a preface and by verses at the end which fix the date by reference to the reconstruction of the citadel and walls of Pisa.[75] Valentinelli indeed infers from these verses that the author was a Pisan of the late fourteenth century, but *nostri cives* would have no point if a Pisan were speaking, and the only others so engaged at Pisa were the Florentines, whose fortification of the city and oppression of the conquered after its final capture[76] are here exactly described. The author is not further indicated, but the name and year can point only to Robert de' Rossi.[77] Freer in style and less indebted to the mediaeval rendering was the more popular Renaissance

---

[72] 'Dignus enim vir ille ut cunctis modis humanitatis auribus insinuetur atque sterilis illa durities quam ad verbum translatio pepererat pro viribus nostris civibus delinienda et demulcenda paulum fuit' [*sic*].   F. 2 v of the MS.

[73] *Wiederbelebung*[3], i. 289, ii. 173.

[74] Parchment, written in a humanistic hand of the fifteenth century. Cf. Valentinelli, *Bibliotheca manuscripta ad S. Marci Venetiarum*, iv. 32.

[76]   Haec ego dum conor nostris aperire Latinis

. . . .

Interea nostri reparabant turribus arcem
Pisanam murisque novis atque aggere cives.

The lines are given in full by Valentinelli.

[76] See *Cronichetta di anonimo pisano*, in Corazzini, *L'assedio di Pisa* (Florence, 1885), p. 75; Matteo Palmieri, in Muratori, *Scriptores*, xix. 194; Morelli, *Cronaca*, p. 338.

[77] The text begins (f. 4): 'Omnis doctrina omnisque disciplina intellectiva ex antea existenti efficitur cognitione. Preclarumque hoc est his qui per cuncta aciem mentis intenderint. Quę enim scientiarum sunt mathematicę per huiusmodi modum acquiruntur atque aliarum etiam quęvis artium. . . .'

version which John Argyropoulos dedicated to Cosmo de' Medici.[78] The Boethian translation, however, persisted in early imprints, corrected and touched up in course of time in ways which still require investigation,[79] but still holding its own by reason of its faithfulness to the text of the master whose words were not to be lightly changed.

[78] It begins: 'Omnis doctrina omnisque disciplina intellectiva ex antecedenti cognitione fieri solet. Id si omnis quo fiunt pacto considerabimus manifestum profecto fiet . . .': MS. Vat. Lat. 2116, f. 49 v. For the author's prefaces in MSS. of the Laurentian, see Bandini, *Catalogus codicum Latinorum*, iii. 4, 350.

[79] The humanistic retouching of the text in the Basel edition and in Migne is obvious but cannot be studied until we have a critical restitution of the mediaeval text. It should, however, be kept in mind that the text of these editions is not, as Grabmann thinks (ii. 72), the same as the version of Argyropoulos; see now the study of Minges in *Philosophisches Jahrbuch*, xxix. 250–263 (1916); and cf. Geyer *ibid.*, xxx. 25–27.

CHAPTER XII

SCIENCE AT THE COURT OF THE EMPEROR FREDERICK II[1]

THE Emperor Frederick II is a subject of perennial interest to the historian. The riddle of his many-sided personality, his place at the centre of one of the great struggles of European politics, the striking anticipation of more modern ideas and practices in his administration, the brilliant and precocious culture of his Sicilian kingdom, have attracted the attention of two generations of scholars without definitive results. We still lack a satisfactory biography and a survey of the governmental system, as well as annals for the later years of the reign, while for its intellectual history nothing has superseded what was written by Amari[2] and Huillard-Bréholles[3] more than half a century ago. As regards vernacular literature, the scantiness of the extant material has so circumscribed the problem that we now understand fairly well the importance of the *Magna Curia* as the cradle of Italian poetry and the origin of particular forms like the sonnet.[4] The Latin literature of the South has been partially explored by Hampe and others, though its relations to intellectual movements in northern

[1] Revised from *The American Historical Review*, xxvii. 669–694 (1922); cf. *Rivista storica italiana*, 1923, pp. 165 ff. The best sketch of Frederick is that of Karl Hampe, "Kaiser Friedrich II," in *Historische Zeitschrift*, lxxxiii. 1–42 (1899). The newer materials for the study of the reign are noted in his *Deutsche Kaisergeschichte* (Leipzig, 1919), pp. 219 ff.; and his *Mittelalterliche Geschichte* (Gotha, 1922), pp. 84 ff. E. Winkelmann's fundamental annals, *Kaiser Friedrich II.* (Leipzig, 1889–97), stop with 1233.

[2] *Storia dei Musulmani di Sicilia* (Florence, 1854–72), iii. 655 ff.

[3] *Historia diplomatica Friderici Secundi* (Paris, 1859–61), introduction, especially pp. dxix–dlv.

[4] See particularly E. F. Langley, "The Extant Repertory of the Sicilian Poets," in *Publications of the Modern Language Association of America*, xxviii. 454–520 (1913); and the important studies of Ernest H. Wilkins on the origin of the *canzone* and the sonnet, *Modern Philology*, xii. 135–166, xiii. 79–110 (1915). For Frederick's relations with Provençal poets, see the studies of de Bartholomaeis, in *Memorie* of the Bologna Academy, i. 69–124 (1911–12); and Bertoni, *I trovatori d'Italia* (Modena, 1915), pp. 25–27.

Italy and elsewhere require further inquiry.[5]   On the scientific side, while much remains to be done with the fragmentary materials, investigation has advanced to a point where it may be worth while to supplement and correct the older writers by a general survey of the present state of our knowledge. If the results do not greatly enlarge our acquaintance with the content of thirteenth-century science, they at least illustrate more fully its methods and the workings of one of the most remarkable minds of the Middle Ages.

The intellectual life of Frederick's court cannot be regarded as an isolated or merely personal phenomenon. Lying between the Middle Ages and the Renaissance, it must be seen against the cosmopolitan background of Norman Sicily, the meeting-point of Greek, Arabic, and Latin culture, central in the history as in the geography of the Mediterranean lands. Frederick was not the first but the second of the "two baptized sultans" on the Sicilian throne,[6] and in intellectual matters as in legislation he followed in the direction of his grandfather Roger. King Roger's chief scientific interest was geography, pursued assiduously throughout the fifteen years of his reign. Finding the Arabian geographies and translations insufficient for his purpose, he called to his court famous travellers from many lands and subjected them to a close examination, accepting only the facts on which they were agreed, and recording the results upon a great silver map and in a volume of descriptive text in Arabic which Edrisi completed in 1154.[7] This method is not unlike that followed by Frederick in consulting experts on falconry, among whom he cites King Roger's falconer, William, who passed as one of the earliest writers on this subject.[8] Under Roger's immediate successors, William I and

---

[5] This is the freshest part of the notable article of the late H. Niese, "Zur Geschichte des geistigen Lebens am Hofe Kaiser Friedrichs II," in *Historische Zeitschrift,* cviii. 473–540 (1912). There are noteworthy essays by F. Novati in his *Freschi e minii del dugento* (Milan, 1908), especially pp. 103–142.

[6] The phrase is Amari's, *Musulmani,* iii. 365.

[7] *L'Italia descritta nel "Libro del Re Ruggero,"* translated by Amari and Schiaparelli (Rome, 1883), pp. 4–8; Edrisi, translated by Reinaud (Paris, 1836), i, pp. xviii–xxii; *Encyclopaedia of Islam,* ii. 451. Pardi has recently argued that the final form of the work must be subsequent to 1154: *Rivista geografica italiana,* xxiv. 380 (1917).   [8] *Infra,* Chapters XIV, XVII.

William II, scientific activity took the form particularly of the translation of Greek works on mathematics and astronomy: the *Data, Optica*, and *Catoptrica* of Euclid, the *Pneumatica* of Hero of Alexandria, the *De motu* of Proclus, even the *Almagest* of Ptolemy. Scientific observation, fed by the *Meteorology* of Aristotle, concerned itself with the phenomena of Etna.[9] At the same time Ptolemy's *Optics* was translated from the Arabic, and the household of William II, as portrayed in the scenes of his death, comprised an Arab physician and an Arab astrologer.[10]

At the court of Frederick II the Greek element is of little significance. Greek versions of his laws were issued, and Italian poets sang his praises in Greek verse, but the influence of Byzantium had declined with the fall of the Greek empire, and we hear little of Greek scholars or Greek translations in this period in the South.[11] On the other hand, Arabic influence was, if anything, stronger under Frederick, especially after his visit to the East, and was maintained by the political and commercial relations with Mohammedan countries, while his imperial interests fostered intercourse with northern Italy, Germany, and Provence. The chronicler who passes by the name of Nicholas of Iamsilla tells us that at Frederick's accession there were few or no scholars in the Sicilian kingdom, and that it was one of his principal tasks by means of liberal rewards to attract masters from various parts of the earth.[12] What scholars were thus drawn to the Sicilian court we know but imperfectly. The loss of the imperial registers, save for a fragment of 1239–40,[13] makes it impossible to reconstruct in detail the organization and personnel of the household, and the scattered documents of the reign tell us almost nothing of the men who aided the emperor in his scientific inquiries. That they were chiefly officials of the *curia* seems alto-

[9] Supra, Chapter IX.

[10] Petrus de Ebulo, *Liber ad honorem Augusti*, plate 3.

[11] Krumbacher, pp. 769 f.; Niese, in *Historische Zeitschrift*, cviii. 490 ff.; cf. Bresslau, *Urkundenlehre* (1915), ii. 380 ff. Further investigation is needed respecting Greek in the South in the thirteenth century.

[12] Muratori, viii. 496.

[13] On which see the recent studies of Niese, in *Archiv für Urkundenforschung*, v. 1–20 (1913); and Sthamer, in Berlin *Sitzungsberichte*, 1920, pp. 584 ff.

gether likely. Several of the Sicilian school of poets held official positions as notaries, judges, or falconers,[14] and we are not surprised to find Frederick's astrologer, Theodore, engaged in the same year in casting horoscopes, going on missions, making confectionery, drafting letters, and translating an Arabic work on falconry. In this busy court science, like literature, would seem to have been a matter for leisure hours, and its votaries could be no narrow specialists.

Two of Frederick's courtiers seem to have borne the official title of 'philosopher,' and in an age when philosophy and science were inseparable these two were naturally the chief advisers of the emperor in scientific matters. The more famous of them, Michael Scot,[15] who hailed originally from Scotland, came to Sicily with a reputation gained chiefly in the schools of Spain. Appearing at Toledo as early as 1217, Michael there distinguished himself by translating al-Bitrogi *On the Sphere* and Aristotle *On Animals*, as well as the *De caelo* and the *De anima* with the commentaries of Averroës thereon. By 1220 he is in Italy, and from 1224 to 1227 he enjoys the favor of the pope and the grant of benefices in England and Scotland; but soon thereafter he is found in the emperor's service, in which, though not mentioned in any surviving official documents, he remained until his death, which occurred before 1236. His official position was that of court astrologer, but he made for the emperor a Latin summary of Avicenna's *De animalibus* and busied himself with a series of writings on astrology, meteorology, and physiognomy, all dedicated to Frederick. These show acquaintance with medicine, music, and alchemy, as well as with the Aristotelian philosophy in general. We are told that he knew Hebrew as well as Arabic, but his linguistic attainments are the occasion of unfavorable comment on the part of Roger Bacon. Scot had a respectable knowledge of the Arabian astronomy and its applications, and prided himself on the accuracy of his observations and calculations. His faith in astrology does not, in his age, militate against his stand-

---

[14] See Langley's list in *Publications of the Modern Language Association*, xxviii. 468 ff., and the references there cited, especially the researches of Scandone in *Studî di letteratura italiana*, v, vi.

[15] See the following chapter.

ing as a scientist, but his own writings show him to have been pretentious and boastful, with no clear sense of the limits of his knowledge, and with a tendency to overstep the line, if line there be, between astrology and necromancy. At the same time he had an experimental habit of mind, and a final judgment as to his scientific attainments must await the more careful sifting of his extensive treatises on astrology, the *Liber introductorius* and the *Liber particularis*.

If Michael Scot represented the learning of Moorish Spain and Western Christendom, Master Theodore 'the philosopher' seems to have maintained relations particularly with the East.[16] Greek, or perhaps Jewish,[17] by name, he is said to have been sent to Frederick by the Great Caliph, probably the sultan of Egypt, some time before 1236.[18] If we may believe the prologue to the French romance of *Sidrach*, Theodore, here called "Todre li phylosophes," came from Antioch and remained in relations with its Latin patriarch; while Abulfaragius makes him a Jacobite Christian of Antioch who studied at Mosul and Bagdad and enjoyed the favor of the sultan.[19] In the autumn of 1238, at the siege of Brescia, he appears in the Dominican annals as silencing the friars in philosophical disputes until, challenged to public debate on any subject of philosophy with the doughty Roland of Cremona, he is triumphantly confuted, to the great glory of the

---

[16] See, in general, Amari, *Musulmani*, iii. 692–695; Steinschneider, *E. U.*, no. 116; Sudhoff, in *Archiv für die Geschichte der Medizin*, ix. 1–9 (1915); Suter, in the Erlangen *Abhandlungen zur Geschichte der Naturwissenschaften*, iv. 7 f. (1922).

[17] Renan, in *Histoire littéraire de la France*, xxxi. 290.

[18] 'Explicit liber novem iudicum quem missit soldanus Babilonie imperatori Federico tempore quo et magnus chalif misit magistrum Theodorum eidem imperatori Federico': British Museum, Royal MS. 12 G. VIII; cf. French version in Langlois, *La connaissance de la nature au moyen âge* (1911), p. 191; Amari, iii. 694. The *Liber novem iudicum* is cited by Michael Scot in his *Liber introductorius* (Munich, cod. lat. 10268, f. 128), and must thus have reached Sicily before 1236. The phrase 'magnus chalif' does not strengthen our faith in this colophon.

The references to Theodore in the writings of Leonard of Pisa may well be earlier, but the answers to Theodore's questions look like later additions to the original text of Leonard's *Flos* and *Liber quadratorum*, so that they cannot be dated with certainty.

[19] H. L. D. Ward, *Catalogue of Romances in the British Museum*, i. 904 ff.; *Histoire littéraire*, xxxi. 288–290; Langlois, p. 204; Erlangen *Abhandlungen*, iv. 8.

order.[20]   Probably succeeding Scot as court astrologer, Theodore
casts the imperial horoscope at Padua in 1239, where he is ridi-
culed by the local chronicler for seeking a favorable conjunction
impossible at the time and failing to search in Scorpio for the im-
pending failure of the expedition.[21]   In the register of 1239–40 he
is found drafting the emperor's Arabic letters to the king of Tunis
and acting as his trusty messenger.   In this same year he is busy
compounding syrups and sugar of violet for the emperor and his
household, with free credit in money and costly sugar for this
purpose, and a box of the violet sugar is sent to Piero della Vigna
during his recovery from an illness.[22]   In 1240–41 the emperor
corrects his translation from the Arabic.[23]   No further dates are
known in Theodore's career, but he continued to enjoy imperial
favor until his death not long before November, 1250, when
Frederick regranted the extensive domains which "the late Theodore our philosopher held so long as he lived." [24]

While the biographical data are somewhat fuller in the case of
Theodore than in that of Michael Scot, the evidence of his literary
activity is much less.   Apart from a doubtful connection with the
transmission of the philosophical romance of *Sidrach*, Theodore
is known only as the author of a treatise on hygiene extracted for
the emperor's benefit from the *Secretum secretorum* of the Pseudo-
Aristotle,[25] and a Latin version of the work of Moamyn on the
care of falcons and dogs.[26]   His preface to this shows acquaintance

[20] Quétif and Échard, *Scriptores Ordinis Praedicatorum*, i. 126, col. 2.   On Roland
of Cremona see now Ehrle, in the anniversary *Miscellanea Dominicana* (Rome,
1923), pp. 85–134, especially p. 94.

[21] Rolandino, in Muratori, viii. 228 (new edition, viii. 66); and in *M. G. H., Scriptores*, xix. 73.

[22] Huillard-Bréholles, *Historia diplomatica*, v. 556, 630, 727, 745, 750 ff.; idem,
Pierre de la Vigne, p. 347.

[23] Infra, Chapter XIV, n. 122.

[24] Original charter published by Schneider in *Quellen und Forschungen aus itali-
enischen Archiven*, xvi. 51 (1913); cf. the inquest of the Angevin period published
by Scandone in *Studi di letteratura italiana*, v. 308 (1903).   Theodore may well have
been one of the astrologers lost in the defeat before Parma in 1248: Hartwig, in
Centralblatt jür Bibliothekswesen, iii. 183.   The account of Thadhûri of Antioch in
Abulfaragius makes him take poison after flight from the emperor: Suter, no. 345;
Z. M. Ph., xxxi, sup., pp. 107 f.; Erlangen *Abhandlungen*, iv. 8.

[25] Ed. Sudhoff, in *Archiv für die Geschichte der Medizin*, ix. 4 (1915).

[26] Chapter XIV, n. 122.

with Aristotle, including the *Ethics* and the *Rhetoric*, such as a court philosopher should have, while he also exhibits medical knowledge. Mathematician as well as astrologer, he puts problems to Leonard of Pisa, and is addressed by him as "the supreme philosopher of the imperial court," whose cosmopolitan culture he well represents.[27]

Another court philosopher, John of Palermo, mentioned by Leonard of Pisa in 1225, is probably identical with the Master John the notary who acts as confidential agent of the emperor in 1240, but we know nothing of his scientific tastes beyond his interest in mathematics.[28] A Master Dominicus, perhaps a Spaniard, appears in the same connection.[29] The Sicilian Moslem who tutored Frederick in logic during his crusade remains anonymous,[30] with many other scholars who must have attended the court. One of these, for example, appears in correspondence on mathematical subjects with a learned Jew of Spain.[31]

The more literary members of the *Magna Curia*, such as Piero della Vigna, are silent respecting their scientific associates, save for such an exchange of compliments and sugar plums as has been cited. The interests of Piero, as of the other members of the Capuan school, were primarily literary, and his letters would not have become models of Latin style for the thirteenth century[32] had he not been first and foremost a phrasemaker who spoke "obscurely and in the grand manner."[33] The extant collections of correspondence which pass under his name were preserved for rhetorical rather than historical purposes, and there was no occasion for retaining in them whatever of the scientific life of the court the originals might have reflected. Nevertheless, some of his phrases suggest its other intellectual interests, as when he borrows the language of the current cosmogony in the preface to

---

[27] *Scritti di Leonardo Pisano*, ed. Boncompagni (Rome, 1857–62), ii. 247, 279.

[28] *Ibid.*, ii. 227, 253; Huillard-Bréholles, ii. 185, v. 726 ff., 745, 928.

[29] Leonardo, *Scritti*, ii. 1, 253; Cantor, ii. 35 ff., 41.

[30] Amari, *Biblioteca Arabo-Sicula*, ii. 254.

[31] Steinschneider, *H. U.*, p. 3.

[32] Critical edition lacking. See Huillard-Bréholles, *Pierre de la Vigne*, pp. 249 ff.; Hanauer, in *Mitteilungen des Instituts für oesterreichische Geschichtsforschung*, xxi. 527–536 (1900).

[33] So Odofredus characterizes him, *Mitteilungen des Instituts*, xxx. 653, n. 1.

the emperor's *Constitutions*,[34] or refers to the preoccupation of
the friars with the form of the globe, the course of the sun in the
zodiac, the squaring of the circle, or the conversion of triangles
into quadrangles.[35] Piero's correspondence with the masters of
Bologna and Naples and the *dictatores* of his native Campania
runs parallel to the scientific correspondence of Frederick and his
philosophers with scholars in Italy and Mohammedan lands.

So far as Italy is concerned, the outstanding scientific genius of
the thirteenth century is undoubtedly the mathematician Leonard of Pisa.[36] Beyond the fact of his African education, and his
"sovereign possession of the whole mathematical knowledge of
his own and every preceding generation," [37] his personal history
is unknown; but though he resided at Pisa, he was well known to
Frederick and the philosophers of his court, to whom his extant
works are in large measure dedicated. It is Michael Scot who in
1228 receives from Leonard's hands the revised edition of his
epoch-making treatise on the *Abacus*, first issued in 1202.[38] Already Master John of Palermo had accompanied Leonard into
the emperor's presence and proposed questions involving quad-
ratic and cubic equations, the answers to which are found in the
Flos and *Liber quadratorum*.[39] Like the solutions of various problems submitted to Leonard by Master Theodore, these are designed to illustrate method rather than to form a systematic
treatise. The *Liber quadratorum* is directed to the emperor, who
has himself deigned to read the treatise on the *Abacus* and to hear
the discussion of subtle problems of arithmetic and geometry,
such as those once propounded in his presence by Master John.[40]

[34] Niese, in *Historische Zeitschrift*, cviii. 501, 523. Those who doubt Piero's
authorship of the original constitutions admit his influence on their style as we have
them: e. g., Garufi, in *Studt medioevali*, ii. 105, note.

[35] Poem printed by Huillard-Bréholles, *Pierre de la Vigne*, p. 414.

[36] Cantor, ii, cc. 41, 42; S. Günther, *Geschichte der Mathematik* (Leipzig, 1908),
i, c. 15.

[37] Günther, p. 258.   [38] *Scritti*, i. 1.

[39] *Scritti*, ii. 227–283. The date 1225 which heads the *Liber quadratorum* has
perplexed historians, since Frederick first visited Pisa in the following year. Eneström has tried to reconcile the difficulties by placing the first meeting elsewhere:
B. M., ix. 72 (1908).

[40] *Scritti*, ii. 253.

Relations with other scholars of northern Italy seem to have concerned chiefly matters of law or literature, as Niese has well brought out,[41] but we should not overlook the treatise on the hygiene of a crusading army dedicated to Frederick by Adam, chanter of Cremona, in 1227 and recently brought to light by Sudhoff.[42]

It is characteristic of Frederick's strongly personal policy that the intellectual life of his kingdom centres in his court rather than in universities, and that the southern universities in his reign show little vigor of life and leadership. His absolute and paternal ideas of government left no place for independent corporations of masters and students living the free and turbulent life of the northern *studia*. So Salerno, which had grown to eminence as a school of medicine without the aid of prince or pope, found itself tied down by royal statute in 1231 as part of a comprehensive regulation of the practice of medicine, surgery, and pharmacy throughout the kingdom of Sicily, issued in the interests of bureaucratic administration rather than of university development. The course of study is laid down by law, and royal officers are to be present at the examinations.[43] A similar bureaucratic purpose runs through the statutes establishing the University of Naples in 1224 and reforming it in 1234 and 1239. Frederick needed trained public servants, and he preferred to have them brought up in his own kingdom rather than in Bologna and other Guelfic cities of the North. Although the new university was to comprise all the fields of study then current, its strength lay in law and rhetorical composition, and it is no accident that the masters whose names have reached us are chiefly jurists and grammarians, closely connected with the judges and clerks of the royal *curia*.[44]

[41] *Historische Zeitschrift*, cviii. 513 ff.

[42] F. Hönger, *Aertzliche Verhaltungsmassregeln auf dem Heerzug ins heilige Land für Kaiser Friedrich II. geschrieben von Adam von Cremona* (Leipzig diss., 1913).

[43] Constitutions in Huillard-Bréholles, iv. 150 ff., 235; Greek text, ed. Sudhoff, in *Mitteilungen zur Geschichte der Medizin*, xiii. 180 (1914). See Rashdall, *Universities*, i. 83 ff.; and the commentary of A. Bäumer, *Die Aertztegesetzgebung Kaiser Friedrichs II* (Leipzig, 1911).

[44] See the principal documents concerning the beginnings of the university in Huillard-Bréholles, ii. 450, iv. 497, v. 493–496; and the discussion in Denifle, *Die Universitäten*, i. 452–456. A much-needed study of its early history is promised by

Nevertheless we read of a professor of natural philosophy, Master Arnold the Catalan, who taught the courses of the stars and the nature of the elements but was unable to predict his own sudden death, which occurred "as he was lecturing on the soul," very likely in the midst of a commentary on the *De anima* of Aristotle.[45] No less a person than Thomas Aquinas began his study of natural philosophy at Naples, under an Irish master, one Petrus de Hibernia, who is later found holding a disputation at King Manfred's court.[46]

Frederick's patronage of learning was not limited to Christian scholars. The Jewish translator of the logical commentary of Averroës and Ptolemy's *Almagest*, Jacob Anatoli, praises this "friend of wisdom and its votaries" for pecuniary support, and even hopes the Messiah may come in this reign; his versions into Hebrew, begun in Provence, were continued at Naples in 1232 and brought him into relations with Michael Scot as well as the emperor.[47] A Spanish Jew, the encyclopedist Jehuda ben Solomon

E. Sthamer. Two masters connected with the university in this period are the subjects of recent monographs: G. Ferretti, "Roffredo Epifanio da Benevento," in *Studi medioevali*, iii. 230–275 (1909); and F. Torraca, "Maestro Terrisio di Atina," in *Archivio storico napoletano*, xxxvi. 231–253 (1911). Another professor of grammar, Walter of Ascoli, has left an etymological cyclopaedia entitled *Dedignomion*, or *Summa derivationum*, or *Speculum artis grammatice*, based on Isidore and Hugutio. I have used MS. 449 at Laon and MS. Vat. lat. 1500 of the Vatican, both ca. 1300; there is a later copy at the University of Bologna, MS. 1515 (2832). The Laon manuscript was ascribed to Walter, archbishop of Palermo in the twelfth century (*Catalogue*, p. 238), but 'Gualterius Hesculanus' appears clearly in the preface, and a further sentence printed by Morelli, *Codices MSS. Latini bibliothecae Nanianae* (Venice, 1726), p. 160, states that the book was begun at Bologna in 1229 and afterward completed at Naples. Walter is probably the 'Magister G[ualterius] grammaticus,' professor at Naples, whose death is lamented in a letter of Piero della Vigna (*Epp.*, iv, no. 8; Huillard-Bréholles, *Pierre de la Vigne*, p. 394). In the Laon MS. the *Dedignomion* is followed by the notes of another southern grammarian, Anellus de Gaieta.

[45] See the letter of condolence of Master Terrisio, published by Paolucci in the *Atti* of the Palermo Academy, iv. 44 (1896); and by Torraca in the article just cited, p. 247.

[46] Denifle, *Universitäten*, i. 456 ff.; Baeumker, "Petrus de Hibernia," in Munich *Sitzungsberichte*, 1920; Grabmann, in *Philosophisches Jahrbuch*, xxxiii. 347–362 (1920); infra, n. 138.

[47] Renan, in *Histoire littéraire*, xxvii. 580–589; Steinschneider, *H. U.*, pp. 58–61, 523; Huillard-Bréholles, iv. 382, n.

Cohen, was in correspondence with one of the court philosophers at the age of eighteen, coming later to Italy, where he met the emperor and is found in Tuscany in 1247.[48] Through these or others Frederick had some knowledge of Maimonides, whose *Guide for the Perplexed* seems to have been translated into Latin in southern Italy in this period.[49]

Whether eminent Mohammedan scholars actually resided at Frederick's court, is a question which cannot be answered from the information at our disposal. His colony of Saracens at Lucera [50] and his well known tolerance of the infidel combined with the environment of his youth and his semi-oriental habits of life to spread stories that he preferred to surround himself with Moslem rather than Christian influences, in learning as in everything else.[51] That he was friendly to the learning of Islam appears from the various questionnaires which, as we shall see, he sent out to Mohammedan rulers, partly as puzzles, partly in a real search for knowledge. His crusade led to political and commercial relations with the sultan of Egypt which lasted throughout his reign, while the commercial treaty of 1231 with the ruler of Tunis was followed by the establishment of a Sicilian consulate at Tunis and a series of diplomatic missions of various sorts.[52] Such missions were regularly the occasion of an exchange of presents, and it was well understood that the emperor valued a book,

[48] Steinschneider, *H. U.*, pp. 1–3, 164, 507; idem, *Verzeichniss der hebräischen Handschriften der königlichen Bibliothek zu Berlin*, ii. 121–126; and in *Z. M. Ph.*, xxxi, 2, pp. 106 ff. On Jewish culture under Frederick, see M. Güdemann, *Geschichte des Erziehungswesens der Juden in Italien* (Vienna, 1884), pp. 101–107, 268 ff.; R. Straus, *Die Juden im Königreich Sizilien* (Heidelberg, 1910), pp. 79–91.

[49] Amari, iii. 705 ff.; Steinschneider, in *Hebräische Bibliographie*, vii. 62–66 (1864); idem, *H. U.*, p. 433; infra, Chapter XIII, n. 63.

[50] On which see now Egidi, in *Archivio storico napoletano*, xxxvi–xxxix.

[51] Current views of Frederick's relations with the Saracen world are illustrated by Matthew Paris, *Chronica majora*, iii. 520; iv. 268, 526, 567 ff., 635; v. 60 ff., 217.

[52] See, in general, Amari, *Musulmani*, iii. 621–655; A. Schaube, *Handelsgeschichte der romanischen Völker*, pp. 185, 302–304; Huillard-Bréholles, introduction, ch. 5; Mas Latrie, *Traités de paix avec les Arabes de l'Afrique septentrionale*, introduction, pp. 82 ff., 122–124; Blochet, "Les relations diplomatiques des Hohenstaufen avec les Sultans d'Égypte," in *Revue historique*, lxxx. 51–64 (1902); and, under the several Mohammedan rulers, the indexes to the *Regesta Imperii* and Winkelmann, *Kaiser Friedrich II.*

a rare bird, or a cunning piece of workmanship more highly than mere objects of luxury. Thus in 1232 al-Ashraf, sultan of Damascus, sent him a wonderful *planetarium*, with figures of the sun and moon marking the hours on their appointed rounds; valued at 20,000 marks, this was kept with the royal treasure at Venosa.[53] Frederick gave in return a white bear and a white peacock which astonished the Oriental chroniclers, much as their western contemporaries were impressed by "the marvellous beasts, such as the West had not seen or known," which Frederick had earlier received from Egypt.[54]

At the end of a series of such costly exchanges, Frederick, his treasury exhausted, propounded to the sultan problems of mathematics and philosophy, the solutions of which, due to a famous scholar of Egypt,[55] came back in the sultan's own hand. While in the East Frederick asked an interview with some one learned in astronomy, and in response Sultan Malik al-Kamil sent him a most learned astronomer and mathematician surnamed al-Hanifi.[56] It will be recalled that Theodore the philosopher is said to have been first sent to the emperor by the 'caliph,' and it is he who drafts the Arabic letters to the ruler of Tunis.[57] There can be no doubt of the impression which Frederick made on the scholars of the East as one well versed in philosophy, mathematics, and the natural sciences in general;[58] but such reports, transmitted through later Arabic compilers, are too vague to throw much light on his relation to specific fields of science.

The list of scholars with whom Frederick was in contact fades

---

[53] *Chronica Regia Coloniensis* (ed. Waitz, 1880), p. 263; Huillard-Bréholles, iv. 369; cf. Winkelmann, *Kaiser Friedrich II.*, ii. 399 ff.; Wiedemann, in *Archiv für Kulturgeschichte*, xi. 485 (1914).

[54] *M. G. H., Scriptores*, xxviii. 61. Cf. the white Indian psitacus sent by the sultan: *De arte*, i, c. 23.

[55] *Revue historique*, lxxx. 60; infra, note 122.

[56] Tarih Mansuri, in *Archivio storico siciliano*, ix. 119.

[57] See notes 18, 22, above.

[58] See the passages cited by Röhricht, *Beiträge zur Geschichte der Kreuzzüge* (Berlin, 1874), i. 73 ff.; Winkelmann, *Kaiser Friedrich II*, ii. 137, n. 3. Frederick's fame in the East is further illustrated by the eulogy of Theodore Lascaris: Pappadopoulos, *Théodore II Lascaris* (Paris, 1908), pp. 183–189; Βυζαντίς, ii. 404–413 (1912).

into a penumbra of mythical attributions and romantic tales, interesting at least as showing the reputation which the emperor and his court acquired in the field of learning and literature.[59] Thus *Le régime du corps* of Aldebrandino of Siena, written in 1256 for Countess Beatrice of Provence, appears in certain later manuscripts as translated in 1234 "from Greek into Latin and from Latin into French" at the request of "Frederick formerly emperor of Rome." [60] The famous letter of Prester John concerning the marvels of the East, which in the Latin original is sent to the Greek emperor Manuel, is in its French form addressed to "Fedri l'empereour de Rome," [61] as the mythical account of Alexander's conquests in Central Asia is directed to his philosopher Theodore.[62] The French prophecies of Merlin profess to have been compiled at the desire of Frederick and then turned into Arabic as a present to the Sultan of Egypt,[63] while the romance of *Sidrach* purports to have been brought from Tunis for Frederick and turned into Latin by Friar Roger of Palermo.[64] A medical treatise is said to have been translated for the emperor in 1212 with the aid of Gerard of Cremona, who died twenty-five years earlier.[65]

The nature of the scientific interests of Frederick's court has by this time become in some measure apparent. For one thing, he was deeply interested in all kinds of animals, collecting a menagerie which followed him about Italy and even into Germany. In November, 1231, he came to Ravenna "with many animals unknown to Italy: elephants, dromedaries, camels, panthers, gerfalcons, lions, leopards, white falcons, and bearded

[59] Cf. Langlois, *La connaissance de la nature au moyen âge*, p. 191.

[60] *Le régime du corps de Maître Aldebrandin de Sienne*, ed. L. Landouzy and R. Pépin (Paris, 1911), pp. xxxii, lv.

[61] See, for the Latin text, the various studies of F. Zarncke; and, for the French version, Ruteboeuf, ed. Jubinal (1875), iii. 355; P. Meyer, in *Romania*, xv. 177, xxxix. 271. The reference may be to Frederick Barbarossa: R. Köhler, *Romania*, v. 76; supra, Chapter X, n. 173. On Frederick II and Prester John see the *Cento Novelle Antiche*, no. 1.

[62] Sudhoff, in *Archiv für die Geschichte der Medizin*, ix. 9; Steinschneider, in *Hebräische Bibliographie*, viii. 41.

[63] H. L. D. Ward, *Catalogue of Romances in the British Museum*, i. 371 ff., 905.

[64] *Ibid.*, i. 904; *Histoire littéraire*, xxxi. 288; Langlois, p. 204.

[65] Steinschneider, *H. U.*, p. 793.

owls." [66] Five years later a similar procession passed through
Parma, to the delight of a boy of fifteen later known as Salimbene.[67] The elephant, a present from the sultan, stayed in
Ghibelline Cremona, where he was put through his paces for the
Earl of Cornwall [68] and died thirteen years later "full of humors,"
amid the popular expectation that his bones would ultimately
turn into ivory.[69] In 1245 the monks of Santo Zeno at Verona, in
extending their hospitality to the emperor, had to entertain with
him an elephant, five leopards, and twenty-four camels.[70] The
camels were used for transport and were even taken over the
Alps, with monkeys and leopards, to the wonder of the un-
travelled Germans.[71] Another marvel of the collection was a
giraffe from the sultan, the first to appear in mediaeval Europe.[72]
Throughout runs the motif of ivory, apes, and peacocks from
the East, as old as Nineveh and Tyre and as new as the modern
'Zoo,' with the touch of the thirteenth century seen in the elephant which Matthew Paris thought rare enough to preserve
in a special drawing in his history,[73] and the lion which Villard
de Honnecourt saw on his travels and carefully labelled in his
sketchbook, "drawn from life"! [74]

Frederick's menagerie illustrates various sides of his nature —
his delight in magnificence and display, his fondness for the unusual and the exotic, his joy in hunting, for which he used cours-
ing leopards [75] and panthers as well as hawks and falcons and the

[66] Scheffer-Boichorst, *Zur Geschichte des XII. und XIII. Jahrhunderts* (Berlin,
1897), pp. 282, 286.

[67] *Cronica*, ed. Holder-Egger, pp. 92 ff.

[68] Matthew Paris, *Chronica majora*, iv. 166 ff.

[69] *Chronicon Placentinum*, ed. Huillard-Bréholles (Paris, 1856), p. 215.

[70] *Nuovo archivio veneto*, vi. 129.

[71] Annals of Colmar, *M. G. H., Scriptores*, xvii. 189; Böhmer-Ficker, nos. 2098 a,
2973, 3475 a.

[72] Albertus Magnus, *De animalibus*, ed. Stadler, p. 1417; Michaud, *Bibliothèque
des Croisades*, iv. 436.

[73] *Chronica majora*, iv. 166, v. 489.

[74] "Et bien sacies que cis lions fu contrefais al vif." *Album de Villard de Honnecourt*, plates 47, 48; cf. 52, 53 (facsimile edition published by the Bibliothèque
Nationale).

[75] Böhmer-Ficker, nos. 2661, 2783, 2883, 3029. Cf. the three leopards sent to
Henry III: Matthew Paris, *M. G. H., Scriptores*, xxviii. 131, 407, 409.

humbler companions of the chase — but it also fed a genuine scientific interest in animals and their habits. His *De arte venandi cum avibus*, of which more will be said below, not only deals comprehensively with all the practical phases of the art, but begins with a systematic and careful discussion of the species, structure, and habits of birds, for which the author utilizes the *De animalibus* of Aristotle, such previous treatises as he could find on the subject, and the results of his own observation and inquiry.[76] A similar interest appears in the case of horses, to whose breeding the emperor gave special attention and concerning whose diseases he ordered one of his marshals, the Calabrian knight Giordano Ruffo, to prepare under imperial supervision a treatise, which was not completed until after Frederick's death. The first western manual of the veterinary art, this was widely popular, especially in Italy, being translated into many languages and imitated by the writers of the next generation.[77] Frederick's reputation as a hunter, if not his personal inspiration to authorship, may also be seen in the little treatise on hunting of a certain Guicennas, "master in every kind of hunting by the testimony of the hunters of Lord Frederick, emperor of the Romans." [78]

The medical interests of the court are well attested, though they are not known to have produced notable additions to medi-

[76] Infra, Chapter XIV.

[77] Edited by Molin (Padua, 1818). For manuscripts and translations, see L. Moulé, *Histoire de la médecine vétérinaire* (Paris, 1898), ii. 25–30, where some account will be found of the later Italian treatises. There are four copies at Naples, MSS. viii. D. 65–67 *bis*. See further Huillard-Bréholles, introduction, p. dxxxvi; *Romania*, xxiii. 350, xl. 353; Steinschneider, *H. U.*, p. 985. This author is probably the Jordanus de Calabria who was made castellan of Ceseno in 1239 (Richard of San Germano, *ad annum*).

[78] '*Incipit liber Guicennatis de arte bersandi*. Si quis scire desideret de arte bersandi, in hoc tractatu cognoscere poterit magistratum. Huius autem artis liber vocatur Guicennas et rationabiliter vocatur Guicennas nomine cuiusdam militis Teotonici qui appellabatur Guicennas qui huius artis et libri prebuit materiam. Iste vero dominus Guicennas Teotonicus fuit magister in omni venatione et insuper summus omnium venatorum et specialiter in arte bersandi, sicut testificabantur magni barones et principes de Allemannia et maxime venatores excellentis viri domini Frederici Romanorum imperatoris. . . .' Vatican, MS. Vat. lat. 5366, ff. 75 v–78 v (ca. 1300); MS. Reg. lat. 1227, ff. 66 v–70) (fifteenth century). Guicennas, who is cited by writers on falconry, is identified with Avicenna by Werth but without any reasons given (*Zeitschrift für romanische Philologie*, xiii. 10).

cal knowledge. Thus Pietro da Eboli, early in the reign, dedicated to Frederick his poem on the baths of Pozzuoli,[79] whose
healing qualities the emperor was to put to proof after his illness
in 1227.[80] The treatise of Adam of Cremona on the hygiene of
the crusading army has already been mentioned, as has also the
series of hygienic precepts formulated for the emperor by Master
Theodore,[81] while a similar treatise purports to be dedicated to
Frederick by his 'alumnus,' Petrus Hispanus, who claims Theodore as his master. Frederick seems to have shown some anxiety
concerning paralysis, and a marvellous powder was current in
his name, efficacious against many "chronic ailments of the
head and the stomach." [82] An incantation for the healing of
wounds was also ascribed to him.[83] Frederick gave careful attention to personal hygiene in such matters as blood-letting,[84] diet,
and bathing; indeed his Sunday bath was a cause of much scan-
dal to good Christians.[85] One is reminded of the slander on the
Middle Ages as a thousand years without a bath!

Without astrologers Frederick's court would not have been an
Italian court of the thirteenth century, when even the universities had their professors of astrology.[86] Guido of Montefeltro
kept in his employ one of the most distinguished and successful

[79] For a discussion of the questions concerning this poem, see Ries, in *Mitteilungen des Instituts für oesterreichische Geschichtsforschung*, xxxii. 576–593 (1911),
and the works there cited.

[80] Winkelmann, i. 333.

[81] See notes 25 and 42, above, and for Petrus Hispanus, Harleian MS. 5218, f. 1;
P. Pansier, *Collectio ophtalmologica* (Paris, 1908), vi. 108 f.; and Thorndike, ch. 58.
In the Rossi MSS. recently acquired by the Vatican there are (MS. XI. 7) a series
of 953 prescriptions in the name of "Maestro Bene medico dellomperadore Fede-
rigo"; and a *Libro de consegli de poveri infermi* ascribed to Michael Scot (MS.
XI. 144).

[82] Ed. Sudhoff, in *Archiv für die Geschichte der Medizin*, ix. 6, note. Cf. the
'pills of King Roger,' Worcester cathedral, MS. Q. 60, f. 88 v (*Catalogue of MSS.*,
p. 141).

[83] Huillard-Bréholles, introduction, p. dxxxviii.

[84] Chapter XIII, n. 108.

[85] John of Winterthur, ed. Wyss (Zurich, 1856), p. 8.

[86] Cf. T. O. Wedel, "The Mediaeval Attitude toward Astrology," *Yale Studies
in English*, lx, ch. 5; Novati, *Freschi e minii*, pp. 129–134; Thorndike, ii, especially ch. 67. Gerard of Sabionetta has left a register of his consultations, 1256–
60: B. Boncompagni, in *Atti dell' Accademia Pontificia*, iv. 458 ff. (1851).

of mediaeval astrologers, Guido Bonatti, who is said to have directed his master's military expeditions from a campanile with the precision of a fire alarm: first bell, to arms; second, to horse; third, off to battle.[87]   Ezzelino da Romano also had Bonatti among his many astrologers, along with Master Salio, canon of Padua, Riprandino of Verona, and "a long-bearded Saracen named Paul, who came from Baldach on the confines of the far East, and by his origin, appearance, and actions deserved the name of a second Balaam." [88]   There is no certain evidence that Guido Bonatti resided at Frederick's court, but he tells us that he discovered the conspiracy of 1246 by the stars at Forlì and sent timely word to the emperor at Grosseto.[89] Of the emperor's astrologers we know by name only Michael Scot and Theodore, but his enemies exulted over the troop of astrologers and magicians which this devotee of Beelzebub, Ashtaroth, and other demons lost in the great defeat before Parma.[90]   It is plain that much reliance was placed on such advice, even in quite personal matters.[91]   Scot prided himself on his successful predictions of campaigns and the avoidance of unfavorable seasons; [92] another astrologer guided the emperor through a breach in the wall at Vicenza in 1236; [93] and Theodore stood on the tower of Padua in 1239 seeking a fortunate conjunction for an expedition which was ultimately turned back by an eclipse.[94]   Indeed the story ran that Frederick avoided Florence because of an astrologer's prediction,

---

[87] Boncompagni, *Della vita e delle opere di Guido Bonatti* (Rome, 1851), pp. 6 ff.; cf. Thorndike, ii. 825–835.

[88] Boncompagni, *op. cit.*, pp. 29–32; Muratori, viii. 344, 705, xiv. 930.   On Salio, see Steinschneider, *E. U.*, no. 107; Thorndike, ii. 221.

[89] Boncompagni, *Guido*, p. 24; Guido Bonatti, *Decem libri de astronomia*, tractatus iv, cons. 58. I have used the Venice edition of 1506 in the Boston Public Library. The *Census of Fifteenth Century Books owned in America* seems to be in error in listing the Augsburg edition of 1491 (Hain, 3461*), as at Brown University. On the conspiracy of 1246, see Böhmer-Ficker, no. 3547 a.

[90] Albert of Behaim, ed. Höfler, pp. 126, 128. On Frederick's devotion to astrology, see also Saba Malaspina, in Muratori, viii. 788.

[91] Matthew Paris, in *M. G. H., Scriptores*, xxviii. 131; cf. Scot's *Physiognomy*.

[92] Infra, Chapter XIII, nn. 107, 108. Cf. Salimbene, ed. Holder-Egger, pp. 353, 360, 512, 530; *Forschungen zur deutschen Geschichte*, xviii. 486.

[93] Antonio Godi, in Muratori, viii. 83.

[94] Muratori, viii. 228 ff.

and recognized when it was too late that the obscure Fiorentino would be the scene of his death.[95]   The literary output of the *Magna Curia* in this field is represented by Scot's three treatises, the *Physiognomy*, *Liber introductorius*, and *Liber particularis*, all dedicated to the emperor, the *Physiognomy* being designed to aid him directly in his judgment of men.   Indeed Scot speaks of 'the new astrology' as proudly as writers now speak of the new chemistry or the new history.[96]

With astrology there naturally went a considerable amount of astronomy, for astrology is only applied astronomy, wrongly applied as we now believe, but a thoroughly practical subject in the eyes of the later Middle Ages.   The works of Michael Scot show familiarity with Ptolemy and the principal Arabic writers on astronomy, already translated in the twelfth century; and the Hebrew versions of Ptolemy and his abbreviators by Jacob Anatoli are further evidence of attention to this science.   The mathematical interests of the court reach their highest expression in the relations with Leonard of Pisa, in which, it will be remembered, the emperor himself took an active part.   Frederick's own work shows an acquaintance with the fundamentals of geometry, and while in the East he sought out the company of mathematicians and astronomers.[97]   His castles show much interest in architecture, the towers at Capua being designed with his own hand; [98] indeed we are told that he was "skilled in all mechanical arts to which he gave himself." [99]   No direct contributions to mathematical literature have, however, been connected with the Sicilian court.

To what extent studies in alchemy were pursued at Frederick's court, it is impossible to say with our present loose knowledge of the alchemical literature of the thirteenth century.   The alchemical treatises ascribed to Michael Scot are uncertain enough, as we shall see in the next chapter, and the attribution of others

[95] Muratori, viii. 788.
[96] 'Qui vero hos duos libros plene noverit ac sciverit operari nomen novi astrologi optinebit': *Liber particularis*, Bodleian, MS. Canon. Misc. 555, f. 1 v.
[97] Chapter XIV, n. 107; *Archivio storico siciliano*, ix. 119.
[98] Richard of San Germano, *M. G. H., Scriptores*, xix. 372.
[99] Muratori, ix. 132, 661.

to Friar Elias may be entirely mythical;[100] yet there seems enough basis of fact in the case of Scot's writings to indicate some activity in this direction.

The philosophical interests of the court were strongly marked. Frederick was well trained in logic, even taking a master of dialectic with him on the crusade, and his *De arte* shows familiarity with scholastic terminology and classification. His mind, however, was in no sense formal but actively questioning, and the range of his inquiries touched far-reaching problems of the universe and the human soul, as we shall see from his questionnaires. The doctrines of Averroës were well known and often discussed at his court, so that Mohammedan writers considered him no Christian at heart;[101] and many European contemporaries shook their heads over the current stories of his scepticism and unbelief.[102]

How far the scientific life of Frederick's court was fed by new versions of the works of Aristotle and his commentators, it is not easy to say. By 1215 western Europe knew not only the logical treatises, but the *Metaphysics*, the *Ethics*, and the principal writings on natural philosophy. New versions, often with the commentaries of Averroës and Avicenna, continued to appear in the course of the thirteenth century, but few of these can be specifically connected with Sicily.[103] Roger Bacon, it is true, speaks of the appearance of Michael Scot ca. 1230, bearing "certain parts of the natural philosophy and metaphysics with the authentic commentaries," as constituting a turning-point in Aristotelian

---

[100] Thorndike, ii. 308, 335. The Vatican MS. Reg. lat. 1242, a modern MS. of 11 folios, contains 'Liber patris Rev^{mi} Elie generalis ordinis Minorum ad Fredericum imperatorem.'

[101] Amari, *Biblioteca Arabo-Sicula*, ii. 254; Michaud, *Histoire des croisades*, vii. 810; Röhricht, *Beiträge*, i. 73 ff.

[102] E. g., Matthew Paris, *M. G. H., Scriptores*, xxviii. 147, 230, 416; Salimbene, p. 349.

[103] See, in general, Jourdain; and M. Grabmann, *Forschungen über die lateinischen Aristotelesübersetzungen des XIII. Jahrhunderts* (Münster, 1916). For the *Logic*, see Chapter XI, supra; for the *Ethics*, A. Pelzer, "Les versions latines des ouvrages de morale conservés sous le nom d'Aristote," in *Revue néo-scolastique*, xxiii, 316–341, 378–400 (1921); for the *Metaphysics*, Geyer, in *Philosophisches Jahrbuch*, xxx. 392–415 (1917); F. Pelster, in *Festgabe Baeumker*, pp. 89–118 (1923).

studies;[104] but this seems to be one of the occasions when the friar is speaking loosely. The only work of Aristotle first translated by Scot was the *De animalibus*, in a version made before he joined the Sicilian court, and the only new versions of texts already known which are certainly by him are the *De caelo* and *De anima*, with the commentary of Averroës.[105] To these should be added Scot's Latin abbreviation of Avicenna's commentary on the *De animalibus*, which is dedicated to the emperor before 1232,[106] and the Hebrew versions of Averroës's commentary on the *Logic* made by Jacob Anatoli for Frederick in or about that year.[107] At the same time other works of the Stagyrite were freely used at court. Thus Scot quotes the *Ethics* and draws largely on the *Meteorology*,[108] while Theodore the philosopher cites the *Rhetoric* and *Ethics*, as well as the *Secretum secretorum*.[109] The emperor himself, in the *De arte venandi*, draws on the pseudo-Aristotelian *Mechanics* as well as on the *De animalibus*.[110] Nevertheless what was new in all this was Averroës rather than Aristotle, nor can we be certain, as investigation now stands, that the Sicilian school did more than give wider currency to treatises and doctrines of Averroës which had already begun to spread from Spain.

Frederick has been called "an unrestrained admirer of Aristotle," [111] but his own writings are far from bearing this out. We

[104] *Opus majus*, ed. Bridges, i. 55, iii. 66; *M. G. H.*, *Scriptores*, xxviii. 571.

[105] Besides Grabmann, see below, Chapter XIII.

[106] J. Wood Brown, *Michael Scot*, pp. 53 ff., corrected in Chapter XIII. The University of Michigan has a copy of the printed text of this version.

[107] See note 47, above.

[108] Chapter XIII, n. 78; *Revue néo-scolastique*, xxiii. 326, n. 2.

[109] Chapter XIV, n. 124; *Archiv für die Geschichte der Medizin*, ix. 4–8. On the new version of the *Secretum secretorum* attributed to Philip of Tripoli, see Steele, *Opera hactenus inedita Rogeri Baconi*, v, pp. xviii–xxii; and Chapter VII, supra.

[110] Chapter XIV, n. 113.

[111] Biehringer, *Kaiser Friedrich II*. (Berlin, 1912), p. 244. Frederick's devotion to Aristotle has been argued from a letter ascribed to him which transmits new versions of Aristotle's work to some university, but I agree with most recent scholars in assigning this letter to Manfred and connecting it with the translations of the *Magna moralia* and various pseudo-Aristotelian treatises made by his direction. See Jourdain, p. 156, with French translation; Huillard-Bréholles, *Historia diplomatica*, iv. 383; Denifle and Chatelain, *Chartularium Universitatis Parisiensis*, i,

have, he says in the preface to the *De arte*, followed the prince of philosophers where required, but not in all things, for we have learned by experience that at several points he deviates from the truth. Aristotle relies too much on hearsay, and has evidently "rarely or never had experience of falconry, which we have loved and practised all our life." More than once he must be directly corrected from the emperor's observation — *non sic se habet.*

It is this experimental habit of mind, the emperor's restless desire to see and know for himself, which lies behind those *superstitiones et curiositates* at which the good Salimbene holds up his hands.[112] There is the story of the man whom Frederick shut up in a wine-cask to prove that the soul died with the body, and the two men whom he disembowelled in order to show the respective effects of sleep and exercise on digestion. There were the children whom he caused to be brought up in silence in order to settle the question "whether they would speak Hebrew, which was the first language, or Greek or Latin or Arabic or at least the language of their parents; but he labored in vain, for the children all died." There was the diver, Nicholas, surnamed the Fish, hero of Schiller's *Der Taucher*, whom he sent repeatedly to explore the watery fastnesses of Scylla and Charybdis, and the memory of whose exploits was handed on by the Friars Minor of Messina,[113] not to mention the "other superstitions and curiosities and maledictions and incredulities and perversities and abuses" which the friar of Parma had set down in another chronicle now lost.[114] Such again was the story of the great pike brought to the Elector Palatine in 1497, in its gills a copper ring placed there by Frederick to test the longevity of fish, and still bearing the inscription in Greek, "I am that fish which Emperor Frederick II placed in

---

no. 394; Böhmer-Ficker, *Regesta*, no. 4750; Schirrmacher, *Die letzten Hohenstaufen* (Göttingen, 1871), p. 624; Grabmann, *Aristotelesübersetzungen*, pp. 200–204, 237 ff.; Helene M. Arndt, *Studien zur inneren Regierungsgeschichte Manfreds* (Heidelberg, 1911), p. 149; Pelzer, in *Revue néo-scolastique*, xxiii. 319 ff.

[112] Ed. Holder-Egger, pp. 350–353.

[113] The story appears also in Francesco Pippini (Muratori, ix. 669), Riccobaldo of Ferrara (*ibid.*, ix. 248), and Jacopo d'Acqui (*Neues Archiv*, xvii. 500).

[114] Salimbene, ed. Holder-Egger, p. 351. On Frederick's insatiable curiosity, see also Malaspina, in Muratori, viii. 788.

this lake with his own hand the fifth day of October, 1230." [115]
On another occasion Frederick is said to have sent messengers to
Norway in order to verify the existence of a spring which turned
to stone garments and other objects immersed therein.[116] According to Albertus Magnus, Frederick had a magnet which instead of attracting iron was drawn to it.[117]

Whatever value these tales may have, the emperor's scientific
habit of mind is seen best of all in his own writings. His treatise
on falconry, De arte venandi cum avibus,[118] is compact of personal
observation of the habits of birds, especially falcons, carried on
throughout a busy life of sport and study, and verified by birds
and falconers brought from distant lands. Indeed, his systematic
use for such inquiries of the resources of his royal administration
constitutes an interesting example of the pursuit of research by
governmental agencies. "Not without great expense," he tells
us, "did we call to ourselves from afar those who were expert in
this art, extracting from them whatever they knew best and
committing to memory their sayings and practices." "When we
crossed the sea we saw the Arabs using a hood in falconry, and
their kings sent us those most skilled in this art, with many
species of falcons." The emperor not only tested the artificial
incubation of hens' eggs, but, on hearing that ostrich eggs were
hatched by the sun in Egypt, he had eggs and experts brought to
Apulia that he might test the matter for himself. The fable that
barnacle geese were hatched from barnacles he exploded by sending north for such barnacles, concluding that the story arose from
ignorance of the actual nesting-places of the geese. Whether
vultures find their food by sight or by smell he ascertained by
seeling their eyes while their nostrils remained open. Nests,
eggs, and birds were repeatedly brought to him for observation
and note, and the minute accuracy of his descriptions attests the

[115] A. Hauber, "Kaiser Friedrich der Staufer und der langlebige Fisch," in
Archiv für Geschichte der Naturwissenschaften, iii. 315–329 (1911), brings together
the various reports but shows that the date 1230 is impossible.

[116] The original has 'in regione Armenie Norwegie.' Extract from mediaeval
encyclopaedia published by Delisle, in Notices et extraits des MSS., xxxii, 1, p. 48;
M. G. H., Scriptores, xxviii. 571.

[117] De mineralibus, cited by Thorndike, ii. 525, n.   [118] See Chapter XIV.

fidelity with which his observations were made. The whole of the practical portion of his *De arte* is a setting down in systematic form of the results of actual practice of the art. The author's statements are supported by facts rather than by authority or mere personal opinion, and if information is lacking no conclusion is drawn. One who reads the *De arte* through gets inevitably the impression of the work of a first-rate mind, open, inquiring, realistic, trying to see things as they are without *parti pris*, and working throughout on the basis of systematized experience. To follow this up by a course of reading in the confused and pretentious astrology of Michael Scot is to realize how far the emperor was intellectually superior to those about him.

Observation and experiment on a large scale Frederick supplemented by the questionnaire, applied not only to the scholars of his court and the experts who came at his summons, but to savants of other lands whom he could not interrogate personally. The method seems to have been to draw up a list of questions upon which the emperor could get no final or satisfactory response at home, and to send them to other rulers, most naturally the Mohammedan princes, requesting that they be submitted to the leading local scholars for answer, a procedure which assumes autocratic governments like that which Frederick himself utilized to satisfy intellectual curiosity. Such was the practice followed in the most famous instance, the so-called 'Sicilian questions' published by Amari many years ago.[119] According to the response which has reached us, Frederick, not long before 1242, sent a series of questions to be answered by Mohammedan philosophers in Egypt, Syria, Irak, Asia Minor, and Yemen, and later to the Almohad caliph of Morocco, ar-Rashid, by whom they were forwarded, with a sum of money as the emperor's reward, to ibn Sabin, a Spanish philosopher then living at Ceuta. Refusing the money, ibn Sabin answers at some length in terms of Mohammedan orthodoxy, expressing some contempt for Frederick's attain-

---

[119] M. Amari, "Questions philosophiques adressées aux savants musulmans par l'Empereur Frédéric II," in *Journal Asiatique*, fifth ser., i. 240–274 (1853); idem, *Biblioteca Arabo-Sicula*, ii. 414–419; more fully by A. F. Mehren, in *Journal Asiatique*, seventh ser., xiv. 341–454 (1879). Cf. the problems proposed by Chosroës, published by Quicherat, in *Bibliothèque de l'Ecole des Chartes*, xiv. 248–263 (1853).

ments as seen in his untechnical phraseology, and offering to set him right in a personal interview. The emperor's questions, as they are here cited in refutation, cover the eternity of matter and the immortality of the soul, the end and foundations of theology, and the number and nature of the categories — demanding always the proofs of the opinions advanced in reply. Thus: "Aristotle the sage in all his writings declares clearly the existence of the world from all eternity. If he demonstrates this, what are his arguments, and if not, what is the nature of his reasoning on this matter?" Plainly Frederick was familiar with the Aristotelian doctrines which agitated the Christian and Mohammedan worlds in the thirteenth century, indeed there was a legend that Averroës had lived at his court.[120] The very suggestion of doubt respecting immortality was enough to justify the current belief that Frederick was one of those Epicurean heretics "who make the soul die with the body."

We hear also of geometrical and astronomical problems such as the squaring of a circle's segment, solved for the emperor at Mosul; and we have another series of geometrical questions sent by one of Frederick's philosophers, in Arabic, to the young Jehuda ben Solomon Cohen in Toledo, together with the replies, at which the emperor expressed much satisfaction.[121] Again we learn that in the time of al-Malik al-Kamil, sultan of Egypt (1218–38), the emperor set seven hard problems in order to test Moslem scholars. Three of these, which concern optics, have been preserved with their answers: Why do objects partly covered by water appear bent? Why does Canopus appear bigger when near the horizon, whereas the absence of moisture in the southern deserts precludes moisture as an explanation? What is the cause of the illusion of spots before the eyes? [122]

[120] Renan, *Averroès* (1869), pp. 254, 291.

[121] Steinschneider, in *Z. M. Ph.*, xxxi, 2, pp. 106 ff. (1886); idem, *H. U.*, p. 3; idem, *Verzeichniss der hebräischen Handschriften der königlichen Bibliothek zu Berlin*, ii. 126 (1897); Suter, "Beiträge zu den Beziehungen Kaiser Friedrichs II. zu zeitgenössichen Gelehrten . . . insbesondere zu Kemâl ed-din ibn Jûnis," in *Abhandlungen zur Geschichte der Naturwissenschaften* (Erlangen, 1922), iv. 1–8.

[122] E. Wiedemann, "Fragen aus dem Gebiet der Naturwissenschaften, gestellt von Friedrich II," in *Archiv für Kulturgeschichte*, xi. 483–485 (1914).

Another and a less technical questionnaire has been handed down to us by Michael Scot; and as it does not appear to have been hitherto published or even cited by others, it may not be uninteresting to translate it as it stands in the manuscripts: [123]

"When Frederick, emperor of Rome and always enlarger of the empire, had long meditated according to the order which he had established concerning the various things which are and appear to be on the earth, above, within, and beneath it, on a certain occasion he privately summoned me, Michael Scot, faithful to him among all astrologers, and secretly put to me at his pleasure a series of questions concerning the foundations of the earth and the marvels within it, as follows:

"My dearest master, we have often and in divers ways listened to questions and solutions from one and another concerning the heavenly bodies, that is the sun, moon, and fixed stars, the elements, the soul of the world, peoples pagan and Christian, and other creatures above and on the earth, such as plants and metals; yet we have heard nothing respecting those secrets which pertain to the delight of the spirit and the wisdom thereof, such as paradise, purgatory, hell, and the foundations and marvels of the earth. Wherefore we pray you, by your love of knowledge and the reverence you bear our crown, explain to us the foundations of the earth, that is to say how it is established over the abyss and how the abyss stands beneath the earth, and whether there is anything else than air and water which supports the earth, and whether it stands of itself or rests on the heavens beneath it. Also how many heavens there are and who are their rulers and principal inhabitants, and exactly how far one heaven is from another, and by how much one is greater than another, and what is beyond the last heaven if there are several; and in which heaven God is in the person of His divine majesty and how He sits on His throne, and how He is accompanied by angels and saints, and what these continually do before God. Tell us also how many abysses there are and the names of the spirits that dwell therein, and just where are hell, purgatory, and the heavenly paradise, whether under or on or above the earth [or above or in the abysses, and what is the difference between the souls who are daily borne thither and the spirits which fell from heaven; and whether one soul in the next world knows another and whether one can return to this life to speak and show one's self; and how many are the pains of hell]. Tell us also the measure of this earth by thickness and length, and the distance from the earth to the highest heaven and to the abyss, and whether there is one abyss or several; and if several how far one is from another; and whether the earth has empty spaces or is a solid body like a living stone; and how far it is from the surface of the earth down to the lower heaven.

"Likewise tell us how it happens that the waters of the sea are so bitter and the waters are salt in many places and some waters away from the sea are sweet although they all come from the living sea. Tell us too concerning the sweet waters how they continually gush forth from the earth and some-

[123] For the Latin text, see below, pp. 292–294.

times from stones and trees, as from vines when they are pruned in the springtime, where they have their source and how it is that certain waters come forth sweet and fresh, some clear, others turbid, others thick and gummy; for we greatly wonder at these things, knowing already that all waters come from the sea and passing through divers lands and cavities return to the sea, which is the bed and receptacle of all running waters. Hence we should like to know whether there is one place by itself which has sweet water only and one with salt water only, or whether there is one place for both kinds, and in this case how the two kinds of water are so unlike, since by reason of difference of color, taste, and movement there would seem to be two places. So, if there are two places for these waters, we wish to be informed which is the greater and which the smaller, and how the running waters in all parts of the world seem to pour forth of their superabundance continually from their source, and although their flow is copious yet they do not increase as if more were added beyond the common measure but remain constant at a flow which is uniform or nearly so. We should like to know further whence come the salt and bitter waters which gush forth in some places, and the fetid waters in many baths and pools, whether they come of themselves or from elsewhere; likewise concerning those waters which come forth warm or hot or boiling as if in a caldron on a blazing fire, whence they come and how it is that some of them are always muddy and some always clear. Also we should like to know concerning the wind which issues from many parts of the earth, and the fire which bursts from plains as well as from mountains, and likewise what produces the smoke which appears now in one place and now in another, and what causes its blasts, as is seen in the region of Sicily and Messina, as Etna, Vulcano, Lipari, and Stromboli. How comes it that a flaming fire appears not only from the earth but also in certain parts of the sea of India?

["And how is it that the soul of a living man which has passed away to another life than ours cannot be induced to return by first love or even by hate, just as if it had been nothing, nor does it seem to care at all for what it has left behind whether it be saved or lost?"]

A notable series of questions this, in spite of a certain amount of confusion and repetition which may be due to the less clear medium of Michael Scot through which they have been transmitted. Besides the previous discussions which they assume respecting astronomy, geography, and natural history, they cut to the heart of the current cosmology, which readers of Dante will recognize, with an insistent demand for exact and definite information. Just where are heaven and hell and purgatory; exactly how far is one heaven or one abyss from another; what is the structure of the earth and the explanation of its fires and waters — questions that might easily have cost Michael Scot his reputation, in spite of his boastful promise to answer them all, and may well have led him to seek to measure the distance to heaven by means of a church tower with an apparent exactness which seems to have imposed on the emperor.[124]   Astronomy and cosmology cannot avoid theology: In which heaven is God to be found, and where are the souls of the departed, and why do they not communicate with us for love or even hate? "Or even hate" — a very human touch which shows us Frederick's own passion in the midst of the eternal riddles and reminds us of that hatred for Viterbo which he would come back from Paradise to assuage.[125] And here as in the stories of Moslem writers we recognize the note of scepticism, the trace of that Epicurean heretic whose lurid figure haunts one of the thousand fiery tombs of the tenth canto of the *Inferno*.

The nature of Frederick's ultimate religious opinions lies beyond the ken of the historian, for we have no direct statements of his own beyond his general assertions of orthodoxy, against many highly colored stories from his enemies.   When, however, Gregory IX accuses him of declaring that one should believe only in what is proved by the force and reason of nature,[126] the assertion falls in entirely with what we know of Frederick's habit of mind.   Profoundly rationalistic, he applied the test of reason and experience to affairs of state as well as to matters of science, as the body of his Sicilian legislation abundantly testifies.   When he abolishes the ordeal, his reason is that it is not in accord with nature and does not lead to truth.[127]   In matters of commercial policy, "he was the first mediaeval ruler to use consistent economic principles as his standards."[128]   *Immutator mirabilis*, he has none of the mediaeval horror of change.   Yet it is scarcely historical to call him a modern, for he looks in both directions. He harks back to King Roger and the Mohammedan East, while

---

[124] See the passage printed below, Chapter XIII, n. 110.

[125] *Historische Zeitschrift*, lxxxiii. 30.

[126] Encyclical of July 1, 1239, in Huillard-Bréholles, v. 340; Böhmer-Ficker, no. 7245; Potthast, no. 10766.   Frederick's reply is in Huillard-Bréholles, v. 348 (Böhmer-Ficker, nos. 2454, 2455); see also the examination of his orthodoxy in 1246, *ibid.*, vi. 426, 615 (Böhmer-Ficker, no. 3543).

[127] Hampe, in *Historische Zeitschrift*, lxxxiii. 14.

[128] Jastrow-Winter, *Deutsche Geschichte im Zeitalter der Hohenstaufen*, ii. 549.

in his many-sided patronage of learning and his free and critical spirit of inquiry he belongs rather to the Italian Renaissance. Only in part does he belong to the thirteenth century, and he was in no sense its type. He was above all an individual, *stupor mundi* to his own age, and a marvel still to ours.

Frederick's favorite son, Manfred, appears linked with his father in Dante's mention of the two illustrious heroes who, while fortune lasted, despised the merely brutal and followed humane pursuits.[129] Certainly Manfred inherited many of his father's tastes and something of the same habit of mind, and his court continued much of the scientific activity of the earlier reign.[130] He tells us that the masters of his father's court [131] taught him the nature of the world and the properties of both the transient and the eternal. At the age of twenty-five he fortified himself during a severe illness with the teachings of the treatise *De pomo*,[132] then ascribed to Aristotle, and on his recovery had it translated from Hebrew into Latin. Latin versions of the *Magna moralia* and pseudo-Aristotelian works, apparently those sent by the king to the students of Paris,[133] were made directly from the Greek by an official translator, Bartholomew of Messina, who also translated at Manfred's command the veterinary treatise of Hierocles.[134] Translation from the Arabic is represented by an

---

[129] *De vulgari eloquentia*, i, c. 12.

[130] See, in general, Schirrmacher, *Die letzten Hohenstaufen*, pp. 209–216; Capasso, *Historia diplomatica regni Siciliae*, pp. 324 ff.; Helene M. Arndt, *Studien zur inneren Regierungsgeschichte Manfreds*, c. 4; O. Cartellieri, "König Manfred," in *Centenario Michele Amari* (Palermo, 1910), i. 116–138.

[131] The arguments of Hampe, *Neues Archiv*, xxxvi. 231 ff., and Arndt, pp. 146 ff., that Manfred was a student at Bologna and Paris, are to me unconvincing.

[132] Preface in Huillard-Bréholles, *Monuments de la maison de Souabe*, p. 169; Schirrmacher, p. 622; Capasso, p. 112, note; Böhmer-Ficker, no. 4653. Cf. Steinschneider, *H. U.*, p. 268, who thinks it unlikely that the king himself was the translator. A copy of this version in the Biblioteca Colombina at Seville purports to have been made 'de greco in latinum' (MS. 7-6-2).

[133] Supra, note 111.

[134] MSS. of Hierocles at Pisa and Bologna: *Studi italiani di filologia classica*, viii. 395, xvii. 76; *Rheinisches Museum*, n. s., xlvi. 377 (1891). For the pseudo-Aristotle see Grabmann, pp. 201 ff.; Foerster, *De translatione Latina Physiognomicorum* (Kiel, 1884); particularly the evidence of MS. xvii. 370 of the Biblioteca Antoniana at Padua. Another translator, Nicholas of Sicily, may belong to this group: Grabmann, p. 203.

astrological treatise, the *Centiloquium Hermetis*, turned into Latin by Stephen of Messina and also dedicated to the king,[135] and by a set of astronomical and astrological tables translated by John 'de Dumpno' and preserved in a fine codex at Madrid.[136] Manfred's knowledge of philosophy and mathematics, especially Euclid, as well as of languages, is praised by an Egyptian visitor, who dedicated to him a work on logic,[137] and a further illustration of his philosophical tastes is found in a disputation in which he asks whether members exist because of their functions or functions because of their members, the final 'determination' of this scholastic dispute being made by that *gemma magistrorum et laurea morum*, Master Petrus de Hibernia.[138]

Like his father, Manfred had his menagerie, including a giraffe from the East,[139] and he also shared his father's devotion to astrology [140] and to sportsmanship. The *De arte venandi*, originally dedicated to Manfred, has come down to us as he revised it, with certain additions from his own observations but primarily with the aim of filling blanks in the original by the aid of his father's notes, reading and rereading the book with filial piety that he might obtain the full fruits of its science and that no scribal errors might be left to frustrate the author's purpose.[141] This was only one of the numerous books by many hands which filled the presses of the royal library,[142] including philosophical and mathematical works in Greek and Arabic, certain of which are believed

---

[135] Steinschneider, *E. U.*, no. 114; Thorndike, ii. 221. Many MSS., e. g., Madrid, MS. 10009, f. 225.

[136] Biblioteca Nacional, MS. 10023, ff. 1–23: 'Perfectus est interpretatio et translatio istarum portarum de arabico in latinum per Iohannem de Dumpno filium Philippi de Dumpno in civitate Panormi anno a nativitate domini nostri Iesu Christi 1262, sub laude et gloria omnipotentis Dei feliciter amen.'

[137] Djemal-Edin, in Michaud, *Bibliothèque des Croisades*, vii. 367; *Revue historique*, lxxx. 64; Suter, no. 380.

[138] Text published by Baeumker, "Petrus de Hibernia," in Munich *Sitzungsberichte*, 1920. See also Pelzer, in *Revue néo-scolastique*, 1922, pp. 355 f.

[139] Röhricht, *Beiträge*, i. 74.

[140] Huillard-Bréholles, introduction, p. dxxxii; Arndt, p. 151.

[141] Chapter XIV, p. 304.

[142] 'Librorum ergo volumina, quorum multifarie multisque modis distincta cyrographa diviciarum nostrarum armaria locupletant': *Chartularium Universitatis Parisiensis*, i, no. 394.

to have gone as a present to the Pope from the victorious Charles of Anjou,[143] and thus served to hand on something of the scientific interests of Manfred and of Frederick to a later age. At best, however, Manfred's court is but an echo of that of Frederick, and under the Angevins the intellectual history of Sicilian royalty enters upon a new and different period.[144]

[143] Chapter IX, n. 35.

[144] On translations under Charles of Anjou, see Amari, *La guerra del Vespro Siciliano*, edition of 1886, iii. 483–489; Hartwig, in *Centralblatt für Bibliothekswesen*, iii. 185–188; Steinschneider, *E. U.*, nos. 39, 86; *Hermes*, viii. 339; de Renzi, *Collectio Salernitana*, i. 336; Thorndike, ii. 757.

CHAPTER XIII

MICHAEL SCOT[1]

In any judgment respecting the scientific activity of the court of Frederick II, much depends upon the opinion formed of Michael Scot, the emperor's astrologer, whose writings form a large part of the scientific and philosophic product of the *Magna Curia*. Condemned by Roger Bacon as "ignorant of the sciences and languages," Scot is praised by Gregory IX for his knowledge of Hebrew and Arabic, and addressed as *summe philosophe* by Leonard of Pisa, the most eminent mathematical genius of his time. Naturally enough for an astrologer, Scot early became a subject of legend, and the small body of fragmentary fact has not yet been winnowed from the mass of tradition. The elaborate biography by James Wood Brown[2] contains far too much of pleasing conjecture, and its insecure chronology has misled more than one subsequent writer. It may help investigation if we try to set down the ascertainable events of Scot's life and to group his works in some chronological order, as a preliminary to an examination of his treatises on astrology and his intellectual relations with the emperor.

Concerning the place and date of Scot's birth no evidence has reached us. We may, however, be sure that when Master Michael calls himself Scot[3] he means a native of Scotland and not an Irishman, as the name frequently signifies in mediaeval usage. Not only did he hold benefices in Scotland,[4] but he refused a most lucrative appointment, the archbishopric of Cashel, because he

---

[1] Revised from *Isis*, iv. 250–275 (1922). Cf. *American Geographical Review*, xiii. 141 f. (1923); *Mitteilungen zur Geschichte der Medizin*, xxii. 4.

[2] *An Enquiry into the Life and Legend of Michael Scot* (Edinburgh, 1897), followed closely in the article in the *Dictionary of National Biography*, and by Comrie in *Edinburgh Medical Journal*, July 1920; Thorndike, ii, ch. 51, is more independent.

[3] 'Cui ego Michael Scottus tanquam scottatus a multis et a diversis': Bodleian, MS. Canon. Misc. 555, f. 45; infra, p. 294. 'Ego Michael Scotus': Jourdain, pp. 127–129; MS. Pisa 11, n. 10, below.

[4] Bliss, *Calendar of Papal Letters*, i. 102.

was ignorant of the Irish tongue.[5] That he knew English might be inferred from a list of Anglo-Saxon names of months which he inserts in his *Liber introductorius*, did not a similar list appear in Bede.[6] The facts of his career place his birth somewhere in the late years of the twelfth century. Of his education we know nothing, the statements concerning his studies at Durham, Oxford, Paris, and Bologna, being mere guesses of modern writers.[7] All that we can say is that his writings show a knowledge of the elements of Latin culture — the Bible, Augustine, the writers on the *trivium* and *quadrivium* — and that this was probably gained before he went to Spain for more special studies.

We must likewise dismiss as entirely baseless Brown's chapter which makes Scot tutor of the young Frederick II and author of various works composed in Sicily in 1209 and 1210. The sole foundation for this elaborate construction is the misreading as 'MCCX' of the 'MCC etc.' of a Vatican codex of the *Abbreviatio Avicenne*;[8] and there is no evidence connecting Scot with Sicily until many years later.

The first specific date in Scot's career is 18 August 1217, when he completed at Toledo his translation of al-Bitrogi (Alpetragius) *On the Sphere*.[9] He had plainly been for some time in Spain and

---

[5] *Ibid.*, i. 98.

[6] '*Nomina mensium secundum Anglicos.* Primus mensis anni Anglorum est giuli, id est januarius; 2. est solmonant, id est februarius; 3, est heredemonath, id est martius; 4. est turmonath, id est aprilis; 5. est thrumlei, id est maius; 6. est lidan; 7. est lydi; 8. est vendmonath; 9. est aligmonanth; 10. est gyh. Hee gentes suum annum incipiunt a medianocte nativitatis Domini et quociens sunt kalende mensium tociens solempne pulsant campanas ecclesie maiori post complementum officii matutini cum interpellatione et omnes gentes summa devocione vadunt ad eandem ecclesiam portantes aliquid ad offerendum.' Cod. Lat. Monacensis 10268, f. 71 v. Cf. Bede, *De temporum ratione*, ch. 15.

[7] The story that Michael taught theology at Paris may arise from a confusion with Master Matthew Scot, who appears there in 1218. *Chartularium*, i. 85.

[8] See the facsimile in Brown, p. 55. Monsignore Auguste Pelzer of the Vatican Library informs me, as I had conjectured from the facsimile, that 'MCC etc.' is the necessary reading of the original. I find that Sir John Sandys had also questioned Brown's reading, but without rejecting the inferences from it (*History of Classical Scholarship*[3], i. 566). Thorndike accepts the date. The MS. is Vat. lat. 4428.

[9] Jourdain, p. 133, where one MS. has the Christian and one the Spanish era. This is confirmed by MS. Madrid 10053 (ca. 1300, formerly in the chapter library at Toledo), f. 156 v.: 'Perfectus est liber Avenalpetraug a magistro Michaele

gained something of that acquaintance with Arabic which was to serve him later. The next point in Scot's biography is 21 October 1220, when he appears at Bologna, living in the house of the widow of Alberto Gallo and describing in detail a neighbor's case of calcified fibroid tumor.[10] The sworn note to this effect which he appends to certain copies of the *De animalibus* gives the year as 1221, but the day of the week given shows that he is using the Pisan style, as in his later works.[11] This is his first appearance in Italy, and it should be remarked that Frederick II was in the neighborhood of Bologna at the same time,[12] although we have no evidence that Scot was then in the emperor's service.

From 1224 to 1227 the papal registers show that Scot had the active favor of Pope Honorius III and his successor, Gregory IX. This interesting series of entries begins 16 January 1224 with a letter from Honorius III recommending Scot to the archbishop of Canterbury as a man of eminent learning (*singularis scientia inter alios litteratos*), worthy of a benefice in that province.[13] The church assigned yielded an insufficient income, and 18 March he received permission to hold two benefices,[14] one of which appears from what follows to have been in England. His tenure of these was unaffected by his elevation the following May to the archbishopric of Cashel,[15] but by 20 June he had declined this

Scotto Toleti in decimo octavo die veneris augusti hora tertia cum Abuteo levite anno incarnationis Iesu Christi 1217.' MS. Barberini Lat. 156 of the Vatican, f. 194, has 1221, but with the same day of the week and month. Steinschneider, *E. U.*, no. 84 *i*, gives incorrectly 1267. Cf. MS. Arsenal, 1035, where the date is 1207; Harleian MS. 1, f. 1 (1217).

[10] The note is printed by Dr. M. R. James in the *Catalogue of the Manuscripts in the Library of Gonville and Caius College*, i. 112, from MS. 109; facsimile in *Edinburgh Medical Journal*, 1920, p. 56. It is also found in a thirteenth-century copy of the *De animalibus* in the manuscripts of the Convento S. Caterina at Pisa, MS. 11, f. 133–133 v (cf. *Studî italiani di filologia classica*, viii. 325), where the following is added to Dr. James' text: 'eiecit in octabis sancti Iohannis maiorem post .viii. dies post minorem.'

[11] Below, n. 112.   [12] Böhmer-Ficker, *Regesta imperii*, nos. 1176–94.

[13] Pressutti, *Regesta Honorii Pape III*, no. 4682; *Chartularium Universitatis Parisiensis*, i, no. 48; Brown, p. 275; Bliss, *Calendar of Papal Letters*, i. 94.

[14] Pressutti, no. 4871; Bliss, i. 96.

[15] Pressutti, no. 5025; not in Bliss. A papal letter on the same subject, apparently to Henry III, is printed in my paper on "Two Roman Formularies in Philadelphia," in the *Miscellanea Ehrle* (1924).

preferment because of his ignorance of Irish.[16] 9 May 1225 he
is allowed to hold an additional benefice in England and two in
Scotland.[17] 28 April 1227 Gregory IX, shortly after his accession, urges Michael's claims on the archbishop of Canterbury as
one who had pursued learning since boyhood and added a knowledge of Hebrew and Arabic to his wide familiarity with Latin
learning.[18]

In 1228, or, since we are in Pisa, more probably in 1227, falls
the dedication to Scot of the revised edition of the great treatise
of Leonard of Pisa on the abacus, of which Scot had solicited a
copy from the author.[19] As Leonard was in relations with Frederick II and the philosophers of his entourage as early as 1225 or
1226,[20] Scot may have already become connected with the emperor's court. In any event, Scot disappears from the papal
registers after 28 April 1227, and no long time can have elapsed
before he joined the court of Frederick II, with which he is thereafter identified. Contemporaries call him Frederick's astrologer
and recount various stories of his skill, even to the prediction of
the place of the emperor's death,[21] while Scot himself mentions
instances of his prophesying from the stars the results of Frederick's military operations.[22] Scot's later works are dedicated to
the emperor, and one of them, the *Abbreviatio Avicenne*, was kept
in the emperor's library in 1232. The loss of the imperial registers, save for a fragment of 1239–40, prevents our tracing details
of his activity at the court, except for some indications in Scot's
own writings to which we shall come below. His career is thus
summed up by a poet of the court:

[16] Pressutti, no. 5052; Bliss, i. 98.   [17] Pressutti, no. 5470; Bliss, i. 102.

[18] Auvray, *Registres de Grégoire IX*, no. 61; *Chartularium Universitatis Parisiensis*, i, no. 54; Potthast, *Regesta*, no. 7888; Bliss, i. 117.

[19] Boncompagni, *Scritti di Leonardo Pisano* (Rome, 1857), i. 1; for the date
1228, see Boncompagni in *Atti dei Lincei*, first series, v. 73 f. (1851); Cantor, ii. 7.

[20] *Scritti*, ii. 253. On the chronological difficulties, see Eneström, in *B. M.*, ix.
72 f. (1908).

[21] Salimbene, ed. Holder-Egger, pp. 353, 361, 512, 530; Riccobaldi of Ferrara,
in Muratori, *Scriptores*, ix. 128; Francesco Pipini, *ibid.*, ix. 660, 670.

[22] 'Et ut apercius hec dicta pateant, recordamur duarum questionum inter alias
principis volentis ire super duas civitates sibi rebelles,' followed by the observations, with diagrams, and Scot's deductions: *Liber introductorius*, MS. n. a. lat.
1401, f. 99 v.

Qui fuit astrorum scrutator, qui fuit augur,
Qui fuit ariolus, et qui fuit alter Apollo.[23]

If we could accept the statement of a note which accompanies this prophecy in one manuscript, Scot was at Bologna in 1231, where he was consulted by the *podestà* and notables concerning the fate of the Lombard cities and replied with a famous set of verses predicting the fate of each. The references to the events of 1236 and following are, however, so specific as to indicate that this *Vaticinium* was written subsequently and ascribed to Scot,[24] who was known to have made definite predictions foretelling the emperor's triumph over his enemies.[25]

The date of Scot's own death is apparently fixed by certain verses of Henry of Avranches dedicated to the emperor shortly before his last return to Italy from Germany early in 1236.[26] Scot is here mentioned as one who has passed, apparently recently, into eternal silence, and there is no reason to doubt the testimony of a court poet then in the emperor's following. If we attach any weight to the Paris manuscript of Scot's *Vaticinium*, he was in Germany with the emperor on this journey, and would thus have met his death there.[27] The story ran that he was killed at mass by the falling of a stone, in spite of a metal headpiece by which he had sought to protect himself.[28]

The only reason for seeking to place Scot's death later is connected with the dates of his writings. The manuscripts of his *Liber particularis* bear a title *tempore domini pape Innocentii quarti* (1243–54), and since the preface refers to an event of 1228 this cannot be explained away by Brown as a slip for Innocent III; but, as there is no reference to this pope in the text, we may have no more than the guess of a scribe, itself inconsistent with

---

[23] *Forschungen zur deutschen Geschichte*, xviii. 486.

[24] Holder-Egger, in *Neues Archiv*, xxx. 349–377, where the text of the verses appears as well as in his edition of Salimbene, p. 361. Cf. Winkelmann, *Kaiser Friedrich II*, ii. 323, n. A note in MS. lat. n. a. 1401, f. 124 v., not used by Holder-Egger, states that the verses were recited to the emperor by Scot before the departure from Germany: Delisle, *Catalogue du fonds de La Trémoïlle*, p. 43.

[25] Poem of Henry of Avranches: *Forschungen zur deutschen Geschichte*, xviii. 486.

[26] *Ibid.*

[27] *Catalogue du fonds de La Trémoïlle*, p. 43, cited above.

[28] Pipini in Muratori, ix. 670.

a closing verse of 1256.[29]   The commentary on the *Sphere* of John of Sacrobosco must be subsequent to the date of that work, often stated as 1256, but the facts of Sacrobosco's life have not been sufficiently investigated, and Scot's authorship is too uncertain to permit drawing any decisive conclusion.   I see no reason for identifying him with the clerk Michael of Cornwall, 'dictus Scotus,' who appears at Chartres in 1252–54.[30]

Scot's writings are, with one exception, undated in the form in which they have reached us.   They can, however, be distinguished into two main groups, corresponding to the two chief periods of his activity, the Spanish and the Sicilian.   Speaking broadly, natural philosophy predominates in the earlier period, and astrology in the later.   Let us consider them in this order.

I.   The only dated work is the translation of al-Bitrogi, completed at Toledo 18 August 1217.   This treatise, which develops Aristotle's theory of homocentric spheres against the eccentrics and epicycles of Ptolemy, was of considerable importance as a source of Aristotelian cosmology in the thirteenth century, and Scot's version seems to have been the medium through which it was known to Roger Bacon and others.[31]

Scot's version of Aristotle's *Historia animalium* is in four of the manuscripts dated at Toledo.[32]   His authorship is clear from a memorandum inserted in his own copy and preserved in two extant manuscripts.[33]   This note, dated at Bologna 21 October 1220, shows that the work must have been completed before this date, and thus strengthens the statement that this version belongs to the Toletan period of Scot's life.   As the manuscripts lack a dedication, the words *ad Caesarem* added in current usage would appear to rest on a confusion with the *Abbreviatio Avicenne*. Whether the translation was made from the Hebrew or from the

---

[29] This verse is also found in the *Vaticinium* of John of Toledo: *Neues Archiv* xxx. 353, note.

[30] Clerval, *Les écoles de Chartres*, pp. 350 f.

[31] For the date and manuscripts, see above, n. 9; for the contents, Duhem, iii. 241 ff., 327 f.

[32] Merton College, MS. 278; Cues, MS. 182 (Grabmann, *Aristotelesübersetzungen*, p. 187); Laurentian, Plut. XIII, sin., 9 (Bandini, iv. 109); Cracow, MS. 653.

[33] See above, n. 10.

Arabic has been a matter of dispute;[34] in any event a Jewish interpreter[35] seems to have been used. The version is closely literal, so that it has even been used for reconstructing the Greek original;[36] but there are also numerous errors, which were repeated by Albertus Magnus in using it.[37] Here, as in the usual Arabic tradition of the work, the *Historia animalium* consists of nineteen books, including not only the *De animalibus historia*, with the spurious tenth book, but the *De partibus animalium* and the *De generatione animalium*. For all of these Scot's version was the first and remained in use till the fifteenth century.[38]

In the case of other works of Aristotle the question is complicated by the fact that there was more than one version from the Arabic in circulation in the thirteenth century, as well as by their relation to the accompanying commentary of Averroës. The one entirely clear case is the *De caelo et mundo*, to which Scot has prefixed a preface addressed to Stephen of Provins, doubtless the canon of Reims named by Gregory IX in 1231 as one of the commission to examine and purge the newly translated works of Aristotle on natural science.[39] This version is subsequent to 1217, as it cites Scot's translation of al-Bitrogi. It is altogether likely that Scot is the author of the version of the *De anima* which, with the commentary of Averroës, regularly accompanies his *De caelo*

[34] See especially Wüstenfeld, pp. 101–106 (1877); Steinschneider, *H. U.*, pp. 479–483.

[35] Roger Bacon, *Compendium studii*, ed. Brewer, p. 472.

[36] Rudberg, in *Eranos*, ix. 92 ff.

[37] H. Stadler, *Albertus Magnus de animalibus* (Münster, 1916), i, p. xii; id., in *Archiv für die Geschichte der Naturwissenschaften*, vi. 387–393 (1913); Dittmeyer, *Guilelmi Moerbekensis translatio commentationis Aristotelicae De generatione animalium* (Dillingen programme, 1914).

[38] See in general, Grabmann, *Forschungen über die lateinischen Aristotelesübersetzungen des XIII. Jahrhunderts*, pp. 185–187, and the literature there cited. This version passed quickly into use. Before Albertus Magnus we find it cited by Philip de Grève, 1228–36 (Minges, in *Philosophisches Jahrbuch*, xxvii. 28); and by Bartholomew Anglicus, ca. 1240 (Grabmann, p. 42).

[39] Jourdain, p. 127 f.; Grabmann, p. 175; Renan, *Averroès* (Paris, 1869), p. 206; bull of Gregory in *Chartularium Universitatis Parisiensis*, i, no. 87. Other manuscripts are at Erfurt, F. 351; at Durham, C. I. 17; at the University of Paris, MS. 601 (infra, n. 63); at the Vatican, Vat. lat. 2184, f. 1. On the various persons known as Étienne de Provins in this period, see my paper, "Two Roman Formularies," in the *Miscellanea Ehrle*.

in the manuscripts.[40] Translations of the *Physics*, *Metaphysics*, and *Ethics* have been ascribed to Scot, but without sufficient evidence.[41] The argument is somewhat stronger for certain other commentaries of Averroës, coinciding as they do with Scot's *Questiones Nicolai*,[42] but the matter is not yet clear. In any event Scot's rôle was merely that of translator; it was Averroës *che il gran commento feo!* [43]

Two philosophic treatises of Scot probably belong to the Spanish period. One, a *Divisio philosophica*, or classification of philosophical knowledge, preserved only in fragments by Vincent of Beauvais, is based in considerable measure upon Dominicus Gundisalvi, who worked in Spain in the twelfth century.[44] The other, known in extracts as the *Questiones Nicolai peripatetici*, is definitely assigned by Albertus Magnus to Scot,[45] who here seems to take shelter in anonymity in order to preach strong Averroism.[46]

II. From the Sicilian period of Scot's activity we have, first of all, the *Abbreviatio Avicenne de animalibus*, dedicated to Frederick II as emperor. We have already seen that this cannot be dated 1210,[47] as Brown fondly thought; all that we can say is that it was anterior, and probably not long anterior, to the copy made from the emperor's original by Henry of Cologne at Melfi 9 August 1232.[48] Frederick's keen interest in animals, and especially in birds, is a sufficient explanation of its origin.[49]

[40] Hauréau, *Philosophie scolastique* (1880), ii, 1, p. 125; Grabmann, p. 198.

[41] Jourdain, pp. 128, 141 f., 144; Grabmann, pp. 172, 212, 215, 217. Note that the *Ethics* is cited in the preface to the *Liber introductorius* (see below), and the *Metaphysics* in the commentary on Sacrobosco.

[42] Renan, *Averroès*, p. 205.   [43] Dante, *Inferno*, iv, line 144.

[44] Baur, *Dominicus Gundissalinus De divisione philosophiae* (*Beiträge*, iv, nos. 2–3), pp. 364–367, 398–400; supra, Chapter I.

[45] 'Feda dicta inveniuntur in libro illo qui dicitur *Questiones Nicolai peripatetici*. Consuevi dicere quod Nicolaus non fecit librum illum sed Michael Scotus, qui in rei veritate nescivit naturas nec bene intellexit libros Aristotilis.' *Opera* (ed. Paris, 1890), iv. 697. Birkenmajer is preparing an edition of these *Questiones*.

[46] Hauréau, *Philosophie scolastique* (1880), ii, 1, p. 127; Renan, *Averroès*, pp. 209 f.; Duhem, iii. 245 f., 339, 346 f.

[47] Supra, n. 8.

[48] Huillard-Bréholles, *Historia diplomatica*, iv. 381.

[49] See the next chapter.

The most ambitious of Scot's works belong to this period, the series of treatises on astrology made up of the *Liber introductorius*, the *Liber particularis*, and the *Physionomia*. In their final form these are subsequent to 16 July 1228, since the general preface refers to Francis of Assisi as already a saint.[50] They are dedicated to the emperor, whom they mention in the text, and, as we shall see, contain in part answers to specific questions asked by him.

III. The remaining works attributed to Scot are more or less doubtful. The court of Frederick II became a peg on which to hang all sorts of fictitious attributions,[51] and Scot's popular reputation could easily lead to connecting his name with the works of others.

So of Scot as an alchemist it is hard to speak with any certainty amid the mass of false attributions which accompany the alchemical literature of the later Middle Ages.[52] That he passed as an alchemist is clear from the ascriptions of several manuscripts, notably a list of alchemical writers preserved in a Palermo codex,[53] and his familiarity with alchemical doctrine is seen in the chapter from his own *Liber particularis* printed below.[54] The question is whether he wrote actual treatises on the subject, and, if so, whether any of these can be identified. A definite answer must await the sifting of the confused and uncertain manuscript material. Meanwhile the most promising evidence seems to be afforded by a few pages in the library of Corpus Christi College,

---

[50] 'Quandoque sine vestibus cum alis, ut seraphim ad beatum Franciscum et Michael quando pugnavit cum dracone et quando consignavit in Monte Gargano ecclesiam, propter quod hodie dicitur Mons Angeli qui est prope Romam versus Apuliam': Munich, Cod. lat. 10268, f. 9 v; N. a. lat. 1401, f. 22, omitting what follows 'ecclesiam.'

[51] Ch. V. Langlois, *La connaissance de la nature et du monde au moyen âge* (Paris, 1911), pp. 190–192; supra, Chapter XII, notes 59–65.

[52] See Brown, ch. 4, and the more sceptical pages of Thorndike, ii. 335–337. E. von Lippmann, *Entstehung und Ausbreitung der Alchemie* (Berlin, 1919), does not discuss Scot's alchemical writings.

[53] G. di Marzo, *I MSS. della Biblioteca comunale di Palermo* (1878), iii. 237. This MS. (4Q q. A 10) is cited by Brown, p. 79, as in private hands.

[54] P. 295. The reference of the *Dictionary of National Biography* to Scot's *magisterium* in MS. Bodley 44 is an error.

Oxford.[55]   Here we have not only the attribution of the *explicit* but the frequent mention of Michael in the body of the work, much as in his other works: 'et ego Michael Scotus multociens sum expertus et semper veracem inveni.'[56]   The work purports to be dedicated to Theophilus king of the Saracens, but Friar Elias is mentioned in the second person as Michael's associate in experiments.[57]   Besides the transmarine writers, Hebrew, Arabic, Saracen, Armenian, and other, whom the author has read, he cites specifically Barbaranus the Saracen of Aleppo (Halaph), Theodosius the Saracen of 'Cunusani,' Medibibaz the Saracen of Africa, and Master Jacob the Jew at Catania [58] (?).   He himself has translated a book explaining how to treat salts in alchemy.[59] Besides various eastern substances he mentions alum of Aleppo and gum of Calabria and Montpellier.[60]   The milieu resembles that of Michael Scot, and so does the general style, although the material seems to have been reshaped by another hand.

Similarly the notes appended to the copies of his *De animalibus* at Cambridge and Pisa indicate that Scot observed and treated diseases; [61] but no works of medicine can be certainly identified

[55] MS. 125, ff. 97–100 v (ca. 1400): 'Cum animadverterem nobilem scientiam apud Latinos penitus denegatam vidi quoque neminem pervenire ad perfectionem propter nimiam confusionem in libris philosophorum que reperitur, existimavi secreta nature intelligentibus revelare, incipiens a maiori magisterio et minori que inveni de transformatione metallorum et de permutatione eorum qualiter substantia unius in alterum permutetur. . . . Septem sunt planetarum (f. 97 v) . . . sales qui operantur in solem. Explicit tractatus magistri Michaelis Scoti de alkemia.'

[56] F. 100.

[57] F. 97 v: 'Et que in hac arte sunt necessaria tibi, frater Helya, diligenter et subtiliter enarravi.' F. 98: 'Et ego magister Michael Scotus sic operatus sum solem et docui te, frater Elia, operari et tu mihi sepius retulisti te instabiliter multis vicibus operasse.' F. 98 v: 'Prout Michael predictus probavi et docui, frater Helya.' F. 99 v: 'Sed ego vidi ipsam fieri a fratre Helya et ego multociens sum expertus.'

[58] F. 100: 'Et ego vidi istam operationem fieri apud Cartanam a magistro Iacobo iudeo et ego postea multociens probavi.'

[59] F. 97 v: 'Prout in aliquo libro a me translato dixi quomodo de salibus oportet in arte alkemie operari.'

[60] F. 99: 'Et hoc facit cum alumine de Halaph et cum quadam gumma que in partibus Kalabrie invenitur et in Monte Pessulano.'

[61] James, *Descriptive Catalogue of MSS. in the Library of Gonville and Caius College*, i. 112 f.; Pisa, Convento S. Caterina, MS. 11. See above, n. 10.

beyond the *Physionomia* and the *De urinis* which forms a part of it. For the pills and powders which passed in Scot's name there is no valid authority.[62]

Two versions of Maimonides in a manuscript of the University of Paris [63] are ascribed to Michael Scot by the author of the printed catalogue, but no definite basis for this appears save the fact of their occurrence, in a different hand of the thirteenth century, in the same volume as Michael's translation of the *De caelo*. The second of these [64] is the standard Latin rendering of the *Guide to the Perplexed*, generally supposed to have been made from the Hebrew in southern Italy before ca. 1250. The first [65] discusses parables more fully than the *Guide*, and then the fourteen fundamental classes of precepts and the six hundred and thirteen commandments, but is evidently the work of some adapter, after Maimonides's death, since it is in answer to an inquiry made in the eighth year of the blessed Honorius III (24 July 1223–24). The treatise is directed in an Oriental style to a Roman, or Romanus,[66] and Michael Scot was then high in the Pope's favor and probably at the Curia. Possibly he was already in relation, as later, with Jewish translators,[67] while not concealing the knowledge of Hebrew attributed to him by the Pope.

The commentary on the *Sphera* of John of Holywood has already been mentioned apropos of the date of Scot's death. No manuscript has been cited, and the only basis for ascribing it to

[62] For the medical literature, see Brown, pp. 149–156. The Rossi MSS. now in the Vatican contain (xi. 144) a 'Libro de consegli de poveri infermi e utile per ciascun povero medico segondo che mete Michiel Scoto astrologo del imperador Federico.'

[63] MS. 601. *Catalogue* (1918), p. 150.

[64] Ff. 21–103 v. On this version see Steinschneider, *H. U.*, p. 433; and especially Perles, in *Monatsschrift für Geschichte und Wissenschaft des Judenthums*, xxiv (1875).

[65] Ff. 1–20 v: 'In octavo anno gubernacionis felicis Honorii tercii interrogasti me, potens [MS. poteritis] et humilis Romane (prolonget tibi vitam Deus et augmentet statum), quare mel non adolebatur in sacrificiis et sal valde item (?) parrabatur in eisdem, ut dicitur secundo Levitici circa finem in illo versu [2, 11] . . . vel que removet difficultatem in operando et hoc constituitur (?) consuetudinalis.'

[66] A Romanus was then cardinal of S. Angelo, 1216–35, and later bishop of Porto.

[67] Infra, note 79.

Scot is the title of the printed edition.[68] The preface shows some similarities of phrase to the preface to the *De arte venandi* of Frederick II,[69] and the commentary recalls al-Bitrogi;[70] but there are no references to the emperor in the body of the work, and the scholastic style is quite unlike that of Scot's astrological writings, which are, indeed, professedly popular. The treatise on geomancy ascribed to Scot in a late Munich manuscript is very doubtful;[71] and the *Mensa philosophica*, at times attributed to him,[72] is clearly by another and later hand.

Scot's translations were the occasion of unfavorable judgments on the part of Roger Bacon, who declared that Scot did not really know the languages or the sciences, and that the work was chiefly done by a Jew named Andrew.[73] Help of this sort was usually employed by the Toletan translators;[74] whether Michael was more inaccurate than others is a question which has not been investigated.[75] On the other hand Bacon seems to ascribe too much credit to Scot as the introducer of the natural philosophy of Aristotle,[76] for, as we have seen, only one of these treatises, the

[68] 'Eximii atque excellentissimi physicorum motuum cursusque syderei indagatoris Michaelis Scoti super auctorem sperae cum questionibus diligenter emendatis incipit expositio confecta Illustrissimi Imperatoris Dni. D. Federici precibus.' I have used the Bologna edition of 1495 (Hain, 14555) in the Thatcher collection in the Library of Congress.

[69] 'Causa efficiens est magister Johannes de Sacrobusto et alii compositores. Causa finalis cognitio corporum celestium in se et proprietatum . . . modus agendi est quintuplex, scilicet definitivus, probativus, id est probatitius, exemplorum positivus, ut legitime per se liqueat.' *Ibid.*, f. 1 v. So Frederick considers *intentio*, *utilitas*, and describes the *modus agendi* as 'prosaycus, prohemialis, et executivus, executivus vero multiplex, partim namque divisivus, partim descriptivus, partim convenientiarum et differentiarum assignativus, partim causarum inquisitivus.' Vatican, MS. Pal. lat. 1071, f. 1 v. The preface to Scot's *Liber introductorius* discusses *ars, genus, intentio, utilitas, finis, instrumenta*, etc.: Clm. 10268, f. 16 v; N. a. lat. 1401, f. 35. That of the *Liber luminis luminum* (Brown, pp. 81, 240) has *intentio, causa intentionis, utilitas*. Such terminology appears as early as Gundissalinus, and even in the preface to the *Euclid* of Adelard of Bath (Digby MS. 174, f. 99).

[70] Duhem, iii. 246–248, who accepts Scot's authorship.

[71] Cod. lat. 489, ff. 174–206 v (saec. xvi): *Liber geomantiae Michaelis Scoti*.

[72] As by Querfeld, p. 12.

[73] *Compendium studii*, ed. Brewer, p. 472; *Opus tertium*, ed. Brewer, p. 91.

[74] Rose, in *Hermes*, viii. 332 ff.; supra, Chapter I, n. 57.

[75] Save in the case of the *De animalibus*; supra, n. 37.

[76] 'A tempore Michaelis Scoti qui annis Domini 1230 transactis apparuit de-

De animalibus, was first given to the Latin world by Scot. Bacon's date 1230 has likewise been taken too literally, especially by those who have sought to connect it with the letter recommending the new versions of Aristotle to the universities, a document once ascribed to Frederick II but now generally admitted to come from Manfred and to relate to the translations made at his court.[77]

In general Scot's writings show a respectable education. He quotes the Scriptures freely and refers occasionally to Augustine and Ambrose and more frequently to Boethius, Isidore, and Bede. Classical Latin writers, such as Virgil, Cicero, and Ovid, rarely appear. The citations from Aristotle are fairly numerous; besides the *Meteora* and *De caelo* they include two references to the full text of the *Ethics*, then just coming into use in the West.[78] There is no evidence of any real knowledge of Greek, the etymologies and the Greek names of months, climates, and points of compass being easily available at second hand; indeed it has been pointed out that in mentioning specifically Scot's knowledge of Hebrew and Arabic, Gregory IX would hardly have omitted Greek if Scot had known this language. The extent of Scot's knowledge of Hebrew we are unable to judge, but he seems to have been in relations with Jacob Anatoli, the translator of Averroës and Ptolemy.[79] It may also be noted that the Arabic writers

ferens librorum Aristotilis partes aliquas de naturalibus et metaphysicis cum expositionibus authenticis magnificata est philosophia Aristotilis apud Latinos.' *Opus majus*, ed. Bridges, i. 55, iii. 66.

[77] Document in Huillard-Bréholles, *Historia diplomatica*, iv. 383; *Chartularium Universitatis Parisiensis*, i, no. 394. Cf. Böhmer-Ficker, *Regesta*, no. 4750; Grabmann, pp. 201–203, 237, 249; supra, Chapter XII, n. 111.

[78] 'Ethica est scientia moralis quam reperitur compillavisse Aristotiles, cuius liber sic intitulatur, Ethicorum Nichomachiorum Aristotiles liber primus incipit; et sunt 10. libri cuius primus ita incipit, Omnis ars et omnis doctrina, etc.' Clm. 10268, f. 18 v; N. a. lat. 1401, f. 37. 'Unde Aristotiles in libro Ethicorum: desideratur res propter aliud.' Cod. lat. Mon. 10268, f. 16; MS. lat. n. a. 1401, f. 33 v. The history of the Latin versions of the *Ethics* is treated by Pelzer in the *Revue néoscolastique* for 1921, pp. 316–341, 378–400. Of Grosseteste's version of the commentary of Eustratius there described (pp. 382 ff.) there is a copy in the cathedral library at Seville, MS. Z. 136. 14.

[79] Renan, in *Histoire littéraire*, xxvii. 580–589; Steinschneider, *H. U.*, pp. 58, 61, 523, 553; supra, n. 63. On contemporary Jewish culture in Sicily see further M. Güdemann, *Geschichte des Erziehungswesens der Juden in Italien* (Vienna, 1884), pp. 101–107; R. Straus, *Die Juden im Königreich Sizilien* (Heidelberg, 1910), pp. 79–91.

on astronomy and astrology whom Scot cites freely were in large part available in Latin versions of the twelfth century. His scientific writings show a knowledge of medicine,[80] natural philosophy, and music, as well as a familiarity with the various branches of astronomy and its mediaeval applications. They deserve a closer examination than can here be given in relation to the astronomy and cosmology of his age.

Scot's writings on astrology were the basis of his literary fame in the Middle Ages, and it is by these that his scientific attainments must chiefly be judged today. The three treatises are introduced by a general preface, which he also calls an epilogue and which was hence written after the completion of the series.[81] It is here clear that the three are parts of a single comprehensive work, and cross-references are frequent between the *Liber introductorius* and the *Liber particularis*. This general preface, which is long and diffuse, occupying thirty-eight pages in the principal manuscript, is largely given up to a loose discussion of the Creation — in the course of which the Averroistic doctrine of the eternity of the universe is specifically denied [82] — God, the Trinity, the nature of man, and the various orders of angels and evil spirits. The heavenly bodies are not the cause of the events which they indicate, but only the signs, as the circle before the tavern is only the sign of the wine within; [83] but, granted an accurate knowledge of planets and the zodiac, we may know future events and the right occasions for doing anything.[84] Indeed, we are later told that the astrologer need not err, by God's help.[85] Sound learning (*mathesis*) is carefully distinguished from those magic arts (*matesis*) [86]

---

[80] Cf. also the prescriptions which passed under his name: Brown, pp. 154 f.; supra, n. 62.

[81] Munich, cod. lat. 10268, ff. 1–19 v; Bibliothèque Nationale, MS. n. a. lat. 1401, ff. 11–39; Edinburgh, MS. 132, f. 34. Cf. Boll, *Sphära*, p. 440, n.; Thorndike, ii. 316–322.

[82] 'Ob hanc causam dicunt multi quod mundus sit ab eterno . . . et quod mundus non sit eternus patet aperte.' Clm., f. 1 v; Nal., f. 11 v. Cf. the commentary on Sacrobosco, f. 2.

[83] Clm., f. 1; Nal., f. 11 v.   [84] Clm., f. 15; Nal., f. 32 v.   [85] Clm., f. 118 v.

[86] Clm., ff. 17–17 v. So Roger Bacon, as in the *Secretum secretorum* (ed. Steele), pp. xxviii, 2 f. Cf. Thorndike, ii. 11 f., 158, 580, 668 f.; Webb, *Ioannis Saresberiensis Policraticus*, i. 49; and in *Classical Review*, xxxv. 119 (1921).

which no Christian can rightly practise — geomancy, hydromancy, aeromancy, pyromancy, spatulamancy, necromancy,
divination, auguries, incantations, prestigiation, etc. The examples show that Scot was not unacquainted with these arts, as
when, in the name of the Trinity, he gives an incantation for
summoning evil spirits.[87] The list of magicians includes Simon
Magus, Virgil, Peter Alexandrinus, the *ariolus* of Alexander,
and Peter Abelard; to whom he elsewhere [88] adds Solomon and
Ottonel of Parma. The history of astrology is traced from Zo-
roaster to Gerbert, via Nimrod, whose dialogue with Ioanton,
illustrated with circles and figures, Scot has evidently seen and
indeed uses in the body of the *Liber particularis*.[89] From Egypt,
where it was elaborated by King Ptolemy, astronomical knowledge was carried to Spain by Atlas, all before the birth of Moses,
and from Atlas two French clerks brought the knowledge of the
astrolabe in France to Gerbert, *optimus negrimanticus*, who by
diabolical arts attained the archbishoprics of Reims and Ravenna
and at last the papal see.

The last of the three treatises, the *Physionomia*, or *De secretis
nature*, may be dismissed with a word, as it has long been accessible in print and has been studied by Foerster [90] and more recently by one of Sudhoff's pupils, A. H. Querfeld.[91] Dedicated to
the emperor, whom it professes to guide in his judgments of men,
it contains a treatise on generation and an account of the prog-
nostications from dreams, complexions, and the different parts of
the body. Its indebtedness to the *Physiognomy* of the Pseudo-
Aristotle is limited to the preface; it makes free use of Razi, and
shows some affinities with Trotula and other Salernitan writers.[92]
There is also, possibly through a common Arabic source, some
connection with the contemporary Latin version of the Pseudo-

---

[87] Clm., f. 114 v; not in Nal., so that it may be an interpolation.

[88] Clm., f. 114 v.

[89] See below, Chapter XVI. The figures of the Venetian manuscript of Nimrod
deserve study; cf. n. 99.

[90] *De translatione Latina Physiognomicorum quae feruntur Aristotelis* (Kiel, 1884);
De Aristotelis quae feruntur Secretis secretorum (Kiel, 1888); *Scriptores Physio-
gnomici* (Teubner ed., 1893).

[91] *Michael Scottus und seine Schrift De secretis naturae* (Leipzig diss., 1919).

[92] Foerster, *Scriptores*, pp. xxiii–xxv, clxxix; Querfeld, pp. 20–23, 26.

Aristotelian *Secretum secretorum*.[93] The *Physionomia* was Scot's most popular work, having been printed in a score of incunabula and nearly as many later editions.[94]

The *Liber introductorius*, consisting of four parts or distinctions, is Scot's most ambitious work.[95] It is written in more or less popular fashion (*leviter*) for beginners in' the art of astrology,[96] but is also intended for the convenience of adepts who may not

[93] Foerster, *Scriptores*, p. clxxix; Roger Bacon's *Secretum secretorum*, ed. Steele (Oxford, 1920), pp. xviii–xxi, lxiii; supra, Chapter VII.

[94] Querfeld, pp. 14 f., who has also used the Ambrosian manuscript of 1256. I have used still another printed copy in the Harvard library, ca. 1490 (Reichling, no. 1864), which is omitted from the *Census of Fifteenth Century Books owned in America*. The printed text lacks the chapters on urine, also copied as a separate treatise, which Querfeld prints, pp. 50–60; Italian version at Naples, Biblioteca Nazionale, MS. XV. F. 51.

[95] Munich, cod. lat. 10268, 146 folios, with notable figures, xivth century; Oxford, MS. Bodley 266, a copy of the Munich manuscript (Boll, *Sphära*, p. 444); Paris, Bibliothèque Nationale, Nouv. acq. lat. 1401, ff. 39–128 v, probably copied in 1279 (Delisle, *Catalogue du fonds de La Trémoïlle*, pp. 41–43); Escorial, MS. f. III, 8; modern copy at Munich, cod. lat. 10663. Extracts at the University of Edinburgh, MS. 132 (= Munich MS., ff. 118–146 v); Bibliothèque Nationale, MS. lat. 14070, ff. 112–118 v (= Munich, ff. 86 v–89 v); Vienna, MS. lat. 3124, ff. 206–211, MS. 3394, f. 214 ff. (Saxl, in *Der Islam*, iii. 166); Vatican, MS. Pal. lat. 1363, ff. 90–94; MS. Pal. lat. 1370 (Saxl, in Heidelberg *Sitzungsberichte*, 1915, p. 25); MS. Vat. lat. 4087, ff. 88–99 v; Modena, Estense, MS. lat. 79; Seville, Colombina, MS. 7.7.1, end (saec. xv), with illustrations; Cues, MS. 209, f. 76 v; see also Brown, p. 27. None of these manuscripts seems complete. The Munich and Oxford codices lack the fourth distinction which cross-references show to have contained chapters *De anima* (Munich MS., ff. 15, 88 v), *De arte cyromantie*, and *De elementis* (MS. Canon. Misc. 555, f. 37–37 v); they also contain later additions, as a table of 1320 (Munich, f. 76 v) and a judgment of Bartholomew of Parma in 1287 (f. 125 v). The Paris copy is earlier and considerably briefer, but includes the fourth distinction (ff. 105 v ff., where the elements and the soul are treated). It ends (f. 128 v): 'Librum primum in arte astronomica incepimus in honore ac laude Dei et ad preces domini nostri Frederici Rome imperatoris et semper augusti leviter composuimus propter novicios in arte et pauperes intellectus, et nunc ipsum complevimus suo adiutorio cui sit dignus honor, grandis laus cum actionibus gratiarum, concors amor, una fides, rectus timor, et reverens obedentia cum omni supplicatione humilitatis in preceptis eius per nos et sequentes amen, amen.' The Munich manuscript ends merely: 'Expliciunt iudicia questionum hominum secundum sentenciam Michaelis Scotti grandis astrologi condam imperatoris Frederici de terra Teotonica, Deo gratias amen.' I have used the Munich manuscript, cited as Clm., of which I have a complete rotograph, and the Paris manuscript, cited as Nal. Cf. Thorndike, ii. 322–326, based on the Bodley MS.

[96] Clm., f. 30; cf. ff. 74, 100, and the *explicit* of the preceding note.

have at hand the many works to which the author refers. It is
not well organized, but the early portions are chiefly astronomical
and the later astrological, the various heavenly bodies being taken
up one by one and detailed advice given for the practice of the
astrological art. The calendar is treated at some length, and
there is a certain amount of meteorology, developed more fully
in the *Liber particularis*. Emphasis is laid on the mystical value
of the sevens which rule the world — seven planets, metals, arts,
colors, odors, tones, etc. The music of the spheres leads to a
digression on music, *de notitia totius artis musice*, which gives an
outline of the whole subject, with citations of Boethius and
Guido.[97] The astronomy is based chiefly on al-Fargani, with
occasional citation of the *Almagest*,[98] but the remarkable figures
of the constellations and planets in the Munich and Oxford manuscripts, represent an antique tradition which is ascribed by Boll
to the scholia of Germanicus.[99] Scot uses the Toletan tables,
though he knows those of Arin and others. The astrological
writers cited are the usual ones: Albumasar Jafar, Zael, Hermes,
Dorotheus, Thebit ben Korah, Messehalla, and the *Centiloquium*.[100] In one instance the *Liber novem iudicum* is specially
commended.[101] The author also refers guardedly to more dangerous books: a *Liber perditionis anime et corporis* containing the
names, abodes, and workings of demons; and a *Liber auguriorum,
ymaginum, et prestigiorum* "which we have seen and possessed in
our time, although the Roman church prohibits employing them
or believing in them." [102]

Scot has plainly gone beyond the books and conducted his own

---

[97] Clm., ff. 38 v–43.　　　　　　　[98] E. g., Clm., f. 32 v.

[99] *Sphära*, pp. 441 ff., 540–543; Bruno A. Fuchs, *Die Ikonographie der sieben
Planeten in der Kunst Italiens* (Munich diss., 1909), pp. 24–29 and plates; Saxl, in
Der Islam, iii. 166–168, 175–177, and plate 27; *Catalogus codicum astrologicorum
Graecorum*, v, 1, p. 86. None of these has compared the figures in the Venice manu-
script of the so-called Nimrod (Lat. VIII, 22).

[100] On these and similar authorities see the *Speculum astronomie* usually ascribed to Albertus Magnus (*Opera*, 1891, x. 629), with Steinschneider's commentary
in *Z. M. Ph.*, xvi. 357–396 (1871). For the question of authorship see Mandonnet,
in *Revue néo-scolastique*, xvii. 313; Palitzsch, *Roger Bacons zweite Schrift über die
kritischen Tage* (Leipzig diss., 1918), pp. 12–15; and Thorndike, ii, ch. 62.

[101] Clm., f. 128. Cf. Chapter XII, n. 18.　　　　[102] Clm., ff. 114, 116 v.

experiments, leading at times to new results.[103] That this experimental temper was shared by his imperial patron we know from
Frederick's treatise on falconry,[104] and Scot gives additional illustrations of this side of the emperor's mind. Not only did Frederick, as he himself tells us, have experts brought from Egypt to
Apulia to test the incubation of ostrich eggs by the sun's heat,[105]
but he also experimented with the artificial incubation of hens'
eggs.[106] Scot advised the emperor to seek counsel at the time of
the new moon,[107] and to avoid bloodletting when the moon was in
Gemini, lest the puncture be repeated; but the emperor, wishing
to test this for himself, called his barber at this season. The
barber assured him there was no danger and staked his head upon
it, but after a successful puncture he dropped the lancet acci-
dentally on the emperor's foot, causing a swelling which required
the care of a *cynigus* for a fortnight.[108] Scot also gives his version
of an experiment which is recounted to much the same effect by

[103] 'Nos quidem fecimus multa nostris temporibus nobis et amicis de quibus
vidimus magnam probationem in rebus divinis prout diverse fuerunt instructione
libri ymaginum lune. Verbi gratia quadam vice recipiens semper solis radium per
bussulum magnum in culo totum perforatum ad instar sachi discusiti in ymaginem
quam faciebamus ad valimentum cuiusdam rei future et optate diu.' Clm., f. 114.

[104] See the next chapter.

[105] Infra, p. 311.

[106] 'Et istud fecit probare dominus imperator F. multociens et ita est reperta
veritas eorundem.' Clm., f. 117.

[107] 'Solebamus dicere domino nostro F. imperatori, Domine imperator, si vultis
a sapiente clarum consilium, postulate ipsum crescente luna.' Clm., f. 118.

[108] 'Eligitur purgatio et diminutio sanguinis et proprie manus luna existente in
signo igneo vel aereo, excepto signo Geminorum quod dominatur manibus et brachiis
notando quod tunc geminari solet percussio lanceole. Hoc autem voluit videre
dominus meus F. imperator et sic quadam vice luna existente in signo Geminorum
vocavit suum barberium dicens ei, Est modo tollere sanguinem? Barberius dixit,
Sic domine, quia tempus pulcrum est et quietum, vos autem estis bone sanitatis,
etc. Cui dixit imperator, Magister, timeo ne bis me percutiatis, quod quando
contingit periculosum est, etc. Tunc barberius ait, Domine, volo perdere caput si
plusquam semel vos percussero, etc. Tunc dedit sibi verbum et in uno ictu exivit
rivulus sanguinis. Letatur barberius dicere imperatori, Domine, timebatis de bina
percussione. Habens vero barberius lanceolam in manu apposuit eam sibi in ore,
quam cum sic teneret cecidit super pedem imperatoris et imperator fuit in culpa.
Illa cum carnem tetigisset exivit sanguis cum dolore et inde secutus est tumor unde
locus habuit consilium cynigi 15. diebus. Videns barberius casum et percussam
dixit, Domine, grandis sapientia est in vobis et magna provissio futurorum, etc.'
Clm., f. 114 v.

Salimbene.[109] Frederick had Scot calculate the height of the
starry heavens — whatever that may mean — by the tower of a
certain church, and then had the tower cut off somewhat and
casually brought Scot back to the site. Scot took his observation
and answered that either the heavens were more distant or the
tower had sunk a palm's measure or less into the earth, both of
which were impossible, whereupon the emperor embraced him in
admiration of his skill.[110]

Apart from these mentions of the emperor, there are few references to Italy. Scot tells us he predicted the rising of Aquila in
Italy 20 December.[111] He begins the year in the Pisan style,[112]
and notes that the imperial notaries begin the year at Christmas
and the Venetian notaries with the Lord's incarnation.[113] In the
streets of Messina and Tunis (?) there are fortune-tellers who
follow the Oriental precepts of Alchandrinus and seek out newly
arrived merchants.[114] Among the questions which the astrologer
must be prepared to answer are those concerning the acceptance
of election as *podestà* or the fate of a city in war; [115] indeed the
whole account of the wealth and position of the astrologer and his
mode of life [116] reflect the influence and position of the profession
in the Italy of the thirteenth century.

The *Liber particularis*,[117] also written at Frederick's request, is

---

[109] Ed. Holder-Egger, p. 353.

[110] 'De hoc probavit nos imperator in venatione apud turrim cuiusdam ecclesie
ville. Facta autem ratione per geometriam et arismetricam ei diximus summam
miliariorum et hanc fecit notare in scriptis. Interim fecit latenter truncari turrem
per .i. semissum, iterum conduxit me in venatione per illas partes et cum fuimus
iuxta turrem finxit se non bene recordari de summa numeri mensurationis cacu-
minis turris usque ad celum sydereum et sic secundo petiit rationem fieri a me.
Facta vero ratione sapienter nec invenerim ut prius, dixi, Domine, aut celum superius ascendit quam erat externa die vel turris intravit terram per unum palmum
sive semissum, quod est mihi impossibile credere, et cum non perpenderem detrun-
cationem pedis turris factam latenter ipse imperator amplexatus est me et miratus
est valde de sententia numeri et omnis qui cum eo erat.' Clm., f. 31.

[111] Clm., f. 86 v.   [112] Clm., f. 60.   [113] Clm., f. 71.

[114] 'Et talis modus qualem Alchandrinus ostendit in generali servatur inter
Arabes et aliquos Indorum, ut patet in viis et stratis Messine et Tonisti in quibus
sunt mulieres docte que invitant novos mercatores inquirere de statu illorum, de
domo sua, de fortuna sue mercationis, etc.' Clm., f. 119.

[115] Clm., fols. 133 v, 142 v.   [116] Clm., f. 118 v.

[117] It is found in the Bodleian, MS. Canon. Misc. 555, ff. 1–59, dated 1256 (unless

likewise a popular introduction. Much briefer than the *Liber introductorius*, it seeks to supplement this in certain particulars, as the preface explains:

Incipit liber particularis Michaelis Scotti astrologi domini Frederici Rome imperatoris·et semper augusti quem secundo loco breviter compillavit ad eius preces, in nomine Iesu Christi qui fecit celum et terram in intellectu. Prohemium.

Cum ars astronomie sit grandis sermonibus phylosophorum et quod de ipsa multi multa scripserunt et diversa veluti cognoverunt semel et pluries experimentis celestium et per celestia de terrestribus, idcirco que compendiose sufficiunt scribere novicio in eadem arte ad preces domini nostri Frederici Rome imperatoris et semper augusti iuxta vulgarem in gramatica compillavi ut aliquis novicius hoc opus inveniat quantum per se valeat studere in ipso et de arte astronomie intelligere competenter.[118] . . . Sed quia in precedenti libro tractavimus de hiis que utilia sunt et necessaria omni volentium scire prenominatam artem, in hoc secundo libro adhuc recitamus quedam particularia de arte plenius que vero sunt penitus de necessitate cognoscenda pariter et scienda. Et hec que intendimus dicere in illo non tetigimus quod sciamus. Qui vero hos duos libros plene noverit ac sciverit operari nomen novi astrologi optinebit.[119]

The treatise contains relatively little astrology in the narrower sense, being devoted to the reckoning of time, where the author cites Helperic, Bede, Gerland (?), and modern computists; [120] sun, moon, and stars; the winds and tides; and various meteorological questions, many of which are also touched in the *Liber introductorius*. The whole is a curious mixture of Isidore, Roman

otherwise stated, references below are to this manuscript); the Ambrosian, MS. L. sup. 92, fols. 1–89, where the date 1256 also appears; Bibliothèque Nationale, MS. n. a. lat. 1401, fols. 129–162 v, incomplete at beginning and end, following the *Liber introductorius*; Escorial, MS. e. III. 15, incomplete at the end; Vatican, Rossi MS. ix. 111, of the year 1308 (cf. *Neues Archiv*, xxx. 353 f.); Breslau, MS. f. 21 (Pertz, *Archiv*, xi. 704; Querfeld, p. 14). The extracts in MS. Corpus 221, fols. 2–53 (Coxe, *Catalogi*, p. 88) are probably in part from the *Liber particularis*. Dr. Birkenmajer informs me that there is also a copy at Berlin, Cod. lat. 550.

[118] Here follow a list of writers on astrology, much as in the *Liber introductorius*, and a list of necessary instruments: 'tabule Tolletane vel alie meliores eis ac faciliores si unquam appareant, studiosa compotatio algorismi in suis speciebus, horologium perfectum, astrolabium integrum, quadrans iustum, et spera lignea qua utuntur phylosophi ad oculum cum tractatu regularum Parisiensi, cui spere in nostro magisterio addidimus circulos planetarum sperales quos collocavimus seriatim infra zodiacum cum corporibus planetarum designatis.'

[119] MS. Canon. Misc. 555, f. 1–1 v; MS. Ambrosian L. sup. 92, ff. 1–2.

[120] 'Computiste ecclesie, ut Albericus, Girardus, et Beda,' MS. Canon. Misc. 555, f. 6 v; 'compotiste moderni,' f. 10.

tradition, Aristotle's *Meteorology*, ecclesiastical writers, and bits of Arabic learning. The setting is Italian and in large measure Sicilian, mention being found of the *tramontana* and the oppressive south wind, the Germans in the Romagnola and the march of Ancona,[121] the sulphur baths of Montepulciano, Porretta, and Montegrotto,[122] and the volcanic phenomena of Sicily.

The most interesting part of the *Liber particularis* is the last quarter, consisting of a series of questions of Frederick II on various scientific and quasi-scientific matters, with Michael Scot's answers. Frederick's use of the questionnaire has long been known from the so-called 'Sicilian Questions' directed to the various Saracen rulers and preserved in part through the answers of ibn Sabin of Ceuta analyzed by Amari in 1853.[123] More recently fragments of a set of questions on optics have been recovered by Wiedemann.[124] The series printed below is, so far as I am aware, unknown and doubtless owes its preservation to its incorporation as an addendum to the *Liber particularis*: [125]

Cum diutissime Fredericus imperator Rome et semper augustus oppinatus fuisset per institutum ordinem a semetipso de varietatibus tocius terre que sunt et apparent in ea supra eam inter eam et sub ea, quadam vice me Michaelem Scotum sibi fidelem inter ceteros astrologos domestice advocavit et in occulto fecitque mihi sicut eidem placuit has questiones per ordinem de fundamento terre et de mirrabilibus mundi que infra continentur, sic incipiens verba sua:

Magister mi karissime, frequenter ac multipharie audivimus questiones et solutiones ab uno et a pluribus de corporibus superioribus, scilicet solis et lune ac stellarum fixarum celi, et de elementis, de anima mundi, de gentibus paganis et Christianis, ac de ceteris creaturis que sunt communiter super terram et in terra ut de plantis et metallis. Nundum autem audivimus de

---

[121] 'Idem est de bestiis, verbi gratia gentes Alamanie in asta sunt difficiles gentibus Romaniole ac marchie de Ancona, etc.' MS. Canon., f. 41 v.

[122] 'Ut patet in Pulicano Viterbii, in comitatu Padue ubi dicitur Mons Gotus, etc.' MS. Canon., f. 43 v; see also below.

[123] "Questions philosophiques adressées aux savants musulmans par l'empereur Frédéric II," in *Journal Asiatique*, 5th ser., i. 240–274; 7th ser., xiv. 341.

[124] "Fragen aus dem Gebiet der Naturwissenschaften gestellt von Friedrich II," in *Archiv für Kulturgeschichte*, xi. 483–485 (1914). See above, Chapter XII, nn. 119–122.

[125] MS. Canon. Misc. 555, f. 44 v; Ambrosian MS. L. sup. 92, f. 69; MS. Rossi IX, 111, f. 37; MS. n. a. lat. 1401, f. 156 v, a somewhat different text, briefer at some points but containing the two additional passages printed in the following notes. For an English translation, see above, Chapter XII.

illis secretis que pertinent ad delectum spiritus cum sapientia, ut de para-
diso, purgatorio et inferno ac de fundamento terre et de mirabilibus eius.
Quare te deprecamur amore sapientie ac reverentia nostre corone [126] quatenus
tu exponas nobis fundamentum terre, videlicet quomodo est constancia eius
super habyssum et quomodo stat habyssus sub terra et si est aliud quod
sufferat terram quam aer et aqua, vel stet per se an sit super celos qui sunt
sub ea; quot sint celi et qui sint sui rectores ac in eis principaliter commo-
rentur; et quantum unum celum per veracem mensuram cesset ab alio, et
quod est extra celum ultimum cum sint plures et quanto unum celum est
maius alio; in quo celo Deus est substantialiter, scilicet in divina maiestate,
et qualiter sedet in trono celi; quomodo est associatus ab angelis et a sanctis,
quid angeli et sancti continue faciunt coram Deo. Item dic nobis quot sunt
habyssi et qui sunt spiritus commorantes in eis nomine, ubi sit infernus, pur-
gatorius, et paradisus celestialis, scilicet an sub terra vel in terra vel supra
terram.[127] Item dic nobis quanta est mensura huius corporis terre per gros-
sum et per longum, et quantum est a terra usque in celum altissimum et a terra
usque in habyssum, et si sit una habyssus vel sint plures habyssi, et si sunt
plures quantum cesset una ab alia; et si hec terra habeat loca vacua vel non
ita quod sit corpus solidum ut lapis vivus; et quantum est a facie terre deor-
sum usque ad celum subterius. Item dic nobis quomodo aque maris sunt sic
amare ac fiunt salse in multis locis et quedam sunt dulces extra mare cum
omnes exeant de vivo mari. Item dic nobis de aquis dulcibus quomodo ipse
omni tempore eructuant extra terram, et quandoque de lapidibus et de arboribus ut vitibus velud in vere apparet per putationem, unde veniunt et
surgunt et quomodo est quod earum quedam eructuant dulces et suaves
quedam clare et quedam turbide ac quedam spisse ut gummose, quoniam
mirramur ex eis valde eo quod scimus iamdiu quod omnes aque exeunt de
mari et euntes per diversa loca regionum et venarum adhuc intrent in mare,
et ipsum mare est tantum et tale quod est lectus et receptaculum omnium
aquarum decurrentium. Unde vellemus scire si sit unus locus per se qui
habeat aquam dulcem tantum sicut unus est que habeat aquam salsam, an
sit ambarum aquarum unus locus, et si est unus quomodo iste due aque sunt
sibi tam contrarie cum ratione diversitatis colorum et saporum atque motuum
videatur quod sint duo loca. Unde si sint duo loca aquarum scilicet dulces
et salse, querimus certificari quis eorum sit maior et minor, et quomodo est
quod hee aque decurrentes per orbem terre videantur eructuare omni tempore ex nimia habundancia sui de loco sui lecti, et licet tam copiose habun-
dent illico tamen non multiplicant quasi ultra communem mensuram ratione
tanti additus sed sic stant eructuantes quasi ex una mensura vel ad similitudinem unius mensure. Vellemus etiam scire unde fiunt aque salse et amare
que per loca reperiuntur surgitorie et aque fetide, ut in multis locis balnea-
rum et piscinarum, an ex se ipsis fiant vel aliunde. Similiter iste aque que
per loca eructuant tepide vel bene calide aut ferventes velut essent supra

---

[126] *Ac imperii maiestatis*, the Paris MS. adds.

[127] Here the Paris MS. inserts: 'Et que sit differentia animarum que cotidie
illuc defferuntur et spirituum qui de celo ceciderunt, et si una anima in alia vita
cognoscit aliam et si aliqua potest transire ad hanc vitam causa loquendi et se
demonstrandi alicui, et quot sunt pene inferni.'

ignem ardentem in alliquo vase quomodo sunt ita, unde veniunt et unde sint et quomodo est quod aquarum eructuantium quedam semper fiunt clare quedam turbide. Vellemus etiam scire quomodo est ille ventus qui exit de multis partibus orbis et ignis qui eructuat de terra tam planure quam montis; similiter et fumus apparens modo hic modo illic unde nutritur et quod est illud quod facit ipsum flare, ut patet in partibus Scicilie et Messine sicut in Moncibello, Vulcano, Lippari, et Strongulo. Quomodo etiam est hoc quod flamma ignis ardentis visibiliter apparet non solummodo in terra sed in quibusdam partibus maris Indie.[128]

Then begins Scot's long reply:

Cui ego Michael Scottus tanquam scottatus a multis et a diversis libere spopondi dicere veritatem cum vehementi admirratione tantarum et talium questionum: O bone imperator, per memetipsum oppinor vehementer quod si unquam fuisset homo in hoc mundo qui per suam doctrinam evasisset mortem, tu es ille qui inter ceteros debuisses evadere. Sed mors est talis calix et tam communis quod ex eo bibit et bibet omnis sapiens et insipiens, cum in hoc mundo nihil reperiatur fortius morte. Tamen doctrina sapientum vivorum et mortuorum que in hoc seculo dicitur vel scripta reperitur ad instruendum indoctos et ad memorandum peritoꝫ donec vita permanet proficit multis et in multis, videlicet quantum ad corpus et quantum ad animam, de qua multum curandum est. Et ideo mihi est valde acceptabile duras questiones audire eo quod tunc proficio in scientia multis modis et principaliter dum sunt ipsius scientie qua pocior et glorior inter gentes ac me penes vos video honoratum. Unde sicut constituistis cor vestrum ad has cogitationes questionum quas nunc mihi dilucidastis ordine pretaxato, sic ponite aures vestri capitis ad audiendum et mentem vestram ad intelligendum plenam satisfactionem omnium predictorum que vobis leviter et sine disputatione pandere non pigritabor si Deus voluerit.

This boastful preface, followed by a supplication for divine aid,[129] introduces thirty pages of manuscript which it is unnecessary to reproduce in full. Brief statements concerning hell, purgatory, heaven, and the terrestrial paradise are followed by an account of the marvels of nature — strange lakes and rivers of the East, wondrous metals, stones, plants, drugs, and animals, with their respective virtues. The magnet is mentioned incidentally three times,[130] each time as something well known. The

[128] The Paris MS. adds: 'Et quomodo est hoc quod anima alicuius hominis viventis dum transierit ad aliam vitam quod nec amor primus nec etiam odium dat sibi causam reddeundi tanquam nihil fuisset, nec de remanenti re videtur amplius curare sive sit salvata sive dampnata.'

[129] 'Per meam sapientiam vobis ad tanta et talia non possem veraciter satisfacere nisi esset mihi donum gratie a Deo datum.'

[130] 'Per calamitam scitur ubi est tramontana cum acu, et cognito domino anni

most interesting of these chapters is that on metals, a summary of alchemistic doctrine which can be usefully compared with the alchemical writings attributed to Scot:

Metallum est quedam essentia que dicitur secunde compositionis, cuius species sunt 7, scilicet ferr m, plumbum, stagnum, ramum, cuprum, argentum, et aurum, sciendo quod generantur compositione argenti vivi, sulphuris, et terre. Et secundum unitam materiam eorum quibus componuntur sunt ponderis et coloris. Aurum plus tenet sulphuris quam argenti vivi; argentum tenet plus argenti vivi quam terre et sulphuris; ferrum plus tenet terre quam argenti vivi, etc. Valet quodlibet ad multa ut in compositione sophystica et in aliis virtutibus. Verbi gratia: aurum macinatum valet senibus volentibus vivere sanius et iuniores esse sumptum in cibo, et per eum comparantur multi denarii argenti causa expendendi, fiunt multa monilia, decorantur vasa, et pro eo acquiruntur femine ac multe possessiones. Argentum emit aurum et ex eo multa acquiruntur ut ex auro et fiunt ut denarii, vasa, etc. Stagnum valet ad faciendum vasa et aptandum ferrum laboratum et ramum. Idem dicitur de plumbo ramo etc. Sophysticantur metalla doctrina artis alchimie cum quibusdam additamentis pulverum mediantibus spiritibus quorum species sunt 4, scilicet argentum vivum, sulphur, auripigmentum, et sal ammoniacum. Ex auro cum quibusdam aliis fit plus aurum in apparentia, ex argento et ramo dealbato cum medicina fit plus argentum in apparentia, etc. De argento leviter [fit] azurum. De plumbo leviter fit cerusa. De ramo leviter fit color viridis cum aceto forti et melle. De plumbo et ramo etc. fit aliud metallum. De stagno et ramo fit peltrum cum medicina. Argentum vivum destruit omne metallum ut patet in moneta quam tangit et stagno cuius virgam rumpit tangendo, etc. De plumbo fiunt manubria lime surde quo sonus mortificatur. Argentum vivum interficit edentem et tollit auditum si cadat in aures. Metallorum aqua, ut ferri arsenici vitrioli calcis et virideramini, corodit et frangit calibem. Ex vilibus et muracido ferro fit ferrum andanicum, et ecce mirrabile magnum.[131]

Coming at last to the emperor's penetrating questions concerning the earth, Scot explains that the earth is round like a ball, surrounded with water as the yolk is surrounded in the egg, the waters being held in their place by a secret virtue; but any further knowledge of this is beyond human ken and merit. The distance to the extreme of the waters beneath the earth equals the distance

adequatione tabularum de Tolleto scimus quod futurum est in rebus.' MS. Canon. Misc., f. 48 v. 'Item est lapis qui sua virtute trahit ferrum ad se ut calamita et ostendit locum tramontane septentrionalis. Et est alius lapis generis calamite qui depellit ferrum a se et demonstrat partem tramontane austri.' *Ibid.*, f. 50. 'Calamita reconciliat uxorem ad maritum.' *Ibid.*, f. 50 v. Cf. *Physionomia*, part 1, c. i. On the compass in the thirteenth century see the various studies of Schück (*Isis*, iv. 438) and Günther in *Deutsche Revue*, March, 1914.

[131] MS. Canon., f. 49 v; MS. Ambrosian, f. 76 v; not in Nal.

to the moon. After air ends fire begins, extending from the moon to the eighth sphere, then a multitude of waters and then the ether as far as the ninth sphere, the spheres being fitted one about another like the layers of an onion. The waters of the sea are bitter because they are older and are not exposed to the sun's heat. Waters were created with inexhaustible virtue of pouring forth so long as the world endures, and they move about in the earth like blood in the veins, the quality of the water depending on the earth through which it passes, and its heat coming from dry, hot rocks, especially sulphur. The hot springs of Monte-grotto, Porretta, and Montepulciano and the volcanic outpourings of Etna and the Lipari islands are explained as follows: [132]

Nam illius quod me interrogastis de flammis ignis que visibiliter apparent in multis locis huius mundi ut in partibus Scicilie etc., iam supra diximus intellectum huius in capitulo quod incipit, Tellus Scicilie, etc., et in capitulo alio quod incipit, Queri solet de aquis fluminum.[133] Sed quia de hoc facta est expressa questio iterum studebimus dictam questionem solvere. Unde dicimus quod in ventre terre sunt saxa sulphuris vivi et petre calidissime nature et in eisdem partibus sunt multe vacuitates quas venas appellamus et fistulas. Causa est fervor caloris quo terra grustificatur cessans a sede illius sulphuris, et ventus qui spirat per orbem reperit fixuras terre in extremis partibus et cavernas qui dum intrat in eas non revertitur retrorsum ymo flat antrorsum de vena in venam et de fistula in fistulam et sic tentans loca cavernosa pervenit ad has vacuitates ubi est tanta copia sulphuris et petrarum calidissimarum, et quia ventus est substantia calida et sicca atque subtilissima et se fricat per tales partes magis subtiliatur, et quia est de materia elementali recipit compositionem qua cum exit de locis apertis usque que [134] continuatur illa multitudo sulphuris et petrarum calidissimarum apparet flammabilis vehementer, et a diversis gentibus iudicatur et creditur esse ignis cum habeat omnes condiciones ignis nostri, scilicet motu sintilis figura dumi fumo et cinere in eisdem partibus. Calore vero tali aer in eisdem partibus inflammatur et fit subtilis calidus et sulphure odoriferus. Unde aque calide et bullientes surgunt in eisdem partibus et sunt balnee multe, sicut est Pellicanus apud Viterbium, balneum de Porreta, de Monte Gotto in districtu Padue, etc., sciendo quod ubi habundat calor et sulphur sub terra crescit aurum et nascitur, econtra in contraria parte nascitur plumbum ferrum et argentum utrumque. Sunt etiam aque frigide, lacus magni, nives, etc., unde substantia illius flamme ignis parissibilis in certis locis terre et maris non est aliud quam vapor calidus et siccus violenter inflammatus a maiori calore et

[132] MS. Canon., ff. 56 v–57 v; MS. Ambrosian, ff. 85–86 v; not in Nal.
[133] MS. Canon. Misc. 555, ff. 40, 43, where these topics are more briefly discussed.
[134] I. e., 'usquequaque.'

siccitate, quod totum fit secundum quod prediximus. Et quia ventus non cessat antecedere sive per aerem expeditum ut supra terram sive per cavernas terre prepeditum, aut in exitu loci exit calidus invisibiliter aut inflammabilis visibiliter aut frigidus invisibiliter. Et est sciendum quod si sulphur continuatur producte usque ad exitum venti exit ventus in modum flamme que est magna vel parva secundum quantitatem substantie venti et habundanciam caloris et condictionem aeris quem reperit impeditum ab aliqua impressione vel absolutum, et hoc dico tam de vento invisibili quam visibili et tam de frigido ut in partibus Sclavonie et Alamannie quam in partibus Scicilie, etc. Ut etiam patet per Strongulum montem qui est in medio maris et per Strongulinum, per Vulcanum et Vulcaninum, per Moncibellum et per insulam Lippari in qua sunt omnia genera bonarum arborum et herbarum. Nam Strongulus est mons magnus in mari et de sumitate illius exit continue magna flamma ignis. Similiter exit continue flamma ignis de sumitate montis Strongulini qui est mons minor Strongulo. De monte Vulcani et Vulcanini, Moncibelli et insule Lippari dicimus quod ex eis quandoque exit flamma ignis ut quociens ventus qui dicitur auster spirat et non alias et quando cessat flamma exit fumus maximus. Et est sciendum quod ista flamma ignis cuiuslibet dictorum locorum sepe importat lapides adhustos et quandoque sticiones lignorum et cinerum que cooperit totam terram inde et aerem sepe obcecat ut est in partibus fluminum de arena. Eiciuntur etiam multi igniculi extra in altum cum flamma ardentiores ut ferrum focine fabri sintillans qui descendendo franguntur in multa frustra et magna et parva, et hec reperiuntur esse pomices quibus utuntur scriptores, et has pomices mare portat ad littora et colliguntur a gentibus et inde murantur domus et parificantur ut apud nos de lateribus, quare in eisdem partibus sunt montes et fragmenta ut de lapidibus apud ceteras regiones. Aqua quidem pellagi est inde frigens et sulphurea unde marinarii transeuntes hinc quandoque implent nodos harundinum et catinos de illa aqua que cum est frigida esse sulphur probatur coagulatione, et est sciendum quod quanto plus aqua accedit prope montes ubi bullit tanto magis sulphur est melior. Verum est quod sulphuris alius albus alius niger alius zallus, etc., sciendo quod unusquisque habet certas virtutes magni valoris, ut in alchimia ad commutandum metalla et ad faciendum focum zambanum, unguenta ad scabiem, etc., suffumigatio cuius dealbat setam zallam et folia rose et lilii et cum ardet reddit aerem feculentum. Insuper dicimus quod si illa flamma esset ignis ut noster extingueretur ab aqua que est nostro igni contraria percurrens sub terra in partibus sulphureis in quibus inflammatur, sciendo quod sicut est cursus aquarum super terram et origo fontium lectus fluminum et multitudo lacuum et stagnorum, sic est inter terram. Item dicimus quod si dicte petre tam calide nature essent super terram sicut sunt in ea absconse et sulphur cum eis, iam mundus esset undique consumatus caliditate flatus ventorum inde transeuntium. Sed cum misericordia Dei sit maxima in dispositione constitutionis mundi, hunc sulphurem et hos lapides locavit inter terram propter melius, nolens quod mundus taliter destruatur, unde voluntate Dei flamme dictorum locorum nec mundum destruunt nec loca sibi propinqua, unde super dictos montes sunt domus que ab hominibus inhabitantur et cultus terre quo fructus habentur multi.

Such evil signs have led many to believe that these volcanoes are the entrance to the hell which is vividly described in the vision of St. Paul in prison; but whether the gate to the lower regions is here or in the northern isle seen by St. Brandan, Scot will not decide. Whatever the way in, hell is in the bowels of the earth, and there is no way out.[135]

Scot does not answer all the emperor's questions and his answers are far from satisfactory, yet all is not empty words. He has some acquaintance with the principal sulphur springs and volcanoes of Italy, and, while his knowledge of the Lipari group does not necessarily rest on personal observation, it at least represents inquiry among those who have observed. Although the omission of any special account of Etna is noteworthy, he has in these local matters gone well beyond Aristotle's *Meteorology* and given some real description of volcanic phenomena.[136] Nevertheless, making all allowance for the fact that it is easier to ask questions than to answer them satisfactorily, the emperor's questions show the keener mind and the more penetrating intelligence. They raise real difficulties, and, like those preserved by ibn Sabin, they cut deeply into the current cosmology. That one who can go so far in these directions should at the same time accept implicitly the facile predictions of the court astrologers, is one of the typical contradictions in the intellectual life of the thirteenth century.

[135] The treatise ends: 'Hec autem que breviter et facile diximus nunc ut melius fuit nobis visum, vobis, domine imperator, sufficiant ad presens de recitatione mirrabilium mundi que Deus fecit cum magno delectu ad instar ioculatoris et adhuc facit continue, et de expositione fundamenti terre. Volentes hic finire secundum librum quem incepimus in nomine Dei cui ex parte nostra sit semper grandis laus et gloria benedictio et triumphus in omnibus per infinita secula seculorum amen. Explicit secundus liber Michaelis Scotti qui dicitur liber particularis. Nunc incipit liber physionomie. . . .' MS. Canon. Misc. 555, f. 59; Ambrosian MS. L. sup. 92, fols. 88 v–89.

[136] Cf. *Geographical Review*, xiii. 141 f. (1923).

CHAPTER XIV

THE *DE ARTE VENANDI CUM AVIBUS* OF FREDERICK II[1]

THE reign of the Emperor Frederick II holds an important place in the transition from mediaeval to modern culture. Much has been written of the cosmopolitan intellectual life of his court, of its school of poetry as the cradle of Italian vernacular literature, of the philosophers and translators who linked it with the older world. To many it has seemed that it is under Frederick, "the first modern man upon a throne,"[2] rather than in the days of Petrarch, that the real beginning of the Italian Renaissance is to be sought. In any such discussion much depends upon our judgment of the personality of the emperor, that *stupor mundi* of learning whose *superstitiones et curiositates* scandalized contemporaries. All agree as to the extraordinary activity and extraordinary interest of his mind, yet its principal literary product, his *De arte venandi cum avibus*, has been strangely neglected. Mentioned in rather perfunctory fashion by other historians,[3] its significance has been more fully seen by Karl Hampe, who declares that this book must be studied by all "who wish to learn to know Frederick's method of thinking and working scientifically";[4] yet Hampe devotes but two pages to the treatise, the greater part of which he has not read. The solid volume

---

[1] Revised from *E. H. R.*, xxxvi. 334–355 (1921). Cf. Sudhoff, in *Mitteilungen zur Geschichte der Medizin*, xxi. 41; *Isis*, iv. 203.

[2] J. Burckhardt, *Die Cultur der Renaissance in Italien* (ed. Geiger, Leipzig, 1899), i. 4.

[3] Raumer, *Geschichte der Hohenstaufen* (Leipzig, 1857), iii. 286 f.; Huillard-Bréholles, *Historia diplomatica Friderici Secundi* (Paris, 1859), introduction, pp. dxxxv f.; Ranke, *Weltgeschichte*, viii. 369; Biehringer, *Kaiser Friedrich II* (Berlin, 1912), p. 273; Novati, *Freschi e minii del dugento* (Milan, 1908), pp. 137–143; Paolucci, "Le finanze e la corte di Federico II," in *Atti* of the Palermo Academy, vii. 41–45 (1904); L. Allshorn, *Stupor mundi* (London, 1912), p. 118. The very brief treatment of the *De arte venandi* is a serious gap in the suggestive article of H. Niese, "Zur Geschichte des geistigen Lebens am Hofe Kaiser Friedrichs II," in *Historische Zeitschrift*, cviii. 473–540 (1912).

[4] *Historische Zeitschrift*, lxxxiii. 19 (1899).

required for a complete text would need careful examination by the zoölogist and the falconer, in relation both to its antecedents and to its additions to the store of theoretical and practical bird-lore, and our knowledge of mediaeval zoölogy and of the earlier literature on falconry [5] is still insufficient to permit these specialists to assign the treatise to its final place. Still, a beginning must sooner or later be made, and the fresh use of manuscript material may enable even a layman to draw certain provisional conclusions concerning the sources and composition of the *De arte* and the light it throws on the workings of the emperor's mind.

The chief obstacle to a study of the *De arte venandi cum avibus* is the lack of a complete edition. The treatise contains six books, yet only two have been printed, from an incomplete manuscript then in possession of Joachim Camerarius of Nürnberg, and since supposed lost, but now clearly identifiable with MS. Pal. lat. 1071 of the Vatican. The *editio princeps* of Velser (Augsburg, 1596), reprinted with a valuable zoölogical commentary by J. G. Schneider (Leipzig, 1788–89),[6] not only has *lacunae* which correspond to the considerable *lacunae* and the faint and illegible portions of this codex, but it is in places quite careless, so that it does not furnish a satisfactory edition even of this mutilated copy of the first two books. It became the basis of two translations into German,[7] yet, with all the learning lavished on Frederick II by German writers, no one has published a comparison of the different manuscripts or edited a complete and critical text. There are two principal classes of manuscripts:

[5] The principal study of this material is by Werth, "Altfranzösische Jagdlehr-bücher nebst Handschriftenbibliographie der abendländischen Jagdlitteratur überhaupt," in *Zeitschrift für romanische Philologie*, xii. 146–191, 381–415, xiii. 1–34 (1888–89), who reviews the important mediaeval works on falconry without throwing any new light on the work of Frederick II. He overlooks the Vatican MS., mentioned by Seroux d'Agincourt in 1823, by Huillard-Bréholles in 1859, and by Bethmann in 1874 (Pertz, *Archiv*, xii. 350), and makes no advance in relation to the six-book text, first indicated by Jérôme Pichon in 1863 (*Bulletin du bibliophile*, xvi. 885–900). See also below, Chapter XVII.

[6] In the citations below I have referred to Schneider's text as the more accessible, using the copy at Columbia University, but all such passages have been collated with the Vatican MS.

[7] By Johann Erhard Pacius, Onolzbach, 1756; and by H. Schöpffer, Berlin, 1896.

I. Containing the first two books only, with Manfred's additions:

M. Vatican, MS. Pal. lat. 1071. Parchment, 111 folios, 360 × 250 mm., written not long after the middle of the thirteenth century, with valuable illustrations in a contemporary hand. The chapters are rubricated but not numbered. The rubrics are in red; the initials, in red and blue, are colored only to f. 36 v. The first page, as well as many later pages, has been partly defaced by moisture, and has two holes in the parchment, hence the *lacunae* in the first two pages of the editions. The text breaks off in c. 80 of bk. ii, shortly before the end of the book. As this text contains the additions made by Manfred as king, it falls between his coronation in 1258 and his death in 1266. The considerable *lacuna* between ff. 16 and 17 (bk. i, c. 23), which fills pp. 47–72 of MS. B, existed already in the thirteenth century, since it is found likewise in MS. *m*. (f. 28). The conclusion of bk. ii was probably also missing when the version of *m* was made, for *m* carries the text no further than the last folio of M and rounds out the sentence with a general phrase. On the other hand, the *lacuna* of one folio after f. 58 (ii, 33), not found in *m*, must have been made between ca. 1300 and 1596. On the miniatures, see Seroux d'Agincourt, *L'histoire de l'art* (Paris, 1823), v, pl. 73 and text; Venturi, *Storia dell' arte italiana*, ii, nos. 277 f., iii, nos. 689–698; S. Beissel, *Vaticanische Miniaturen* (Freiburg, 1893), p. 39 and plate xx; Graf zu Erbach-Fürstenau, *Die Manfredbibel* (Leipzig, 1910), c. 2. Those on the second page, one of which is reproduced in the Augsburg edition, evidently represent Frederick II on his throne; that on f. 5 v, on the margin of Manfred's first addition, is plausibly conjectured by Erbach to represent Manfred. The administration of the Vatican library plans a publication of the whole manuscript in facsimile edition. For much of this and for other information and assistance I am specially indebted to Monsignore A. Pelzer.

M 1. Vienna, Nationalbibliothek, MS. 10948. A sixteenth-century copy, apparently from M, omitting the preface and introduction.

m. Paris, Bibliothèque Nationale, MS. Fr. 12400. Parchment, 186 folios, ca. 1300, with illustrations. A French translation, made for Jean de Dampierre and his daughter Isabel, probably ca. 1290–1300. See *Notices et extraits des MSS.*, vi. 404; Pichon, in *Bulletin du bibliophile*, xvi. 894–897 (1863). The text is that of M, including the additions of Manfred; probably the version is based on M itself, for the illustrations of birds in M are followed closely and the same *lacuna* occurs in i. 23; but the text of M had not yet been injured by moisture or by the holes in the first folio. On the miniatures see Vitzthum, *Die*

Pariser Miniaturmalerei des xiii. Jahrhunderts, pp. 228 f. (Leipzig, 1907).

m 1. Geneva, MS. Fr. 170. Parchment, fifteenth century, with similar illustrations. Same translation as *m*. See Senebier, *Catalogue raisonné des MSS.*, pp. 426 f.; Aubert, in *B. E. C.*, lxxii. 307–309.

m 2. Bibliothèque Nationale, MS. Fr. 1296. A different French translation of the second book only. See Pichon, pp. 898 f.

II. Containing the whole six books,[8] without Manfred's additions:

B. Paris, Bibliothèque Mazarine, MS. 3716. Parchment, 589 pages, early fifteenth century, with remnants of a coat of arms of Anjou-Sicily. P. 589: 'Explicit liber falconum cum quibus venantur.' See Pichon, pp. 888–891. I have a complete rotograph of this manuscript. The illuminations, save for the first page, are confined to a few initials and have nothing in common with those of M and its derivatives. In bk. i B contains (pp. 32–37) after c. 15 a passage on the feeding of birds of prey which is lacking in M, and in c. 23 it enables us (pp. 47–72) to fill the important *lacuna* in the M group. At the close of this book (pp. 139 f.) it repeats c. 54 which it has already on p. 120. In bk. ii it omits the last sentence of the prologue and cc. 1–30, resuming with c. 31 on p. 90 of the edition; it fills (pp. 146–149) the *lacuna* in c. 33; inserts (pp. 256 f.) eight lines at the end of c. 76; and finishes (pp. 277–281) the treatment of hooding in c. 80 left incomplete by the break in M.

C. University of Valencia, MS. 402. Parchment, 238 folios, fifteenth century, with the arms of Aragon-Sicily. Attributed in a hand of the eighteenth century to Thomas of Capua (!). Inaccessible when I visited this library in 1913, but now described by Marcelino Gutiérrez del Caño, *Catálogo de los manuscritos existentes en la Biblioteca Universitaria de Valencia* (Valencia, [1915]), i. 154 f., with a facsimile of the first page which shows a text identical with B.

D. Rennes, MS. 227, paper, 404 folios, fifteenth century: 'Liber falconum cum quibus venantur.' With chapter headings throughout and a table of contents at the close, ff. 389–404; text as in B.

E. Bologna, University Library, MS. Lat. 419 (717). The full six books, with some veterinary material at the end.

---

[8] The Bodleian MS. Digby 152 (saec. xiv) contains, ff. 42–54 v, a loose body of extracts comprising a large part of the first half of bk. iii, incorporated as bk. iv of a treatise of which the lost third book dealt with the subject of Frederick's second, even taking over Frederick's reference to his own second book (f. 42 v = MS. B, p. 282). As this manuscript begins with the fourth book of the treatise and breaks off in the middle (= MS. B, p. 323), further comparison is impossible.

F. MS. formerly in possession of Baron Pichon, from whose library it passed in 1869 to M. Giraud de Savine. See *Bulletin du bibliophile*, xvi. 891–893. Closely related to B. Copy executed for Astorre Manfredi of Faenza, probably Astorre II (†1468).

The two families of manuscripts thus correspond to two editions. The first or two-book family is Manfred's edition, with the additional matter which he discovered as well as with notes of his own. The second or six-book family was not thus revised and supplemented, but it fills the *lacunae* in books i and ii. Whether Manfred revised the last four books also is a question which cannot be answered from the manuscripts so far examined. The fact that the French versions likewise contain but two books shows that a two-book text was in circulation in the thirteenth century, and lends probability to Pichon's hypothesis [9] that Manfred's revision did not extend to the later books.

So far as they can be identified, Manfred's additions are of two sorts. One group, consisting of his own practical observations, is brief and relatively unimportant,[10] their brevity not appearing in the edition, where their beginning is marked by 'Rex,' 'Rex Manfredus,' or 'addidit Rex,' but the end of the passage is not indicated. Collation with the text of the second family shows that these are ordinarily but a few lines in length.[11] A good example runs as follows:

Sunt et alie rationes quas Manfridus rex Sicilie, quondam divi Augusti imperatoris huius libri auctoris filius, addendas providit cum librum ipsum coram se legi mandavit. Cum aves omnes tam aquatice et medie quam terrestres tantum laborent pro acquirendo cibo, eundo redeundo et stando super pedes fatigantur valde, sed, nocte veniente qua quiescere consueverunt, cum stando pedes quiescere volunt vicissim aliquando super uno pede aliquando

---

[9] *Bulletin du bibliophile*, xvi. 887.

[10] They are less important than is supposed by Helene M. Arndt, *Studien zur inneren Regierungsgeschichte Manfreds* (Heidelberg, 1911), pp. 152 f.

[11] Besides those given above in the text, Manfred's glosses are in the edition as follows: i. 4 'Causa . . . rationabiliter' (26 lines); i. 53 'Inter modos . . . semper in aquis' (18 lines); i. 54 'Preterea aves . . . ut dicit Philosophus in libro celi et mundi' (8 lines); ii. 15 'Necessitas . . . pascuntur' (6 lines); ii. 53 'Amplius . . . falconum' (10 lines); ii. 59 'Et si in hoc . . . inquietat se' (18 lines); ii. 69 'Dimittens falconum . . . portandus' (3 lines). The following also appears in the Vatican text (f. 40 v), but not in the edition: 'REX. Nam tunc . . . motu' (i, c. 54, ed. Schneider, p. 60).

super alio quiescunt, sicut accidit fixis animalibus ambulabilibus dum quiescere volunt stando super pedes, quandoque super uno pede quandoque super
altero quiescunt.[12]

A more important class of additions is found in two passages
where Manfred uses indications or material left by his father.
One of these is c. 60 of book ii, a long chapter which, beginning
as follows, shows that the original, or rather the copy then surviving, contained marginal directions for later additions:

REX: Cum non contineretur in hoc libro qualiter falco deciliatus poni
debeat ad sedendum in pertica et levari ab ea et de diverberationibus et
lesuris que possent in ea contingere, sed esset in margine eius scriptum quod
addi deberet presens capitulum, tanquam necessarium prelibatis documentis de falconibus editis, prout melius expedire vidimus duximus inseren-
dum.[13]

A longer passage in ii. 18, explaining the insertion of ii. 1–30,
shows that the emperor's codex left spaces blank, and that loose
notes and drafts were also left by the author:

REX: Cum sepe legeremus et relegeremus hunc librum ut fructum
scientie caperemus et ne vitio scriptoris aliquid remanserit corrigendum,
finito prohemio invenimus quod dominus pater noster subsequenter ordina-
verat capitulum istud primo inter alia capitula, videlicet de modis quibus
habentur falcones; tamen inter capitulum istud et prohemium erant carte
non scripte, quibus repertis existimavimus aliquod aliud capitulum obmis-
sum fuisse quod scribi debebat in eis. Post spatium vero temporis, dum
quereremus quaternos et notulas libri istius, eo quod videbamus ipsum ratione scriptoris correctione egere, invenimus in quibusdam cartulis quoddam
capitulum intitulatum de plumagio falconum, quo capitulo docebantur differentie falconum per membra et plumagia ipsorum. Nos autem rememo-
rantes dubietatis quam habuimus cum perlegendo librum pervenimus ad
capitulum predictum quod prohemium sequebatur, ubi credebamus aliquem
fuisse defectum propterea quod cartas non scriptas videramus ibidem, visum
fuit nobis quod capitulum de forma membrorum et plumagio falconum illic
locari debebat, eo quod capitulum de cognoscendis falconibus capitulum de
habendis ipsis precedere debet et quod ignota et incognita, si querantur,
reperiri non possunt (quia quod est incognitum qualiter reperitur?), et si
accidit inveniri, non est ratione scientie sed fortune. Propter quod, ut inven-
toris intentio non frustretur et avem unius speciei loco alterius non acquirat,
vidimus preponendum esse capitulum quo docetur qualiter cognoscantur
falcones et in quibus conveniant et differant ratione plumagii et membrorum,
capitulo quo docetur qualiter habeantur.[14]

[12] MS. M, f. 8 v; Schneider, p. 13.   [13] MS. M, f. 90 v; Schneider, p. 140.
[14] MS. M, f. 52 v; Schneider, p. 82; translated by Pichon, p. 890, who (p. 898)
also gives the text of *m* and *m* 2.

Another important addition to the text of the *De arte* has been ascribed to Manfred, namely the remarkable illustrations found in the two-book family, but absent from all manuscripts of the second family so far found. This attribution is perhaps strengthened if we accept Erbach's identification of Manfred with a figure in the Vatican codex, and the close parallelism which he finds with the illuminations of the Manfred Bible.[15] Nevertheless, while the figures in their present form date, like the earliest manuscript, from Manfred's time, I do not believe that he first introduced them into the margin of the text, which it appears from his own words he scrupulously respected as his father's work. Indeed the emperor's book captured in 1248 already had notable marginal illustrations.[16] We know from Richard of San Germano that Frederick could draw, designing with his own hands the towers of Capua,[17] and it is probable that he at least gave the directions for these illustrations which are almost a part of the text and plainly go back to a common original. Probably they were omitted from the unrevised archetype of the six-book family. These illustrations constitute a document of the very first importance for the scientific observation and the artistic skill of their age. They must be studied in the Vatican codex,[18] save where others of the same family supply missing or injured figures,[19] and few pages lack such embellishments. The figures of the seated emperor and of one who is probably Manfred are Byzantine in pose and treatment, and the background of architecture and landscape shows little advance on the art of the *Exultet* rolls; but while the grouping is conventional and quite lacking in perspective, the drawing of birds is extraordinarily

---

[15] *Die Manfredbibel*, c. 2.   [16] See the letter published below, n. 36.

[17] 'Quod ipse manu propria consignavit': *M. G. H., Scriptores*, xix. 372; cf. E. Bertaux, *L'art dans l'Italie méridionale*, i. 717; H. W. Schulz, *Denkmäler der Kunst in Unteritalien* (Dresden, 1860), ii. 167.

[18] For references to reproductions, without colors, see p. 301 above. Venturi, *Storia dell'arte italiana*, iii. 758–768, gives some account of the coloring, which stops at f. 93 v. The water is regularly a striated blue or bluish green, the land green, streams blue, flowers generally red, buildings red, blue, brown, etc. Clothing shows some variety, but the greatest effort to reproduce differences of color is seen in the case of birds.

[19] As on f. 96 of *m*, which corresponds to the *lacuna* between ff. 58 and 59 of M.

lifelike. There are in all more than nine hundred figures of individual birds, not only falcons in various positions, with their attendants and the instruments of the art, but a great variety of other birds to illustrate the general matter of the first book. Brilliant in coloring, the work is accurate and minute, even to details of plumage, while the representation of birds in flight has an almost photographic quality which suggests similar subjects in modern Japanese art. Saracen influence has been offered as an explanation,[20] but in any case these illustrations rest upon a close and faithful study of bird life, and thus form an essential part of the work which they accompany.

Whatever the occasion for the separate preservation of the first two books, the six books of the *De arte* form a unit. After an introductory chapter on falconry as the noblest of arts, a subject for elaborate debate on the part of later writers,[21] the first book is a general treatise on the habits and structure of birds. Book ii then deals with birds of prey, their capture and training. The third book explains the different kinds of lures and their uses. The three remaining books describe, in parallel fashion, the practice of hunting cranes with gerfalcons (iv), herons with the sacred falcon (v), and water birds with smaller types of falcons (vi). The style and manner of treatment are the same throughout. There are also several cross references. Thus the first book refers to the second and others,[22] the second to those which follow.[23] The preface to the second gives the plan of the

[20] Venturi suggests the influence not only of Saracen art but of the Vienna MS. of Dioscorides (facsimile edition, Leyden, 1906), but its drawings of birds (ff. 474–483 v) show no close resemblance to those in the Vatican codex. Erbach, *Die Manfredbibel*, pp. 1, 47–52, finds parallels with the illuminations of the Manfred Bible. In the face of the close agreement of the illustrations in M and *m*, the difference of treatment noted by Erbach in his figures 14 and 15 does not seem to me sufficient to indicate the derivation of *m* from another original than M. The 'gallina de India,' correctly described in the text (i, c. 23; MS. M, f. 19), had evidently not been seen by the illustrator. See A. Thomas, "La pintade (poule d'Inde) dans les textes du moyen âge," in *Comptes-rendus de l'Académie des Inscriptions*, 1917, pp. 40 ff.

[21] Cf. Werth, *Zeitschrift für romanische Philologie*, xii. 391 f.

[22] 'De horum autem falconum et accipitrum modis plenius et evidentius manifestatur in secundo tractatu et aliis in quibus nostra intentio per se super eos descendit,' MS. B, pp. 34 f.

[23] 'In hoc tractatu secundo et in ceteris accedemus,' MS. M, f. 45 v; MS. B,

later books.[24]   Book ii, 71 refers forward to the book on gerfalcons.[25]   The opening of the third book refers to the preface.[26]
Book iv refers back to book i,[27] and repeats an interesting observation already made in the earlier book.[28]   Book v refers also to
book i.[29]

Nevertheless it is also apparent that we have not the complete work as the author planned it, probably not even as he
executed it.   Besides the subjects actually treated in the following books, the preface to book ii promises an account of the
care of birds during moulting and of the treatment of their
diseases.[30]   None of this is found in the six-book text, although
it was common in works on falconry.   There are also specific
references in the text [31] to a subsequent discussion of moulting which does not appear.   Moreover the author three times
promises a book on hawks, which was evidently to be a separate work.[32]   Now Albertus Magnus cites the *experta Frederici
imperatoris* on the care of hawks,[33] as well as a passage on black

p. 140; the edition (Schneider, p. 69) omits 'et in ceteris.' *Liber* is regularly used
of the work as a whole, and *tractatus* of the individual books which compose it; but
MS. B, p. 282, has 'ut in 2° libro huius operis diximus.'

[24] MS. M, f. 46 v; MS. B, pp. 142 f.; ed. Schneider, p. 70.

[25] 'Dicitur plene in tractatu de venatione girofalconis ad grues,' MS. M, f. 98;
MS. B, p. 241; ed. Schneider, p. 152. Note that this remains in the two-book text.

[26] 'Intentio nostra ita ut in principio diximus est docere venationes quas faciunt
homines cum avibus rapacibus ad predandum non rapaces,' MS. B, p. 281.

[27] 'Ut dictum est in capitulo de reditu avium,' MS. B, p. 359. Cf. the reference
to bk. iv on cranes in i, 55 (MS. M, f. 42; ed. Schneider, p. 64).

[28] MS. B, pp. 54 f., repeated p. 361.   See the passage below, p. 312.

[29] 'Nidificant autem in canetis paludum et in arboribus prope aquas ut in primo
tractatu dictum est,' MS. B, p. 440, where the reference is to the treatment of nest-
ing on pp. 60 ff., where there is a *lacuna* in M and the editions.

[30] 'Quedam in conservando sanas etiam quando iam mutant pennas, ut domuncula que dicitur muta, et plumas et multe medicinarum, quedam in curando egrotas
ut ipse medicine et vasa necessaria ad dandum ipsas medicinas; de singulis horum
instrumentorum dicetur ubi conveniet,' MS. M, f. 46 v; MS. B, p. 143; ed.
Schneider, p. 70.

[31] MS. M, f. 45; MS. B, p. 138; ed. Schneider, p. 68. Also the following from
bk. iii: 'dicemus infra quando dicemus de muta et de omni eo quod convenit muta-
tioni,' MS. B, p. 324.

[32] MS. M, ff. 49, 57, 58 v; ed. Schneider, pp. 75, 89, 92.

[33] *De animalibus*, xxiii, c. 40, par. 20 (*Opera*, ed. Paris, 1891, xii. 477), for which
we should now consult Stadler's edition from the original Cologne MS. (*Beiträge*, xv-

falcons[34] which cannot be found in the present text, and in each case he refers at the same time to the *dicta* of King Roger's falconer, William, of whom we shall have more to say. A separate treatise on other forms of hunting which he promised after the completion of this[35] may not have been written, if indeed it was ever begun.

On all these questions interesting light is thrown by a letter addressed to Charles of Anjou in 1264 or 1265 by a certain Guilielmus Bottatus of Milan, of which the original is preserved in the Archives des Bouches-du-Rhône at Marseilles: [36]

Magnifico et glorioso domino K. filio regis Francie Andegav[ie] Provintie et Forc[alquerii] illustri comiti et marchioni Provintie, Guilielmus Bottatus Mediolanensis salutem et paratum devotionis et famulatus obsequium. Quia de magnifice serenitatis vestre prestantia et egregiis liberalitatis strenuitatis prudentie benignitatis et virtutum omnium ac nobilitatum titulis quibus inter cunctos seculi principes vos excellentissime prepolere fama predicat totus mundus testatur et opera laudis argumento certiori declarant, qualibet pretiosi prerogativa decorari preeminentia vestra singulari meretur privilegio. Ego quamvis inter maiestatis vestre subditos per obsequiorum exhibitionem ignotus totis tamen cordis affectibus et ex tota possibilitate devotus ad honoris vestri cumulum, iuxta morem evangelice vidue minutum meum quod mihi contulit facultas offerre cupiens, quoddam in meis facultatibus pretiosum solis excellentibus dignum dominationi vestre tradere preelegi, nobilem scilicet librum de avibus et canibus bone recordationis olim domini FR. gloriosi Romanorum imperatoris quem pre ceteris placidis habere noscebatur precipuum, cuius pulcritudinis et valoris admirationem lingua prorsus non sufficeret enarare; auri enim et argenti decore artificiose politus et imperatorie maiestatis effigie decoratus in psalteriorum duorum voluminis spatio, per compositam capitulorum distinctionem docet ancipitrum, falconum, ierofalconum, asturum, et ceterarum nobilium avium

xvi, 1916–21), p. 1481. On the dates of Albert's works, see F. Pelster, *Kritische Studien* (Freiburg, 1920); and his note on Albert's recently discovered *Questiones super libris de animalibus* of 1258 (*Zeitschrift für katholische Theologie*, xlvi. 332, 1922).

[34] *Loc. cit.*, par. 10, ed. Stadler, p. 1465; infra, Chapter XVII, n. 17.

[35] 'De reliquis vero venationibus precipue de illis in quibus nobiles delectantur vita comite post complementum huius operis dicetur a nobis,' MS. M, f. 3; ed. Schneider, p. 4.

[36] B 365, for a photograph of which I am indebted to the archivist, M. R. Busquet. Extracts, omitting the most significant portions, in Papon, *Histoire de Provence* (Paris, 1778), ii, preuves, p. lxxxv. The date must fall in or about 1264, when Charles had entered into relations with Lombardy but had not yet taken the title of king in 1265; cf. Sternfeld, *Karl von Anjou als Graf von Provence* (Berlin, 1888), p. 218.

et canum omnium cognitionem, nutrituram, eruditionem, et eorum omnium
[infir]mitates et earum causas, signa, et curationes similiter earumdem; illic
etiam ostenditur quomodo si [quis ab] aucupe fugerit possit et debeat mirabiliter rehaberi; venationes insuper describit et quomodo versari venator se
debeat ad perfectionem artis venatorie demonstratur. Ad decus etiam et
utilitatem operis in margine libri ingeniosissime depicti sunt canes et aves,
egritudines eorum et earum signa, cure, et eruditiones, et universa sicut per
litteram denotantur. Quem a quodam ad cuius manus incasu quem memo-
ratus imperator sustinuit in castris Victorie penes Parmam pervenerat blanda
et ingeniosa collatione munerum adquisivi et eum nisi prolixitas itineris et
viarum turbassent discrimina celsitudini vestre dudum fueram oblaturus.
Quo circa excellentiam vestram reverenter propulsare duxi presentibus
quatinus, si dominationi vestre memorati libri placet iocunditas, devotioni
mee benignitas vestra dignetur rescribere quid de ipso per me iusseritis
faciendum. Quia paratus sum librum ipsum sicut et ubi decreveritis trans-
mittere et cunctis beneplacidis vestris liberaliter exponere me et mea.
Valete.

Et ut plenius libri ipsi[us] qualitatem et intentionem vestra comprehendere possit industria, libri ipsius capitula que ob eorum prolixitatem incongruum est literis contexcre in cedula per ordinem sicut in ipso seriatim
habentur duxi cum presentibus vestre preeminencie destinare, cuius reor
in toto orbe similem vel exemplum nisi penes me vobis devotissimum re-
pereri[s].

The accompanying table of contents has unfortunately long
since disappeared, but the description given in the body of the
letter is sufficient to show that the two large volumes thus offered
do not correspond to any known work, whether the *De arte* or
other contemporary treatise. Covering as they did dogs as well
as falcons, hunting as well as hawking, and the diseases of such
animals as well, they cannot be identified with the *De arte* in its
present form nor with the brief treatises of Moamyn or Yatrib described below.[37] Conceivably they might have contained a collection of materials on all these topics for the emperor's use, but
the gold and silver adornment and the size of the work clearly
point to an *édition de luxe* and not a mere set of *documents pour
servir*. Moreover, the 'imperial effigy' still meets us on the first
page of the Vatican manuscript. I believe we here have described Frederick's own copy of the *De arte*, with illustrations
throughout such as the two-book text has preserved, and comprising the lost portions to which he refers in the early books.
Captured with his crown and all his treasure in the defeat before

[37] Pp. 318-320.

Parma in February, 1248,[38] it would seem to have ultimately disappeared with the rest of the scattered loot of the camp. With the completed work thus lost to the enemy, there would be left only such drafts and notes as Manfred describes, very likely kept in some Apulian castle. Indeed if a final official copy had been preserved in the South, Manfred would hardly have undertaken his search for such scattered material.

That Frederick himself was the author can no longer be doubted. Apart from the citations by Albertus Magnus [39] and the specific mention by the so-called Nicholas of Iamsilla,[40] we have the explicit words of Manfred mentioning *dominus pater noster* as the author, as well as the reference to himself in the third person as *imperatoris huius libri autoris filius*.[41] Furthermore, Frederick appears as the author in the preface, as printed below, and in the further prefatory matter.[42] If he did not actually write the book with his own hand, he at least directed its composition and dictated the greater part of its substance.

That the *De arte* belongs to the later years of Frederick's reign is also clear. He tells us in the preface that he had had it in mind for about thirty years, and had completed it at the urgent request of Manfred, to whom it is dedicated.[43] Manfred, born in 1232,[44] could hardly have been much interested in such a book before the age of, say, twelve, which would bring us to 1244, even

---

[38] On the capture of Frederick's treasure at Vittoria, see Böhmer-Ficker, *Regesta imperii*, nos. 3666 a, 13649 f.; Salimbene, ed. Holder-Egger, pp. 203 f., 342 f.

[39] *De animalibus*, xxiii, c. 40, pars. 10, 20, ed. Stadler, pp. 1465, 1478, 1481.

[40] 'Ipse quoque imperator de ingenti sui perspicacitate, que precipue circa scientiam naturalem vigebat, librum composuit de natura et cura avium in quo manifeste patet in quantum ipse imperator studiosus fuerit philosophie,' Muratori, *Scriptores*, viii. 496.

[41] Supra, pp. 303 f.

[42] 'Actor est vir inquisitor et sapientie amator divus Augustus Fredericus secundus Romanorum imperator Ierusalem et Sicilie rex. . . . Libri titulus talis est, Liber divi Augusti Frederici secundi Romanorum imperatoris Ierusalem et Sicilie regis de arte venandi cum avibus,' MS. M, f. 1 v; ed. Schneider, p. 2.

[43] See the preface printed below, p. 312.

[44] On Manfred's youth see Böhmer-Ficker, *Regesta imperii*, nos. 4632 b-h, and A. Karst, *Geschichte Manfreds* (Berlin, 1897), p. 1, who discuss the question of his legitimacy. If his formal legitimation could be established and dated. it might perhaps furnish a *terminus post quem* for the dedication.

if we allow that Frederick's own precocity [45] might have started the idea of the book in his own mind some years before 1214, when he reached the age of twenty. In 1241 the author was still gathering material, as we see from the translation in that year, under his supervision, of the Arabic treatise of the falconer Moamyn rendered into Latin by Theodore the interpreter.[46] The *De arte* can safely be assigned to the period ca. 1244–50. A date before 1248 has been suggested,[47] because of the troubles of the following years; if we are correct in the conclusion that Frederick's personal copy was captured in February of that year, this would be the latest limit.

The local allusions refer almost wholly to Apulia, where the emperor's correspondence shows that, many of his falcons were kept.[48] It must be said that such allusions are rare: the form of the treatise is general and scientific, with little illustrative detail and no hunting stories. Only twice does he mention his experiences in the East, once in connection with the flight of Syrian doves,[49] and again apropos of the Arabian methods of hooding falcons which he introduced into the West under the guidance of oriental falconers.[50] When he wants to test the incubation of ostrich eggs by the sun's heat, he has experts brought from Egypt to Apulia:

> Et hoc vidimus et fieri fecimus in Apulia, vocavimus namque ad nos de Egipto peritos et expertos in hac rc.[51]

Pelicans are called *cofani* in Apulia.[52] Young birds should be protected especially against the south winds,[53] a precaution

---

[45] See the letter describing him as a youth ca. 1207 published by Hampe, *Mitteilungen des Instituts für österreichische Geschichtsforschung*, xxii. 597.

[46] See below, n. 122.   [47] Pichon, *op. cit.*, p. 886.

[48] Böhmer-Ficker, *Regesta*, nos. 2589, 2668, 2705, 2749, 2801, 2807, 2814. See below, p. 324.

[49] MS. M, f. 39; MS. B, p. 124; ed. Schneider, p. 60. It is not expressly stated that the emperor saw these in the East, but this seems probable.

[50] MS. M, f. 104 v; MS. B, p. 258; ed. Schneider, pp. 162 f.; infra, p. 320.

[51] MS. B, p. 67; lacking in M and the editions. Cf. the experiments with hens' eggs, Chapter XIII, n. 106.

[52] 'Pellicani qui ab Apuliensibus dicuntur cofani,' MS. M, f. 3 v; ed. Schneider, p. 6. 'Pellicani quos quidam in Ytalia dicunt cofanos,' MS. M, f. 6; ed. Schneider, p. 9.

[53] MS. M, f. 58 v; ed. Schneider, p. 92. Cf. Moamyn (MS. Corpus 287, f. 48 v): 'Domus non sit aperta a parte austri.'

necessary in Frederick's dominions only in the land of the sirocco. One passage brings us more specifically to that region of the Capitanata where Frederick's favorite castles lay:

In quadam regione Apulie plane que dicitur Capitanata in tempore reditus gruum capte sunt iam grues cum girofalcis, falconibus et aliis avibus rapacibus, que erant sanguinolente in plumis et pennis sub alis et in lateribus et erant adeo debiles quod vix poterant volare et alique de talibus iam fuerunt capte manibus hominum, cuius rei simile non audivimus in aliis regionibus visum fuisse.[54]

The purpose and method of the treatise can best be seen from the preface, where, planning the first comprehensive and finished work on the subject, he declares his independence of Aristotle on the ground that the philosopher had little or no practice in falconry, and indicates his own reliance on experience and the results of long inquiry among experts brought from a distance. Fragmentary and corrupt in the edition, the preface reads as follows: [55]

Liber divi Augusti Frederici secundi Romanorum imperatoris, Ierusalem et Sicilie regis, de arte venandi cum avibus [56]

Pre[sens opus ag]gredi[57] nos induxit et [58] insta[ns tua pe]titio, fili karissi[me Man]fride,[59] et ut removeremus errorem plurium circa presens negocium qui sine arte hiis[60] que artis erant in eodem negocio abutebantur imitando[61] quorundam libros mendaces et insufficienter compositos de ipso, et ut relinqueremus posteris artificiosam traditionem de materia huius libri. Nos tamen, licet proposuissemus ex multo tem[pore ante] componere presens [opus, dis]tulimus fere per trigi[nta a]nnos propositum in scripto redigere, quoniam non

---

[54] MS. B, p. 361; repeated from pp. 54 f.

[55] The text is based on MS. M, with the portions in brackets filled in from B, C, and D. I have not included the introductory matter which follows, since it appears sufficiently in the editions.

[56] There is no heading in the manuscripts, but the title is given in the introductory matter which follows the preface proper: 'Libri titulus talis est, Liber divi Augusti Frederici secundi Romanorum imperatoris, Ierusalem et Sicilie regis, de arte venandi cum avibus divisivus et inquisitivus ad manifestationem operationum nature in venatione que fit per aves.' So M, f. 1 v. The edition omits all after 'avibus.' B and D omit 'de arte venandi cum avibus.' C has further at the end of i, c. 1, 'Divi Augusti Federici secundi Romanorum imperatoris, Ierusalem et Sicilie regis, super librum de avibus et aucupando prologus explicit.'

[57] *agendi*, B C D.   [58] Om. B C D.

[59] *vir clarissime M.S.*, B C D, the last letter blotted in C. *m* has *Tres chiers fils Manfroi*. The edition omits everything to this point.

[60] *habentes*, ed.   [61] *in imitando*, C.

putabamus nos extunc sufficere neque [l]egeramus umquam aliquem preces-
sisse qui huius libri materiam complete tractasset,[62] particule vero aliquot
ab aliquibus per solum visum scite [63] erant et inartificialiter tradite. Ideo [64]
multis temporibus cum sollicitudine [65] diligenter [66] inquisivimus ea que huius
artis erant, exercitantes nos mente [67] et opere in [eadem] ut tandem suffi-
ceremus redig[ere in librum] quicquit nostra [experientia aut ali]orum didi-
cerat.[68] [quosque[69] erant ex]perti circa [praticam huius artis] non sine magn[is
dispendi]is ad nos vocavimus [70] de longinquo vocatosque [undecumque]
nobiscum habuimus, deflorando [71] quicquid melius noverant [72] eorumque
dicta [et facta] memorie [73] commendando. Qui quamvis arduis et inexplica-
bilibus fere nego[ciis perse]pe prepediti essemus circa regnorum et imperii
regimina, tamen hanc nostram intencionem [predi]ctis [74] negociis non post-
posuimus. [In scri]bendo etiam [75] Aristotilem [76] [ubi oportui]t secuti sumus,
in [77] pluribus enim sicut experientia didicimus maxime [78] in naturis a[vium
quarundam [79] discrepa]re a veritate [videtur. Propter hoc] non sequi[mur
principem philosophorum in omnibus, raro namque aut nunquam] vena-
tion[es avium exercuit], sed nos semper [dileximus] et exercuimus. De multis
vero que narrat in libro animalium dicit quosdam sic dixisse, sed id quod
quidam sic dixerunt nec ipse forsan vidit nec dicentes viderunt, fides enim [80]
certa non proven[it] ex auditu. Quod vero multi multos [libros] scripserunt
et non nisi q[uedam de arte], signum est artem ipsam pluri[mum esse diffi-]
cilem et [81] ad[huc diffusam]. Et dicimus quod aliqui nobiles minus negociosi
nobis si huic arti attente ope[ram exhibebunt cum adiu]torio huius libri
[poterunt meliorem com]ponere, assidue siquidem nova et difficilia emergunt
circa negocia huius artis. Rogamus autem unumquemque nobilem huic
libro ex sola sua [82] nobilitate intendere debentem [83] quod [84] ab aliquo scientiarum perito ipsum legi faciat et exponi, minus benedictis indulgens. Nam
cum ars habeat sua vocabula [85] propria quemadmodum et cetere artium et
nos non inveniremus in gramatica Latinorum verba convenientia in omnibus, [app]osuimus illa que magis videbantur esse propinqua [86] per que
intelligi possit [87] intentio nostra.

For the composition of the *De arte* three kinds of sources were
available: systematic works on natural history and related fields

[62] *complere tentasset,* ed.

[63] So B C D. *Sicut,* M.

[64] So B C D. *Immo,* M. *Et pour ce, m.*

[65] *et studio,* insert B C D.

[66] *diligenti,* B C D.

[67] *in ea,* ed.

[68] *diderat,* B D.

[69] *quos quod,* ed.

[70] *venientes,* ed.

[71] *denolundo,* ed.

[72] *noverint,* ed.

[73] *memoriter,* ed.

[74] *predicta,* M. *presentis negocii,* B C D.

[75] *contra,* B C D.

[76] *artem,* ed.

[77] Om. ed.

[78] *maximorum,* ed.

[79] *quarundam avium,* B C D.

[80] *est,* B C D.

[81] Om. C.

[82] *sua sola,* C.

[83] Here the facsimile of C ends.

[84] *qui,* ed.

[85] Om. D.

[86] *propinqua esse,* B D.

[87] *posset,* B D.

of science, notably Aristotle's *De animalibus historia*; practical treatises on falconry; and the direct observation and personal inquiries of the author. Let us examine them in this order:

1. Aristotle, says the preface, is followed where required (*ubi oportuit*). He is frequently cited in the first or general book, sometimes by name only,[88] sometimes specifically as the author of the *Liber animalium*.[89] Once the reference is merely to a *Liber animalium* which seems to be Avicenna's commentary on Aristotle.[90] In the Arabic tradition of the Middle Ages the *Liber animalium* comprised the three Aristotelian treatises, *De animalibus historia*, *De partibus animalium*, and *De generatione animalium*, in all nineteen books. Translations of the Arabic text and of Avicenna's commentary had been recently made by Michael Scot,[91] and it is probably in this form that the emperor was acquainted with Aristotle's writings on natural history, for while his references can ordinarily be identified in the *De animalibus historia*,[92] not all of them can be made to square with the Greek text.[93] Doubtless Aristotle was used in other places where he is not cited, but Frederick's treatment is independent, and is

---

[88] Ed. Schneider, pp. 5 f., 8, 13, 16, 24, 25, 31, 72 f.; infra, n. 98.

[89] *Ibid.*, pp. 5, 6, 8, 43.

[90] 'Oculi sunt instrumenta visus, de quibus quare sint duo, quare in prora capitis locati, et quare altius instrumentis aliorum sensuum, et quomodo constant ex tribus humoribus septem tunicis, dictum est in libro animalium,' MS. M, f. 19; ed. Schneider, p. 29, who points out (i, p. xvi; ii. 17) that this is not found in Aristotle. A long passage deals with these matters in Michael Scot's translation of Avicenna, *De animalibus*, xiii, c. 8, f. 32 r of the printed text (Hain 2220*; copy in the Library of the University of Michigan); cf. the *Canon* of Avicenna, iii, 3, 1, 1, whence the passage is taken by Albertus Magnus, *De animalibus*, i, 2, 7 (ed. Stadler, p. 73). In general the *De arte* has little in common with Michael Scot's version of Avicenna.

[91] Jourdain, pp. 129–134, 327–349; Steinschneider, *H. U.*, pp. 478–483; J. Wood Brown, *Michael Scot* (Edinburgh, 1897), c. 3; Dittmeyer, preface to Teubner edition of the *De animalibus* (1907), pp. xix–xxi; G. Rudberg, in *Eranos*, viii. 151–160, ix. 92–128; H. Stadler, *Albertus Magnus de Animalibus*, i, p. xii; Grabmann, pp. 185 ff.; supra, Chapter XIII, nn. 37 ff.

[92] Thus p. 5 in Schneider's edition = *H. A.*, viii, 2; p. 6 = viii, 12; p. 13 = i, 1; p. 16 = ix, 34; p. 24 = viii, 12; p. 25 = ix, 10.

[93] Thus in the passage printed below, p. 321, Aristotle is made to say that no one has seen a vulture's nest (*Hist. animal.*, ix, 11); but he elsewhere says specifically that nests have been seen (vi, 5). Nor does Aristotle say (ix, 10) that the leader of cranes is permanent, as the *De arte* asserts (p. 25). I have not been able to compare the text of Michael Scot's translation.

much fuller than it could be made by the amplest use of ancient authorities, including Pliny, who is mentioned by name but once.[94] Thus one may compare the brief treatment of migration by Aristotle [95] with the account in the first book of the *De arte*,[96] which uses Aristotle but treats the subject far more amply with the aid of personal observation. Schneider, the learned commentator of Aristotle and Frederick II, declares that the emperor's description of down and feathers is the most careful he knows,[97] and one has only to read the first book to see that much of it rests upon minute and varied observation. As a matter of fact, Aristotle is cited mainly where the author disagrees with him and seeks to correct him from personal experience: *non sic se habet.*[98] The Stagyrite is evidently viewed as a man of books, to whom the reader may be referred for learned detail,[99] but who has little or no practical knowledge of falcons and relies too much on hearsay.[100] To the author he is plainly not 'the master of them that know' birds. Nowhere does Frederick's emancipation from tradition and authority stand out more clearly than in his attitude towards Aristotle.[101]

With the exception of Aristotle there are few specific citations,

[94] Schneider, p. 73.   [95] *Hist. animal.*, viii, 12.

[96] Cc. 16–23, ed. Schneider, pp. 19–26, with the following *lacuna* filled in from MS. B, pp. 47–56.

[97] *Reliqua librorum Friderici II*, ii. 41.

[98] 'Quod ergo Aristotiles dicit in libro animalium, aves uncorum unguium idem sunt quod aves rapaces, non sic se habet,' MS. M, f. 28 v; ed. Schneider, p. 43. 'Non est ergo verisimile quod scribitur ab Aristotile,' MS. M, f. 16 v; ed. Schneider, p. 25. 'Non . . . ut dicit Aristotiles,' MS. M, f. 15; ed. Schneider, p. 24. 'Quamvis Aristotiles dicat contrarium,' MS. M, f. 20; ed. Schneider, p. 31. 'Licet dixerit Aristotiles,' MS. M, f. 47 v; ed. Schneider, p. 72.

[99] 'Quomodo autem generatur pullus in ovo et que membra ipsius prius apparent et formantur et quod tempus est aptius cubationi et per quantum tempus cubant aves et reliqua constantia circa hec pretermittimus, eo quod sufficienter dictum est in libro animalium (*H. A.*, 6, 1–9) nec spectat ad nostrum propositum, quod est de perfectis avibus rapacibus qualiter docentur rapere aves non rapaces iam exclusas de ovibus et perfectas,' MS. B, p. 67. Cf. MS. M, f. 3 v; ed. Schneider, p. 5: 'Reliqua vero omnia que pretermittimus de naturis avium in libro Aristotilis de animalibus requirantur.'

[100] See the preface, supra, p. 313.

[101] Yet Biehringer (*Kaiser Friedrich II*, p. 244) can speak of the emperor as 'ein bedingungsloser Bewunderer des Aristoteles.'

and an examination of the literary sources would require a wide range of reading, especially in the scientific literature of the Arabs. As regards general scientific knowledge, the author follows the traditional division into climates, the third, fourth, fifth, and sixth climates being called *nostre regiones*.[102] Outside the Mediterranean he mentions *Britannia que vocatur Anglia*,[103] and Iceland, the home of the gerfalcon, between Norway and Greenland.[104] The *Aphorisms* of Hippocrates are cited in one passage.[105] In mathematics he is acquainted with the nature of tangents [106] and the *figura quam geumetre dicunt piramidalem*.[107] He fixes his seasons specifically by the progress of the sun through the zodiac.[108] His terminology and arrangement, as in the introductory matter and the prologue to the second book,[109] show training in the philosophical methods of the age. *Legitur in pluribus libris philosophorum*, we read at the beginning of the chapter on the relative size of male and female birds (ii, 2), but its discussion of humors and complexions shows the influence, not merely, as Niese says,[110] of the physiognomic writers, but of the whole physiological tradition of the period; certainly the physiognomic element is not sufficient to support Niese's conjecture of the collaboration of Michael Scot, who died probably before 1236,[111] and whose *Liber phisionomie*, dedicated to the emperor,[112] shows no parallelisms with the *De arte*. At one point[113] there is a citation of the pseudo-Aristotelian *Mechanics*,

---

[102] 'In nostris regionibus, scilicet sexti climatis quinti quarti et tertii,' MS. B, p. 515.   [103] Infra, p. 323.

[104] 'In quadam insula que est inter Noroegiam et Gallandiam et vocatur theutonice Yslandia et latine interpretatur contrata seu regio glaciei,' MS. M, f. 49 v; ed. Schneider, p. 75. Moamyn has 'nascuntur in partibus frigidis ut in Dacia et Norodia' (MS. Corpus 287, f. 45 v).

[105] MS. M, f. 60; ed. Schneider, p. 94.   [108] MS. B, pp. 52, 440–443.

[106] MS. M, f. 27; ed. Schneider, p. 42.   [109] Ed. Schneider, pp. 2, 69 f.

[107] MS. M, f. 75; ed. Schneider, p. 117.   [110] *Historische Zeitschrift*, cviii. 510, n.

[111] Henry d'Avranches, in *Forschungen zur deutschen Geschichte*, xviii. 482 ff.; supra, Chapter XIII, n. 26.

[112] Various editions; I have used Hain 14546*, in possession of Dr. E. C. Streeter of Boston, and a copy in the Harvard library (Reichling, no. 1864). Cf. Chapter XIII, n. 94.

[113] 'Portiones circuli quas faciunt singule penne sunt de circumferentiis equidistantibus, et illa que facit portionem maioris ambitus et magis distat a corpore avis

which has not hitherto been noted in a mediaeval version either Arabic or Latin.[114]

2. Existing works on the art of falconry, Frederick characterizes as incorrect and badly written (*mendaces et insufficienter compositos*), at best dealing in rude fashion with certain small portions of the subject (*particule aliquot*).[115] This earlier literature in Latin and the Romance vernaculars[116] is known to us only in fragmentary and confused form: the letters to Ptolemy and Theodosius, the book of the enigmatical King Dancus,[117] the puzzling references made by Frederick's contemporaries, Albertus Magnus and Daude de Pradas,[118] to King Roger's falconer, William,[119] and to the 'book of King Henry of England.'[120] Further study is required before we can venture with confidence into this field. For our present purpose it is sufficient to point out that Frederick draws little or nothing from the known works of these authors, all of them brief and confined to a summary account of the various species of hawks and falcons and to precepts respecting their training and diseases. Even King Roger's falconer, whom Albertus Magnus quotes specifically through the intermediary of Frederick, is not mentioned in the manuscripts of the

iuvat magis sublevari aut impelli et deportari, quod dicit Aristotiles in libro de ingeniis levandi pondera dicens quod magis facit levari pondus maior circulus,' MS. M, fols. 23 v–24; MS. B, p. 89; ed. Schneider, p. 36. See *Mechanica*, ed. Apelt (Leipzig, 1888), especially cc. 1, 3; ed. Bekker, pp. 848–850.

[114] Steinschneider, *H. U.*, pp. 229 f.; idem, in *Centralblatt für Bibliothekswesen*, Beiheft 12, p. 74; Grabmann, *Aristotelesübersetzungen*, pp. 200–204, 248 f., does not mention this among the pseudo-Aristotelian works translated under Manfred.

[115] Preface, supra.

[116] See in general Werth, in *Zeitschrift für romanische Philologie*, xii. 146–171; supplemented by Chapter XVII, below. Which of the Romance languages are reflected in the vocabulary of the *De arte* is a question that must be left to the philologists.

[117] For the MSS. see Chapter XVII, n. 18.

[118] Since Werth wrote, a complete text of Daude de Pradas, *Lo romans dels auzels cassadors*, has been edited from the Barberini MS. by Monaci, in *Studi di filologia romanza*, v. 65–192 (1891).

[119] Infra, pp. 348–350. Werth, xii. 157–159, xiii. 11.

[120] Infra, p. 348. The reference is apparently to a lost work in Provençal, whether prepared under the king's direction or merely dedicated to him does not appear. Werth, xii. 154 f., 166–171, thinks he can identify it as the source of other passages in Daude.

De arte thus far known.   All these writers would have been useful primarily in relation to the treatment of diseases, and this part of Frederick's work has yet to be discovered.

Besides bringing skilled falconers from the East,[121] the emperor also had their writings translated for his own use.   At least one such work has come down to us in numerous copies, the treatise of an Arab falconer, Moamyn, *De scientia venandi per aves*, as turned into Latin by Frederick's interpreter Theodore and corrected by the emperor himself at the siege of Faenza (1240–41).[122] Master Theodore of Antioch, who here styles himself "the least of the emperor's servants," is a characteristic figure of the court.[123] His preface, after an elaborate disquisition on the particular pleasure appropriate to every human act, in the course of which the *De anima*, *Nicomachean Ethics*, and *Rhetoric* of Aristotle are cited,[124] concludes that hunting is the only distinctively royal amusement:

In quantum enim sunt reges non habent propriam delectationem nisi venationem.   Considerans autem dominus noster serenissimus imperator Fredericus secundus semper augustus, Ierusalem et Sicilie rex, istius delec-

---

[121] Preface, supra; also MS. M, f. 104 v (ed. Schneider, p. 163): 'non negleximus ad nos vocare expertos huius rei tam de Arabia quam de regionibus undecumque, ab eo tempore scilicet in quo primitus proposuimus redigere in librum ea que sunt huius artis, et accepimus ab eis quicquid melius noverant, sicut diximus in principio.'

[122] 'Incipit liber magistri Moamini falconerii translatus de arabico in latinum per magistrum Theodorum phisicum domini Federici Romanorum imperatoris, et correptus est per ipsum imperatorem tempore obsidionis Faventie,' Rome, Biblioteca Angelica, MS. 1461, f. 73; see Narducci, *Catalogus codd. MSS.*, p. 628. The mention of correction by Frederick at the siege of Faenza also appears in a manuscript in private hands and in the French translation mentioned below; see Werth, xii. 175–177.   Other manuscripts not mentioned by Werth are: Vatican, Vat. lat. 5366, fols. 1–33 v, 68 v–75 v (saec. xiii); Regina lat. 1446, fols. 31–70 (ca. 1300); 1111; 1227; 1617; Ott. lat. 1811; Urb. lat. 1014; University of Bologna, MS. lat. 164 (153), ff. 33–49 v; Naples, MS. xiv. D. 31; Ambrosian, MS. D. inf. 11.   This would seem to be the 'librum de animalibus traductum a domino Theodoro' which is mentioned in the papal library in 1475: Müntz and Fabre, *La bibliothèque du Vatican au XV⁰ siècle*, p. 271.   [123] See Chapter XII.

[124] 'Operationes quarum principium est per naturam et perfectio per voluntatem et cetere operationes et un[a]queque istarum coniungitur delectationi et tendit ad finem proprium, ut in libro de anima et Nychomachia et rethorica declaratum est': MS. Reg. Lat. 1446, f. 31 v.   The *De anima* was then current, but the known versions of the *Nicomachean Ethics* and *Rhetoric*, made in the thirteenth century, have not hitherto been connected with Sicily; see Grabmann, *Aristotelesübersetzungen*, pp. 204–237, 242 f., 251–256; Pelzer, in *Revue néo-scolastique*, 1921; supra, p. 284.

tationis nobilitatem imperatoribus et regibus appropriandam dumtaxat, et
videns antecessores suos et contemperaneos reges in delectatione a naturali
veritate appropriata sibi et exhibita non sollicitos esse sed potius sompno-
lentos, servorum sui limitis minimo imperavit presentem librum falconarii
transferre de arabico in latinum, ut eorum sit recordatio que sapientium
solertia adinvenit per experimentum et principium inveniendorum inpos-
terum. Ego igitur cum obedientia et devotione debita domini mei dignum
preoccupavi preceptum presens opus tractatu quaternario dividendo, primo
in theoricam huius artis, secundo in medicinas occultarum infirmitatum,
tertio in curas [125] manifestarum infirmitatum, quarto in medicamen rapivo-
rum quadrupedum.[126]

Ordinarily the manuscripts have five books, the last two devoted
to quadrupeds, so that only the first three concern us. Moreover
of these the second and third are confined to diseases and remedies, and there is also much of this in the first book, after the preliminary classification of birds of prey, several of which have only
their oriental names. It will thus be seen that the treatise, which
is mainly a collection of prescriptions, has little in common with
the subject-matter of the *De arte* as we have it, and there is no
indication that the emperor drew upon it.[127] Its popularity is
attested by the numerous surviving manuscripts of the Latin text
and by the French translation made by Daniel of Cremona for
the use of Frederick's son Enzio, which must antedate Enzio's
imprisonment in 1249.[128]

After Moamyn, Daniel of Cremona dedicated to Enzio [129] the

---

[125] MS. 'cuius.'

[126] Vatican, MS. Reg. Lat. 1446, f. 32; cf. Pertz, *Archiv*, xii. 320. This preface
begins: 'Sollicitudo nature gubernans . . .' Other manuscripts have a different
preface, beginning, 'Reges pluribus delectationibus gaudent,' and mentioning
Theodore by name: e. g., Corpus Christi College, Oxford, MS. 287, f. 45. The
treatise itself begins: 'Genera autem volucrum rapidarum quibus sepius utitur gens
aucupando sunt quatuor et xiiii species.' There are important differences between
the Corpus and the Vatican texts.

[127] There are some notes, possibly added at the time of Frederick's revision, e. g.
at the end of bk. i: 'Sed qualiter debeat teneri pugillus secundum diversitatem
avium tacuit auctor' (Corpus MS. 287, f. 50 v).

[128] Ciampoli, *I codici francesi della R. Biblioteca di S. Marco* (Venice, 1897), pp.
112–114; Paul Meyer, in *Atti* of International Congress of History, Rome, iv. 78
(1904).

[129] A Latin work on falconry seems also to have been dedicated to Enzio as king
of Torres and Gallura, 'principi nostro excellentissimo E. Turrensi,' a title which
Enzio seems to have used interchangeably with that of king of Sardinia: E. Besta,
La Sardegna medioevale (Palermo, 1908), i. 207 f. See below, Chapter XVII.

French version of another oriental work, the book of Yatrib, Gatriph, or Tarif, in seventy-five chapters, which he declared had first been compiled in Persian and then turned into Latin.[130] It is not stated that Frederick II had any connection with the Latin translation, but the similarity of the two treatises and the date of the French version make it likely that the Latin text of Yatrib was also due to the emperor's interest in the oriental literature of falconry. Yatrib, whose favorite bird is the sparrow-hawk, gives a mixture of prescriptions and practical maxims, certain of which are attributed to the Great Khan ('Chaycham rex Parthorum') and to 'Bulchassem,' who may have been the author of the Arabic text (ca. 1200).[131] This manual does not appear to have furnished material for the *De arte*.

3. Taken as a whole, the *De arte* gives the impression of being based far less upon books than upon observation and experience, on the part either of the author or his immediate informants.[132] It is a book of the open air, not of the closet. Frederick's eager desire to learn appears from his inquiries of the Arabs both while he was in the East and later:

> Nos quando transivimus mare vidimus quod ipsi Arabes utebantur capello in hac arte. Reges namque Arabum mittebant ad nos falconarios suos peritiores in hac arte cum multis modis falconum, preterea non neglereximus ad nos vocare expertos huius rei tam de Arabia quam de regionibus undecumque, ab eo tempore scilicet in quo primitus proposuimus redigere in librum ea que sunt huius artis, et accepimus ab eis quicquid melius noverant, sicut diximus in principio.[133]

It will be noted here that the emperor not only watched the Saracen falconers, but tried their methods himself and improved on them, just as he himself tested the hatching of eggs by the sun's heat in Apulia.[134] In the following unpublished passages

---

[130] The French translation is found at St. Mark's in the same manuscript with Moamyn (see Ciampoli, *Codici francesi*, p. 113), and the Latin texts also occur together in MS. Angelica 1461, which I have used.

[131] Werth, xii. 173. On falconry at the court of the Great Khan, see Marco Polo, ed. Yule, i. 402–407.

[132] Cf. Theodore's preface to Moamyn, supra: 'que sapientium solertia adinvenit per experimentum.'

[133] MS. M, f. 104 v; MS. B, p. 258; ed. Schneider, pp. 162 f.

[134] Supra, p. 311.

we see the same spirit of observation applied to the nests of herons, cuckoos, and vultures, to the evidence of intelligence in ducks and cranes, and to the popular fable of the hatching of barnacle geese from trees or barnacles, a legend which he ascribes to ignorance of their remote nesting-places:

Quodam enim tempore apportatus fuit nidus ante nos illius avicule que dicitur praenus, et in illo nido erant pulli praeni et una avicula orribilis visu deformis ut nullam fere figuram avis promitteret, ore magno sine pennis pullos multos et longos habens super totum caput usque ad oculos et rostrum. Ut igitur videremus que avis esset illa, cum diligenti custodia nutrivimus illos pullos et illam aliam aviculam et postquam perruerunt vidimus quod erant cuculi, ex quo cognovimus cuculum non facere nidum sed ova sua ponit in alieno nido.[135] . . .

Vidimus tamen aliquando quod quidam ayronum cineratiorum et bisorum nidificant in arboribus altis, ut sunt quercus, fagi, pini, et ulmi, et similes, et etiam super terram, et quando non possunt habere arbores altas et fortes sibi convenientes et sunt ibi salices, tamarisci, aut arbores alie debiles, nunquam nidificabunt in ipsis debilibus, ymo nidificabunt potius in canetis inviis et limosis super cannas, facilius enim est homines et serpentes accedere ad salices et ad huiusmodi arbores parvas quam ad canetas.[136] . . .

Est et aliud genus anserum minorum diversorum colorum albi [137] scilicet in una parte corporis et nigri in alia orbiculariter, que anseres dicuntur berneclee, de quibus nescimus etiam ubi nidificant. Asserit tamen opinio quorundam eas nasci de arbore sicca, dicunt enim quod in regionibus septentrionalibus longinquis sunt ligna navium in quibus lignis de sua putredine nascitur vermis de quo verme fit avis ista pendens per rostrum per lignum siccum donec volare possit. Sed diutius inquisivimus an hec opinio aliquid veritatis continet et misimus illuc plures nuntios nostros et de illis lignis fecimus adferri ad nos et in eis vidimus quasi coquillas adherentes ligno que coquille in nulla sui parte ostendebant aliquam formam avis, et ob hoc non credimus huic opinioni nisi in ea habuerimus congruentius argumentum, sed istorum opinio nascitur, ut nobis videtur, ex hoc quod bernecle nascuntur in tam remotis locis quod homines nescientes ubi nidificant opinantur id quod dictum est.[138] . . .

Vidimus vulturem in nido suo unicum ovum ponere et unicum cubare, cuius rei experientiam pluries habuimus quamvis Aristotiles dicat in libro suo animalium [139] quod nunquam visi fuerunt nidi neque pulli vulturum.[140] . . .

Et iam vidimus de anatibus et aliis pluribus avibus non rapacibus quod [quando] quis appropinquabat nidis suis ipse simulantes se egrotas fingebant

---

[135] MS. B, p. 60.   [136] MS. B, p. 63.   [137] MS. 'alibi.'

[138] MS. B, p. 63. On the fable respecting barnacle geese in this period see Gervase of Tilbury, *Otia imperialia*, iii, c. 123; Liebrecht, *Des Gervasius von Tilbury Otia imperialia* (Hanover, 1856), pp. 163 f.; Carus, *Geschichte der Zoologie*, pp. 190–195. It is accepted by Vincent of Beauvais but denied by Albertus Magnus: Thorndike, ii. 464 f.

[139] *Hist. animal.*, 6, 5; 9, 11.   [140] MS. B, p. 65.

se volare non posse et aliquantulum se cedebant ab ovis aut a pullis et sponte male volabant ut crederentur habere alas lesas aut crura. Ideo fingebant se cadere in terram ut homo sequeretur eas ad capiendum ipsas.[141] . . .

Nos autem, quia vidimus, vituperamus cibum qui fit eis de avibus que comedunt pisces, multo magis reprobamus nutrimentum quod fit de piscibus, aves enim nutrite piscibus erunt mollium carnium et mollium pennarum et malorum humorum.[142] . . .

Astutiam et acumen ingenii gruum experti sumus quandoque tantam quod videns posset credere eas habere rationem. Nam postquam iactaveramus nostrum girofalconem ad eas et ipse iam segregaverat unam a societate illarum et persequebatur segregatam et fortuitu grus videbat vultures stantes in campis, ipsa confugiebat illuc et stabat tuta inter eas, nam girofalcus ex tunc non audebat invadere ipsam, tanquam si grus scivisset quod girofalcus vultures crederet esse aquilas ad quas non audet accedere.[143] . . .

The emperor who insists upon seeing for himself, who investigates legends by sending for the evidence, who seels vultures' eyes to ascertain whether they find food by smell,[144] is clearly the same inquirer who shocked the good Salimbene by bringing up children in isolation to test their speech, and by cutting men open to observe the processes of digestion.[145]   If the facts are not available, he draws no certain conclusion.[146]   *Fides enim certa non provenit ex auditu.*[147]

The last four books are made up of generalized experience, with few particular instances. Elaborate in plan and almost scholastic in subdivision, *divisivus et inquisitivus*, they are severely practical throughout, with little or no speculation and no digressions, but with constant reference to the author's own observation and practice. He approves or disapproves various methods, not

---

[141] MS. B, p. 70.   [142] MS. B, p. 149.   [143] MS. B, p. 401.

[144] 'Non est ergo tenendum quod odoratu sentiant cadaver, ut quidam dicunt, sed potius visu. Quod expertum est per nos pluries, etenim quando vultures erant ex toto ciliati non sentiebant carnes proiectas ante ipsas quamvis odoratum non haberent oppilatum. Experti sumus autem quod non rapiunt aves cum famelici sunt et videntibus proiecimus pullum galline et non capiebant ipsum nec occidebant,' MS. M, f. 11–11 v; MS. B, p. 29; ed. Schneider, p. 17.

[145] *M. G. H., Scriptores,* xxxii. 350–353; supra, Chapter XII, n. 112.

[146] 'De tempore cubationis ovorum avium rapacium certi non sumus pro eo quod plures de avibus rapacibus nidificant in regionibus longinquis et nimis remotis a nobis, de quibus noticiam habere non possumus,' MS. M, f. 51; ed. Schneider, p. 78. Cf. MS. B, p. 70: 'De avibus autem non rapacibus nobis est dubium an prius pascant se an pullos an simul cum pullis; cognoscere difficile videtur.'

[147] Supra, p. 313.

dogmatically, but giving his reasons.[148]   Thus he prefers a lure of cranes' wings,[149] but mentions the use of hens in Spain and southern France, doves in Arabia,[150] and a pig covered with a hare's skin in *insula de Armenia*.[151]   In England hunters do not shout when they lure; he has asked the reason, but can get no explanation save ancient custom: [152]

Quomodo loyrant illi de Anglia.   Illi vero qui habitant Britanniam que vocatur Anglia non loyrant hoc modo quoniam nunquam loyrant equites neque vociferant sed loyrant pedites et loyrum prohiciunt in altum recte et postquam ceciderit in terram iterum prohiciunt in altum, et hoc faciunt donec falco videat loyrum et incipiat venire ad ipsum. Et postquam ille qui prohicit loyrum videt falconem prope venientem stat et dimittit ipsum venire super loyrum, et est causa hec quare non loyrant equites quia non conveniret et difficile esset prohicere loyrum et descendere iterum ad prohiciendum.

Quare non vociferant in loyratione.   De vociferatione vero quesivimus, quare scilicet non vociferant, et nesciunt reddere causam nisi tantummodo quod hoc haberent ex usu; sed opinamur antiquos eorum loyrando non vociferare pro eo scilicet quod falcones quando etiam mittuntur ad hayrones necessarium est vociferare quoniam ayro reddit se frequenter ad aquas timore falconum et cum vocibus perterretur ut surgat ad aerem sepius, et quod falcones gruerii quando in principio venationis sue, hoc est antequam plures aves cepit, iactentur et emittantur ad sedium ad grues, quando inquam falcones sunt prope gruem, oportet vociferare ad grues ut surgant, falco vero audiens, si assuetus fuerit ad loyrum vociferando, credens se re-

[148] 'Nos vero in loyrando habemus hunc modum,' MS. B, p. 290.   'Quod non reprobamus,' p. 310.   'Nos autem in hoc non facimus magnam vim,' p. 462.   'Hic autem modus volandi idcirco non est laudandus,' p. 499.   'Approbavimus et vidimus,' p. 516.   'Diximus de venatione ad grues quam approbavimus girofalconi propter id quod supra dictum est et venatione ayronis quam approbavimus sacro propter id quod similiter dictum est.   Nunc dicamus de venatione que fit ad aves de rivera et specialiter ad anates et sibi similes, et hanc approbamus falconi peregrino,' p. 517 (beginning of bk. vi).   'Nos autem dicimus quod circa mane melius est,' p. 534.   'Hunc morem non multum reprobamus,' p. 540.

[149] MS. B, p. 282.

[150] 'Plures autem gentium in diebus nostris non utebantur loyro quod diximus ad revocandum genera falconum, scilicet [*read* sed?] gallinis vivis ut in Hispania et regionibus eius vicinis occidentalibus, alii columbis vivis ut in Arabia et in ceteris regionibus meridianis et orientalibus; sed nos modum istorum et illorum reprobamus quia non semper de facili possunt haberi aves vive quemadmodum ale avium,' MS. B, p. 285. This passage is also found in what appears to be extracts from bk. iii of the *De arte* in the Bodleian MS. Digby, 152, f. 44.

[151] 'Item homines de insula de Armenia et de regionibus vicinis faciunt traynam leporinam suis sacris zacharis et suis layneriis hoc modo,' MS. B, p. 327.

[152] MS. B, pp. 307 f.; MS. Digby, 152, f. 50 v.

vocari ad loyrum per illas voces dimittet grues et redibit ad vociferantem spe loyri. Propter hoc non vociferant in loyrando, et quoniam ipsi venantur ad ayrones et ad grues plusquam ad alias aves, assuefaciunt falcones ad loyrum non vociferando.

Quod nobis videtur. Nos tamen dicimus quod melius est vociferare loyrando quoniam naturale est falconibus abfugere ab homine sed retrahere ipsum falconem ab hac natura non potest fieri nisi cum accidentali magisterio et convenientibus instrumentis; necessarium est igitur omnia illa ordinare per que possit habitus retineri et si perdatur recuperari et inter ea per que retinetur aut recuperatur propria sunt loyrum et vox. . . .

For his investigations of falcons, Frederick had at his disposal the whole machinery of his bureaucratic administration, and if the registers of his correspondence had been preserved we should perhaps be able to follow in detail some phases of his literary work. As matters stand, the surviving fragment of a register for a few months of 1239–40 has forty entries concerning falcons, mentioning by name more than fifty of the emperor's falconers.[153] Thus in November 1239 he writes from Lodi to his superintendent of buildings in Sicily thanking him for information concerning the haunts and nests of herons, which the emperor longs to see for himself.[154] From Cremona he sends to his falconer Enzio for a report on his falcons, how many there are and in what condition, and especially concerning those captured at Malta and the wild ones taken during the season;[155] he orders another to await him with hawks at Pisa,[156] while he sends to Apulia for two hawks just brought by the emissaries of Michael Comnenus.[157] After Christmas he sends for two sacred falcons, the one called 'Saxo' and another good bird.[158] Although winter is not so good a season

[153] Including Master Walter Anglicus and his son William: Böhmer-Ficker, *Regesta imperii*, nos. 2857, 3082.

[154] 'De sollicitudine et labore quem assumpsisti super inveniendis ayris hayronum et locis ubi degunt te duximus commendandum, quod excellentia nostra satis delectat audire nec minus presentialiter videre peroptat': Huillard-Bréholles, *Historia diplomatica*, v. 510; Böhmer-Ficker, no. 2566. Cf. the *De arte*, MS. B, p. 442: 'In fine vero autumpni et per hyemem magna copia ayronum invenitur in calidis regionis [sic] ad quas confugerunt propter cibum acquirendum sibi et propter frigus . . . et maxime habundant in regionibus Egypti.'

[155] Böhmer-Ficker, no. 2584. Besides the entries concerning falcons, there are many respecting dogs and hunting leopards, e. g. nos. 2661, 2662, 2709, 2751, 2783, 2785, 2811, 2882, 2932, 2944, 3029.

[156] *Ibid.*, no. 2585.   [157] No. 2589.   [158] No. 2668.

for such game,[159] he writes from Gubbio in January to his falconer Sardus that he is taking many fat cranes and keeping the legs as the portion of the absent falconer, who should come at once [160] to that noblest of sports, the hunting of cranes with gerfalcons, which the emperor describes in his fourth book.[161] The next day he sends a valet for training peregrine falcons in the Sicilian kingdom,[162] and two days later sends from Foligno for three falcons and a *turziolus*.[163] Ten days thereafter he sends falcons and dogs back to the south,[164] and various orders provide for wages and equipment of falconers.[165] In February he is concerned with the moulting of falcons, which are distributed among his barons to be kept during that period.[166] In March we read of the training of falcons in the south.[167] In May the emperor, once more in the Capitanata, sends nineteen falconers to Malta for birds,[168] and orders that all the sparrow-hawks in the county of Molise shall be brought together under a special keeper.[169] When he wants live cranes for training falcons, he commands the justiciars of Terra di Lavoro, Bari, and the Capitanata to have as many as possible caught and sent to the justiciar of the Capitanata to be kept at the royal residences.[170]

Such glimpses of the emperor's daily occupations show his passion for falconry, pursued in the midst of more urgent concerns of state and not merely in the intervals of relaxation at his palaces, and illustrate the devotion of the ideal falconer, who is represented in the *De arte* as desiring primarily neither fame nor a plentiful supply for the table, but to have the best falcons. The successful hawker cannot be 'indolent or careless, for this art requires much labor and much study.' [171] Frederick's pride

---

[159] *De arte*, iv (MS. B, pp. 359–361).

[160] Böhmer-Ficker, no. 2745; cf. 2744. The hunting of cranes is also mentioned in no. 2814.

[161] 'Grues sunt famosiores inter omnes aves non rapaces ad quas docentur capiendas aves rapaces, et girofalcus nobilior est avibus rapacibus et est avis que melius capit grues quam alii falcones et que melius volat ad ipsas.' MS. B, p. 282.

[162] Böhmer-Ficker, no. 2749.   [163] *Ibid.*, no. 2753.   [164] No. 2807.

[165] Nos. 2539, 2591, 2680, 2706, 2744, 2814, 2817, 2856 f., 2863, 2907, 2929, 3082.

[166] Nos. 2800, 2855, 2863, 2903.   [167] No. 2907.

[168] No. 3082.   [169] No. 3056.   [170] No. 2801.

[171] MS. M, ff. 68–69; ed. Schneider, pp. 107–109.

in his mastery of the art is illustrated by the story that, when he was ordered to become a subject of the Great Khan and receive an office at the Khan's court, he remarked that he would make a good falconer, for he understood birds very well.[172]   And if we doubt this characteristic tale, we have at least his own prefatory words concerning falconry, *nos semper dileximus et exercuimus*.

Keen sportsman as he was, Frederick II was not the man to lose himself wholly in the mere joy of hawking.   His mind had also to be kept busy, his questions answered, his knowledge extended and put in order.   The lessons of the *De arte (scientia huius libri)* [173] are essential for the falconer, but it is more than a manual of practical instruction.   The first book and the earlier chapters of the second have a systematic and scientific character which give them an important place in the history of mediaeval zoölogy, while the whole treatise is pervaded by the spirit of actual observation and experiment.   While the author uses the ancients, he is not blinded by them, and does not hesitate to correct them when necessary.   So far as the Renaissance is characterized by the spirit of free inquiry and emancipation from authority, the *De arte* lends support to those who would begin the new movement at the court of Frederick II.

[172] Albericus Trium Fontium, *M. G. H., Scriptores*, xxiii. 943.
[173] MS. M, f. 68 v; ed. Schneider, p. 108.

CHAPTER XV

THE ABACUS AND THE EXCHEQUER[1]

A QUESTION of special obscurity respecting the early history of the English exchequer is the origin and introduction of its distinctive system of reckoning, *secundum consuetum cursum scaccarii non legibus arismeticis*.[2] Inasmuch as the exchequer table was merely a peculiar form of the abacus,[3] some light on the problem may be expected from an examination of treatises upon this method of computation, particularly such as can be connected in any way with England and with the king's court. The only compend of this sort which has so far been associated with the English court was written by a royal clerk named Thurkil, and is preserved in a manuscript of the twelfth century in the library of the Vatican. Although it has been in print since 1882,[4] it has not heretofore been studied from this point of view. It begins:

Socio suo Simoni de ROTOL' TURChillus compotista salutem. In his regunculis quas dilectioni tue, venerande amice, super abacum scripsi et obtuli, licet quid quod tibi displiceat forte reperias, non me tamen, more quorundam quibus nulla inest bonitatis soliditas, iniquo dente livoris mordeas, sed si adhuc solite discretionis es, mee impericie pie ignoscas et, si alicubi necesse est, sic et de meo demas et de tuo addas ut eas sapienter corrigas. Non enim usque adeo perverse mei amator sum ut quod ego inveni pro perfecto defendam, cum in humanis inventionibus, ut ait Priscianus, nichil sit perfectum.

---

[1] Revised from my article in *E. H. R.*, xxvii. 101–105 (1912), which was written before the appearance of R. L. Poole, *The Exchequer in the Twelfth Century* (Oxford, 1912).

[2] *Dialogus de Scaccario*, i. 5 (ed. Oxford, 1902, p. 75). On this phase of the origin of the exchequer, see Round, *Commune of London*, pp. 74 f.; the Oxford edition of the *Dialogus*, pp. 42 f.; Petit-Dutaillis' edition of Stubbs, i. 806–808; Poole, *op. cit.*, ch. iii.

[3] It is worth noting that, whereas the analogy of the chessboard is the only argument hitherto adduced for the existence of transverse lines on the exchequer table, such lines are regularly found in the abacus as described in the mediaeval treatises.

[4] Vat. MS. Lat. 3123, ff. 55–63 v, edited by Narducci, in *Bullettino*, xv. 111–154. Cf. Eneström, in *B. M.*, viii. 78 f., 415; and on the Vatican MS. see also Bethmann, in Pertz's *Archiv*, xii. 233–235.

Et si quid in huius inventionis scintillula utilitati tue dilectissime conducibile inveneris, nec mihi nec tibi, cuius gratia hoc specialiter edidi, verum venerabili viro magistro nostro Guillelmo ℞ [et⁵], quem universis calculatoribus hodie viventibus preferre non timeo, ascribas queso. Vale.⁶

The date of the treatise can be approximately fixed by the following sentence:

Ducentę marce sunt inter .ii.dᵈ hidas dividende, que sunt hide totius Eisexie, ut ait Hugo Bocholaudie.⁷

Two men of this name are known in the twelfth century, one of them sheriff of eight counties under Henry I,⁸ the other a tenant in Berkshire in 1166 and sheriff of the same county a few years later.⁹ There is, however, nothing to connect the younger Hugh de Bocland with Essex, which is in other hands throughout the Pipe Rolls of Henry II, whereas the elder Hugh can be traced as sheriff of Essex in 1101 and the years immediately following.¹⁰ He is found in charters as late as 1115,¹¹ but by 1117 his lands are in other hands ¹² and in 1119 he has been succeeded in his principal office, the shrievalty of Berks.¹³ Our treatise is thus anterior to 1117 and may even go back to the reign of William Rufus, under whom Hugh de Bocland, one of this king's 'new *curiales*,' ¹⁴ can be traced as witness to the king's charters ¹⁵ and

⁵ The MS. here has a sign which is apparently meant for &, but which is probably a corruption of an original ℞, the ℞ now in the text having been inserted later above the line.

⁶ P. 135 of the edition. The edition is for the most part careful, but I have made an occasional correction from the MS.

⁷ P. 153. Narducci noted the mention of Hugh de Bocland, but (pp. 128–130) was misled into placing the treatise in the second half of the century by identifying the author with a Thurkil of Essex mentioned in a vision of 1206. Cf. Poole, p. 48, n.

⁸ *Chronicon Monasterii de Abingdon*, ii. 117 *et passim*; Ordericus Vitalis, iv. 164; *E. H. R.*, xxvi. 490; xxxiii. 156; xxxvii. 163.

⁹ *Red Book*, i. 306 f; Eyton, *Itinerary of Henry II*, pp. 313, 337.

¹⁰ Round, *Geoffrey de Mandeville*, p. 328; *Monasticon*, i. 164; vi. 105; *Cartularium S. Iohannis de Colecestria*, i. 22, 24, 27.

¹¹ He is addressed in two charters of Reginald, who became abbot of Ramsey in 1114 (*Cartularium Monasterii de Ramseia*, i. 130, 133); and attests late in 1115 (Farrer, *Itinerary of Henry I*, no. 361).

¹² J. Armitage Robinson, *Gilbert Crispin*, p. 154 f; Farrer, no. 376.

¹³ *Chron. Abingdon*, ii. 160.   ¹⁴ Morris, in *E. H. R.*, xxxiii. 156.

¹⁵ Davis, *Regesta*, nos. 444, 466.

as sheriff of Bedfordshire,[16] Berkshire,[17] and Hertfordshire,[18] the last of which was regularly held with Essex. Indeed a charter of the Red King for Colchester seems to connect him directly with Essex.[19]

Neither Thurkil nor his colleague Simon 'of the rolls,'[20] who must likewise have been an expert with the abacus, has been identified, but both were evidently members of the royal *curia*, since Thurkil says, speaking of ordinary division and division by differences:

> Si quis tamen cur de utroque divisionum genere, cum ut nunc dictum est ad unum utreque redeant, scripsi quesierit, propterea inquam quod ille ad quoslibet, iste vero non nisi ad curiales tantum pertinent.[21]

Their master, 'Guillelmus ℞,' who is mentioned in two other passages,[22] has been sought in vain among the abacists of this period. He is plainly no common teacher or computer, for he has invented a special sign for the *semuncideunx*[23] and is authority for the statement that the conventional figures of the abacus came from the Pythagoreans but their names from the Arabs. The titles *donnus* and *venerabilis vir* would seem to indicate that he was a bishop or an abbot, but I have found no contemporary prelate of this name who would justify Thurkil's characterization, unless it be William, bishop of Syracuse, ca. 1104–15, who is said to have been of Norman origin and whom Adelard of Bath addresses as *omnium mathematicarum artium eruditissime.*[24]

---

[16] *Ibid.*, no. 395.   [17] *Chron. Abingdon*, ii. 43.

[18] The Hertfordshire text of Henry's coronation charter is addressed to him: *Transactions of the Royal Historical Society*, new series, viii. 33, 40; Liebermann, *Gesetze*, i. 521. He is also addressed by William II in a charter concerning Middlesex (Robinson, *Gilbert Crispin*, p. 138, no. 12) and appears as a royal officer in Sussex in *E. H. R.*, xxvii. 103; Davis, no. 416; Haskins, *Norman Institutions*, p. 81.

[19] Davis, no. 471.

[20] Narducci (p. 121) extends 'Rotolandia,' which seems to me much less likely than 'rotolis.'

[21] P. 148.

[22] Pp. 136, 150.

[23] 'Pars illa que est semuncideunx non est in frequenti usu, unde caracterem non habet quo designetur,' says Gerland: St. John's College, Oxford, MS. 17, f. 51 v; British Museum, Add. MS. 22414, f. 7; not in the text as printed in *Bullettino*, x. 603.

[24] *De eodem et diverso*, ed. Willner, p. 3. See Chapter II, n. 8.

If some one must be found who would satisfy also the 'R̴,' we might turn to William de Ros, abbot of Fécamp from 1079 to 1107 and previously canon, dean, and archdeacon of Bayeux and monk of Caen.[25]   The epitaphs and eulogies written after his death celebrate, as is usual, only his Christian virtues,[26] but we learn from Baudri of Dol and the Fécamp annalist that he was a man of much learning.[27]   We hear of the eminence of Fécamp in music in his time,[28] and of the vain efforts of Abbot Thurstin to introduce the chant of a certain William of Fécamp into Glastonbury.[29]   Nothing is said specifically of the mathematical attainments of William de Ros, but, like Thurstin, he was one of the Bayeux clerks of promise whom Bishop Odo sent to study at Liége,[30] then an outstanding centre of mathematical learning.[31]

Besides the treatise on the abacus the Vatican manuscript contains a related tract addressed by Thurkil to a certain Gilbert and explaining the conversion of marks into pounds and vice versa.[32]   That Thurkil was also the author of a work on the ecclesiastical calendar we know from Philip de Thaon, who, writing in 1119, cites six times *Turkils li vaillanz*, along with Bede, Helperic, and Gerland, on such topics as the length of the year and the lunar month, embolisms, epacts, and the date of St. Matthias' day in leap year.[33]   Two of the citations are from the fourth

[25] Ordericus Vitalis, ii. 129, 243 f.; iii. 266; iv. 269–272; cf. *Archaeologia*, xxvii. 26.   A Guillelmus de Ros still appears as canon of Bayeux in 1092–93: *Livre noir*, nos. 22, 23.

[26] Ordericus, iv. 270 f.; Geoffrey of Winchester, in Wright, *Anglo-Latin Poets*, ii. 155; epitaph discovered in 1875 in *Comptes-rendus de l'Académie des Inscriptions*, 1875, pp. 306–309, and *Bulletin des Antiquaires de Normandie*, vii. 497–502.

[27] 'Admodum literatus': *Auctarium Fiscannense*, in Robert of Torigni, ed. Delisle, ii. 149.   'Magna litterarum peritia preditus': Baudri, *Epistola ad Fiscannenses*, in *Neustria pia*, pp. 227–233; Migne, clxvi. 1173–82.   [28] Baudri, *ibid.*

[29] William of Malmesbury, *De antiquitate Glastoniensis ecclesie*, ed. Hearne, p. 114; Carlez, "Le chant de Guillaume de Fécamp," in *Mémoires de l'Académie de Caen*, 1877, pp. 233–251.   The 'Kalendarium Willelmi abbatis' formerly in the Fécamp library (*Catalogue des MSS. des départements*, i, p. xxvi) is apparently merely the sevice-book now at Rouen, MSS. 237–238.

[30] Ordericus, iii. 265 f.   [31] See below, n. 53.

[32] Printed in *Bullettino*, xv. 127 f.   In the MS. (f. 64 v) this is followed without a break by a chapter 'De collectione diei qui dicitur saltus lune,' the beginning of which indicates a continuation: 'Item si scire volueris quot momenta . . .'

[33] *Li cumpoz*, ed. Mall, lines 2080, 2214, 2361, 2399, 2498, 3208.

and ninth chapters of Thurkil's third book,[34] so that identification ought to be easy, but I have not succeeded in discovering the
work cited, which might aid in fixing the author's date and perhaps other facts concerning him. One is tempted to seek this
treatise in the pages which precede the account of the abacus in
the Vatican manuscript [35] and perhaps in the chapter on the
saltus lune which follows, though none of this rather confused
material is divided into books and chapters. The length of the
lunar month is the same as that cited by Philip de Thaon
from Thurkil and Bede,[36] and there are other resemblances but
nothing sufficiently specific to identify the author. The date
is 1102,[37] and Gerland is already quoted as an authority.[38] As
to Thurkil's identity we can only guess, for the name is by no
means unique in the early twelfth century. Perhaps one conjecture may be hazarded, namely the monk Thurkil of Westminster, who appears in 1122 shortly after the abbot Gilbert
Crispin among the deceased members of the convent inserted in
the mortuary roll of Vitalis of Savigny.[39] If this should be our
Thurkil, the Gilbert to whom the tract on the mark is dedicated
may be Abbot Gilbert, himself *doctus quadrivio*,[40] who died in
1117.

In the treatise on the abacus, Thurkil, like other abacists, confines himself to multiplication, division, and fractions, and so

[34]   2399   E Turkils el tierz livre
                   E el nofme chapitle

                   . . . .

            2498   Turkils en sun escrit
                   E enz el quart chapitle
                   Que il fait del tierz livre.

[35] MS. Vat. Lat. 3123, ff. 44 v–55; also in B. N., MS. Lat. 11260, ff. 24–31 v.

[36] 29 days, 12 hours, 29 moments, 348 atoms: *Li cumpoz*, lines 2496 ff.; MS. Vat.
Lat. 3123, f. 50 v; MS. Lat. 11260, f. 28; also in British Museum, Royal MS. 15 B.
iv, f. 141 v (fragments apparently of a related treatise).

[37] MS. Vat. Lat. 3123, f. 46 v; MS. Lat. 11260, f. 25.

[38] MS. Vat. Lat. 3123, f. 54; MS. Lat. 11260, f. 30 v.

[39] Delisle, *Rouleau mortuaire du B. Vital* (Paris, 1909), no. 100; J. A. Robinson,
Gilbert Crispin, p. 27.

[40] Robinson, *op. cit.*, p. 26. If Simon de rotol' be interpreted as Simon of Rutland, it should be remarked that Westminster Abbey held the churches of Rutland
as Alberic the Lotharingian clerk had held them: Davis, *Regesta*, nos. 381, 382, 420;
Round, *Commune of London*, pp. 36–38.

throws no light upon the procedure at the exchequer table, which consisted merely of addition and subtraction. The king's clerks had, however, frequent occasion to multiply and divide, and Thurkil's illustrations are obviously drawn from familiar subject-matter, as in his brief account of the relation of marks to pounds. What is the product when twenty-three knights owe you six marks each? Divide £800,137 among 1009 knights. The most interesting example is the one relating to Essex, which is printed above. A payment of two hundred marks is assessed against a shire and the amount due from each hide is to be determined — just such a case as would arise in levying the *assisa communis* described in the *Dialogus*, and just the amount which Essex pays as *donum* in the early years of Henry II.[41] This coincidence can hardly be accidental, but indicates rather that the *assisa communis*, as a supplement to Danegeld and a corrective to its unequal assessment, goes back to the reign of Henry I, in which case it should probably be identified with the *novo geldo propter hidagium* mentioned between 1100 and 1107 in a charter for Westminster.[42] The hidation which is taken as the dividend, 2500, has already shrunk from the Domesday quota of 2650 [43] but has not yet reached 2364, which is the number of geldant hides in the Pipe Roll of 1130.[44] Moreover, it is reported on the authority of Hugh de Bocland, who as sheriff would know the actual number of hides liable in such a case. A meagre illustration of this sort is especially irritating when we think of what Thurkil might have told us. It may be argued that his failure to mention so interesting a form of the abacus as the exchequer table is an indication that it was not yet in existence; but the answer is that there is no place for this in his treatise,[45] nor

[41] *Dialogus*, i, 8, 11 (ed. Oxford, 1902, pp. 95, 103); *Pipe Roll*, 2–4 Henry II, pp. 18, 133. Cf. Maitland, *Domesday Book and Beyond*, pp. 473–475.

[42] Robinson, *Gilbert Crispin*, p. 141, no. 19.

[43] This is the number given by Maitland, p. 400. Rickwood argues for 2800: *Transactions of the Essex Archaeological Society*, new series, xi. 249.

[44] Pp. 59 f.

[45] 'In multiplicacione et divisione constat hec scientia,' p. 137. 'Huius artis tota pene utilitas in multiplicacione ac divisione constat': Bodleian, MS. Selden supra 25, f. 112 (brief treatise on the abacus).

should we expect an account of its relatively simple operations in a work which had to explain the 'iron process' of division by means of differences. The evidence that royal clerks were familiar with the abacus at the beginning of the twelfth century implies rather that it was already in use for balancing the royal accounts.

That "it was the introduction of this instrument in the form of the Exchequer which made an epoch in the history of the English Treasury" has now been brought out most convincingly by Poole. He argues that Englishmen became acquainted with the abacus in France, probably in the schools of Laon, and calls attention to the fact that Adelard of Bath studied at Laon, wrote on the abacus, and seems to have been in the employment of the court of Henry I.[46] Nevertheless I am inclined to place the introduction of the abacus earlier and to associate it rather with the movement which connected England with the schools of Lorraine. There is nothing as yet to show whether Thurkil's relations were with Laon or Lorraine, but two of his contemporaries mention the abacus in a way that brings it into connection with the *curia regis* at a still earlier date. Robert, who became bishop of Hereford in 1079, is described by William of Malmesbury as *omnium liberalium artium peritissimus, abacum precipue et lunarem compotum et celestium cursum astrorum rimatus.*[47] At his death in 1095 the prior of Winchester, Geoffrey, wrote of him:

> Non tua te mathesis, presul Rodberte, tuetur,
> Non annos aliter dinumerans abacus.[48]

It is not certain that Robert's writings included a treatise on the abacus,[49] but the passages just cited are conclusive as to his

---

[46] *The Exchequer in the Twelfth Century,* pp. 46–57. Note also that a Ralph of Laon witnesses a Bath charter of 1121: *Two Chartularies of Bath Priory,* ed. W. Hunt (1893), i. 51.

[47] *Gesta Pontificum,* p. 300.

[48] Hardy, *Descriptive Catalogue,* ii. 76; Wright, *Anglo-Latin Satirists and Epigrammatists,* ii. 154. It may be observed, in connection with what is said later, that Geoffrey was a native of Cambrai: *Gesta Pontificum,* p. 172.

[49] The mathematical tables ascribed to him by Bale (edition of 1557, ii. 125) may be simply an inference from the phrases of the chroniclers, but the commentary on Marianus Scotus is evidence of his attainments in chronological computation.

special familiarity with this method of reckoning and the fame
it brought him in England. Now Robert was a royal chaplain
before his elevation to the bishopric,[50] and heard pleas in the Red
King's court only a few months before his death.[51] Moreover,
he was a native of Lorraine,[52] which in the eleventh century was
the chief centre for the study of the abacus and produced such
eminent mathematicians as Heriger of Lobbes, Adelbold of
Utrecht, Reginbald of Cologne, and Ralph and Franco of Liége; [53]
and his zeal for the introduction of Lotharingian culture into
England is seen in his importation of the chronicle of Marianus
Scotus and his use of Charlemagne's church at Aachen as the
model for his own cathedral.[54] Robert was, of course, not the
only connecting link with the lands beyond the Scheldt in this
period, for Lotharingian influence had been strong at the court
of Edward the Confessor,[55] and among the prelates of his own
time Walcher of Durham had been a clerk of Liége and Thomas
of York and Samson of Worcester had apparently been at school
there; [56] while Walcher, prior of Malvern, was another Lotharingian abacist, who appears in England by 1091.[57] Still,
Robert's knowledge of the abacus was evidently considered

[50] Annals of Winchester, in *Annales Monastici*, ii. 32.

[51] *Gesta Pontificum*, p. 302; *Vita Wulstani*, in *Anglia Sacra*, ii. 268.

[52] *Gesta Pontificum*, p. 300.

[53] 'Cogis enim et crebris pulsas precibus ut tibi multiformes abaci rationes per-
sequar diligenter. . . . Quod si tibi tedium non esset harum fervore Lotharienses
expetere, quos in his ut cum maxime expertus sum florere. . . .' Bernelinus, in
Olleris, *Oeuvres de Gerbert*, p. 357; and Bubnov, *Gerberti Opera mathematica*, p. 383.
See further the passages cited in Bubnov, p. 205; Tannery and Clerval, *Une corre-
spondance d'écolâtres au XIe siècle*, in *Notices et Extraits des MSS.*, xxxvi. 487-541;
Cantor, i. 872-878, 880-890; Kurth, *Notger de Liège* (Paris, 1905), c. 14, especially
pp. 282-286; Dute, *Die Schulen im Bistum Lüttich im 11. Jahrhundert* (Marburg
Programm, 1882); B. Lefebvre, *Notes d'histoire des mathématiques* (Louvain, 1920),
pp. 93-114; Manitius, *Lateinische Litteratur*, ii. 778-786.

[54] *Gesta Pontificum*, p. 300 f. For the chronological tract in which Robert elaborated the introduction of Marianus, see W. H. Stevenson, *E. H. R.*, xxii. 72 ff.

[55] Freeman, *Norman Conquest*, 3d edition, ii. 81, 455 f., 598-601, 693-698; Stein-
dorff, *Heinrich III*, ii. 67 f.; Pauli, in *Nachrichten* of the Göttingen Gesellschaft der
Wissenschaften, 1879, pp. 324-330; Round, *Commune of London*, pp. 36-38.

[56] Simeon of Durham, i. 9, 105; ii. 195; Ordericus, iii. 265 f.

[57] Supra, Chapter VI, n. 5. A Lotharingian clerk named William appears between 1107 and 1137: Napier and Stevenson, *Crawford Charters*, p. 31.

something new and exceptional in England, and had doubtless been brought from his Lotharingian home. We can at least be sure that the abacus was known to members of the *curia* under William Rufus and, since Robert's promotion dates from 1079, even under the Conqueror, and for light upon its introduction we may well look in the direction of Lorraine.

CHAPTER XVI

NIMROD THE ASTRONOMER[1]

LI CUMPOZ of Philip de Thaon,[2] written in 1119 and important as the earliest monument of Anglo-Norman literature, possesses a special interest for the student of astronomy and chronology as being at once the earliest treatment of the subject in French and one of the latest expositions of the knowledge current in the period just preceding the advent of Arabic astronomy. Of the authorities whom the author cites, three, Bede, Helperic, and Gerland, are the standard writers on these subjects in the earlier Middle Ages,[3] and the citations are sufficiently specific to render easy a comparison with their works. A fourth, Turkils, though unknown to students of *Li cumpoz*, is plainly to be identified with Turchillus compotista, an Anglo-Norman contemporary of Philip who wrote before 1117 a treatise on the abacus which is of much interest for the early history of the English Exchequer; but the quotations are not from this work and are evidently derived from a treatise on chronological computation, consisting of at least three books, which has not yet come to light.[4] There remains a fifth, called Nebrot, Nebrod, Nebroz, Nembroz, or Nembroth, likewise unidentified by the commentators on Philip, who raises a number of interesting problems. Of the five passages in which he appears, the first, at the close of the chapter dealing with Aries, reads:

---

[1] Revised from *The Romanic Review*, v. 203–212 (1914).

[2] E. Mall, *Li Cumpoz Philipe de Thaün mit einer Einleitung* (Strasbourg, 1873); T. Wright, *Popular Treatises on Science* (London, 1841), pp. 20–73; Paul Meyer, "Fragment du Comput de Philippe de Thaon," in *Romania*, xi. 70–76 (1911). Cf. Langlois, *La connaissance de la nature et du monde au moyen âge*, pp. 2–3, 11.

[3] Cf. Chapter V, supra.

[4] See the preceding chapter. G. L. Hamilton, who first suggested the identity of Turkils and Turchillus (*Romanic Review*, iii. 314 (1912)), made the mistake of thinking that Philip cites the treatise on the abacus, which contains nothing on the subjects treated in *Li cumpoz*. That the work of Thurkil here cited comprised at least three books is clear from ll. 2399 and 2500.

> 1249 E ço Helperis dit
> Pur veir en sun escrit
> E Bede e Gerlanz
> E Nebroz, li vaillanz.

At the close of the account of Leo, speaking of the significance of the lion's tail, Philip says:

> 1345 E ço truvum escrit
> Que dans Nebroz le dit.

In the discussion *De saltu lune* we find:

> 2359 De ço trai a guarant
> Maistre Bede e Gerlant,
> Turkil e Helperi
> E Nebrot, ki eissi
> L'unt enquis e guardet.

Apropos of lunations he says:

> 2495 Ço dit Bede e Gerlanz
> E Nebroz, li vaillanz,
> E Helperis le dit,
> Turkils en sun escrit,
> E ens el quart chapitle
> Que il fait del tierz livre.

Finally concerning the septuagesimal term:

> 3341 Eissi cum Gerlanz dit,
> Nebroz en sun escrit.

To Philip, accordingly, Nebroz is an authority on astronomical and chronological matters of the same type as Bede, Helperic, Gerland, and Thurkil. No writer of this name, however, is known to have existed in the Middle Ages, and the form suggests at once the Νεβρώθ of the Septuagint and the Nimrod of modern versions of Genesis, whose name has furnished a fruitful field for the speculations and conjectures of orientalists.[5] The Biblical Nimrod is, of course, no humble chronologer but a king, a mighty one upon the earth, a mighty hunter before the Lord. How can we make an astronomer out of him? An answer to this question would involve studies of the Oriental Nimrod legends which lie beyond the purpose of this article. An astronomer he had certainly be-

---

[5] See Cheyne's article in the *Encyclopaedia Biblica* and the authors there cited.

come in men's minds by the sixth century, when John Malalas makes him king of the Persians and their master in astronomy and astrology,[6] and an astronomer he remained to the men of the Middle Ages. Astronomical tables under his name are known to have been current in Arabic, and his astronomy meets us in the twelfth century, when Philip's contemporary, Hugh of St. Victor, says, *Aiunt quidam Nemrod gigantem summum fuisse astrologum, sub cuius nomine etiam astronomia invenitur.* He is bracketed with Hyginus and Aratus by William of Conches,[7] and in the following century the *Speculum astronomie* says:[8]

Ex libris ergo qui post libros geometricos et arithmeticos invenitur apud nos scripti super his, primus tempore compositionis est liber quem edidit Nemroth gigas ad Iohathonem discipulum suum, qui sic incipit: *Sphera celi,* etc., in quo est parum proficui et falsitates nonnulle; sed nihil est ibi contra fidem, quod sciam.

Contrary to Cumont's opinion,[9] the work of Nimrod the giant is, in its mediaeval form, still extant, in two manuscripts neither of which appears to have been examined in this connection. One, MS. Lat. VIII 22 of the library of St. Mark's at Venice,[10] has the

---

[6] *Chronographia* (ed. Bonn), p. 17: Περσῶν ἐπρώτευσε διδάξας αὐτοὺς ἀστρονομίαν καὶ ἀστρολογίαν, τῇ οὐρανίῳ κινήσει τὰ περὶ τοὺς τικτομένους πάντα δῆθεν σημαίνοντα. Augustine, *De civitate Dei*, 16, 4, 10, 11, knows Nimrod only as the founder of Babylon. So also Gregory of Tours, *Historia Francorum*, i, 6; *De cursu stellarum*, c. 3 (ed. Arndt-Krusch, pp. 36, 858).

[7] Steinschneider, "Zum Speculum astronomicum des Albertus Magnus," in *Z. M. Ph.*, xvi. 380 (1871); and *E. U.*, no. 175 c. The passage of William of Conches will be found under Honorius of Autun, in Migne, clxxii. 59.

[8] *Alberti Magni Opera* (Paris, 1891), x. 629; critical edition of this passage in *Catalogus codicum astrologorum Graecorum*, v. 86; full commentary by Steinschneider, *loc. cit.* The *Speculum* has been generally attributed to Albertus Magnus; Mandonnet's argument for Roger Bacon in *Revue néo-scolastique*, xvii. 313–335 (1910), is discussed by Thorndike, ii, ch. 62.

[9] *Catalogus codicum astrologorum*, v. 86, n.

[10] Classis XI, Cod. 73; Valentinelli, *Bibliotheca manuscripta ad S. Marci Venetiarum*, iv. 255. The MS. is clearly of the thirteenth century, not as the catalogue says of the fifteenth. The treatise extends from f. 1 to the middle of f. 36, where it ends abruptly after the description of Anticanus. The text begins: 'Spera celi quater senis horis dum revolvitur omnes stelle fixe celo quem [sic] cum ea ambiunt circa axem breviores circulos efficiunt. Igitur que polo apparet vicinior inter omnes, tam ei splendor est precipuus, ipsa noctium hor[arum?] computatrix dicitur argumentum eminientum [sic] cardini oppositum. Recta linea si serves luminum intuitu horas noctis nosse potes galli sine vocibus.' Then after a figure of a man

incipit cited in the *Speculum astronomie*; the other, MS. Pal. Lat. 1417 of the Vatican,[11] has a different beginning, but agrees in the body of the treatise. The correspondence between the two is close throughout the first part of the work; in the latter part the Venetian MS. has a fuller treatment of the planets and constellations but lacks the meteorological chapters with which the other concludes. I do not find in either the fable of Taurus mentioned by William of Conches or the account of Leo for which Philip de Thaon cites Nebroz as his source in the only instance where he seems to be directly followed.[12] Evidently there are problems here which require further manuscript evidence.[13]

Both MSS. have, evidently as part of the original text, numerous figures, of which the most notable are the series of constellations in the Venetian codex. At the beginning of the treatise an interesting drawing, much better in the Vatican MS., represents side by side the two kings, Atlas and Nimrod, whom classical and oriental tradition respectively make the founders of astronomy. Atlas is depicted standing on the Pyrenees and bearing on his shoulders the firmament with its stars, while Nimrod stands on the mountain of the Amorites and looks upward while he supports in his hands the heavens without stars. The inscriptions read: *Athlas magnus astrologus rex Ispanensium vegens humeris suis celum inclinatum cum stellis. Nemroth inspector celorum ac rex*

observing the pole, 'Incipit liber de astronomia. De forma celi et quomodo decurrit inclinatum. Celum igitur inclinatum . . .'

[11] The treatise occupies the nineteen folios of the MS., which is written in a clear hand of the twelfth century, with the headings in red. It bears the title in a modern hand, "Ptolomei tractatus ad sciendum horas dierum ac noctis." The introductory matter was evidently lacking in the fifteenth century, when the contents of the volume were thus given at the bottom of f. 1: 'Libellus pulcer Besde de situ et dispositione stellarum et signorum celi; libellus seu tractatus Ptolomei regis ad sciendum horas diei et noctis; tractatus de distinctione climatum mundi et de terminis septem climatum.' On this MS., see now Saxl, in Heidelberg *Sitzungsberichte*, 1915, no. 5, pp. 30 f., plate 21.

[12] Lines 1315-46. Some of these lines reappear in the description of the lion in Philip's *Bestiaire*, ed. Walberg, lines 25 ff.

[13] MS. Ashmole 191, f. 46, of the Bodleian contains only a brief extract from the "Liber responsionum magistri Nemroth ad discipulum Ioaton," beginning, 'Dico enim quod de oriente . . .' An extract appears also in *Archiv für die Geschichte der Medizin*, x. 309 (1917).

Caldeorum vegens manibus celum inclinatum sine stellis. Probably a paragraph on the preceding page, now lost, of the Vatican MS. explained Nimrod, as a quotation from St. Augustine at the top of this page explains Atlas.[14]   The work proper then begins in both MSS.:

De forma [15] *celi et quomodo decurrit inclinatum*

Celum igitur inclinatum volvitur a meridiano usque in septentrionem super terram et de septentrione ad meridianum sub terram et in rotunditatem suam volvens sese inclinatum et quasi [16] eversum [17] videtur, directum [18] per preceptionem creatoris creature.   Ut homo opifex bonus [19] instruens palatium, qui primum mensurat locum et fodit fundamentum et edificat ordinabiliter illud donec adimpleatur [20] edificium suum, ita et Nemroth mensuravit omnem causam celi per suum intellectum et posuit fundamentum super quod edificavit ordinem numeri per capitula superius denominata et [21] dum perlegisset eadem semper in melius construxit.   Et omnia ista capitula se invicem condecorant ut bonus opifex qui edificium suum ordinanter disponit.   Primo in edificio fit [22] fundamentum in [23] terra et primo capitulo expositio minima celo verso sine stellis et post hec apparebit numerus.

ii.   *De una virtute qua dicit Nemroth quia* [24] *sustinet celum*

Et dum recordaretur Nemroth formam celi cognovit quod habuisset creatorem non agnoscens [24a] quis esset.   Et vidit celum volvens in semetipsum [25] non exiens de loco suo et agnovit quod non habuisset [26] de subter [27] quod illud impedisset nec desuper per quod suspenderetur, et in hoc non potuit dicere aliud nisi quod [28] virtus sit que hoc sustinet.   Et eam nominavit fortitudinem sustinentem celum et stantem sub nullo, ut admiranda sit scientia Nemroth quod mensurasset formam celi et cognovit cursus [29] signorum et circulos stellarum et fundamentum terre et non agnovit quod Deus creasset ea.   Sed et hoc [30] cognovit quod [28] desuper creatura fortis et dominatrix est [31] et nominavit eam creatorem, et depinxit et scripsit omnia secundum similitudinem suam, ita ut qui tunc fuerunt voluerunt illum habere ut deum propter suam virtutem et scientiam, dicente illo occulta in compoto astronomie.   Et cognovit Nemroth quod [28] celum fuisset purum et post hoc factus est sol et luna et omnes stelle celi.[32]

[14] *De civitate Dei,* 18, 39 (ed. Hoffmann, ii. 330).

[15] Vat. *fortitudine.*

[16] Ven. *quod;* Vat. om.

[17] Vat. *reversum.*

[18] Vat. *directum est per preceptum creatoris.   Opifex.*

[19] Ven. *bonum.*

[20] Vat. *adimpleat.*

[21] Vat. omits *et . . . construxit.*

[22] Vat. *sit.*

[23] Vat. om.

[24] Vat. *que.*

[24a] Vat. *sed non cognovit.*

[25] Vat. *semetipso.*

[26] Vat. *erat.*

[27] Vat. *subtus.*

[28] Vat. *quia.*

[29] Ven. *cursum.*

[30] Ven. om.

[31] Vat. *sit.*

[32] Vat. omits *celi.*

Chapters follow *De .iiii^{or}. ventis, De duabus fortitudinibus, De .xii. fortitudinibus, De .vii. fortitudinibus,* varied by the insertion, without credit, of the chapters on earthquakes and Etna from Bede's *De naturis rerum.*[33] The more specifically astronomical part of the work then begins with a brief account of the *axis celi* and the zodiac, succeeded by chapters on the planets, the Pleiads, the sun and its eclipses, and the moon and its eclipses. In the midst of the account of the moon there is evidently a lacuna in the Vatican MS.[34] where the Venetian MS. takes up the several planets and their motions. Both then agree in the portions treating of the hours of the day, epacts, concurrents, and days of the week, after which they finally diverge. The Venetian codex devotes the remaining ten pages to a description of the constellations, to the number of forty-three, accompanied by drawings which should have interest for the student of mediaeval astronomy.[35] None of these are found in the Vatican MS., which proceeds to consider the nature of clouds, thunder, lightning, and the rainbow. Save for the quotations from Bede and the section on the constellations, both MSS. maintain throughout the form of a dialogue between Nimrod and Ioathon, who first appears in the fifth chapter. There is very little that could be called astrological, although the concluding chapter, found only in the Vatican MS., seems to presuppose such a treatment:

Quod interrogavit Ioathon magistrum suum et non dedit ei responsum

Et postquam exposuit Nemroth Ioathon discipulo suo quid sit arcus pacis vel unde est, interrogavit eum dicens, Magister, cognovi quod exposuisti mihi quid sit arcus pacis vel unde fit. Tunc prevenit eum infirmitas mortalis et dum vidisset Ioathon magistrum suum Nemroth quia moreretur, venit et cecidit ad pedes eius dicens, Magister, nimis tristis effectus sum quia dum habui patrem efficior orphanus et post divitias multas nunc veniet michi

---

[33] Cc. 49, 50 (Migne, xc. 275–278). C. 51, 'Divisio terre,' also appears on f. 8 of the Vatican MS.

[34] F. 12, where the heading, *De luna .i. usque in .xv. quot punctos luceat donec veniat in potestate noctis,* does not correspond to the text, which assumes a preceding discussion of the planets.

[35] This part of the text begins with the typical description (f. 31 v): 'Helix, Arctus malorum, habet autem in capite stellas obscuras vii., in spatula .i., super pectus .i., in pede .i., in dorso .i., in tibia interiore .ii., super cauda .iii., sunt omnes .xvi.' The treatment is quite different from that of Hyginus.

paupertas et post virtutem quam habui ero debilis. Respondit Nemroth
dicens, Ioathon, fortasse non erit ita ut putas. Respondit Ioathon dicens,
Magister utique ita erit. Numquid quod a te didici non est veritas? Et si
verus est compotus quem ostendisti mihi pro infirmo, ipse significavit mihi
mortem meam. Ait illi Nemroth, Ioathon, omnia que docui te vera sunt et
compotus qui est super infirmum non erit tibi in aliquo error. Ego autem
vadam ad patres meos et tu venies postea et ego ad te non revertar, quia ita
hoc est quod nemo potest transgredi; et si habes aliquod ad interrogandum
unde tibi cure sit interroga velociter antequam inebreetur anima de potu
calicis mortis et antequam colligatur lingua et quietudine cursus sanguinis
tollatur sensus per fortitudinem magni pavoris cum victus exieris de termino vite ad potestatem mortis. Respondit Ioathon dicens, Magister bone,
de omnibus que ostendisti mihi aliquit cognovi, de vento autem aperte non
exposuisti michi. . . . Usque huc interrogavit Ioathon Nemroth magistrum suum et non dedit illi responsum et dum interrogat de vento insufflavit
in eum ventus mortis et non respondit ei ullum verbum et dimisit doctrinam
suam aliis.

It is plain, merely from the extracts here given, that the author
of the treatise does not speak in the name of Nimrod but bases
his work upon a dialogue between Nimrod and Ioathon which he
supplements and modifies. He refers to *alii doctores qui fuerunt
post Nemroth*,[36] and in two passages cites a certain Alexander.[37]
The Oriental touch is apparent, but there is no trace of Arabic
terms or of the Arabic astronomy, so that the work is plainly
anterior to the introduction of Saracen learning into Latin
Europe. Words like *planetes* and *sinodus* and the passage (gloss?)
on the Pleiads [38] show a certain amount of Greek influence,[39] but

[36] 'Et alii doctores qui fuerunt post Nemroth et Ioathon exposuerunt obscurita-
tem que apparet in luna. Nos autem modo exponimus subterius in loco oportuno.'
Vat. MS., f. 6 v.

[37] Vat. MS., f. 2 v: 'Nam quod ipse dixit quia discurrunt inter signa disposuit
Alexander dicens quia iste fortitudines quas ait ipse Nemroth ipse sunt quas exposuit
superius.' F. 10 (= MS. Venice, f. 12 v): '*In quo signo currit luna ut exposuit
Alexander.* Exposuimus superius in quo signo currat luna, nunc ostende mihi sicut
Alexander exposuit qui mensuravit et coequavit numero astronomie.'

[38] MS. Vat., f. 10 v: 'Pliades vii stelle splendide que post vere exoriuntur vel
Pliades a pluralitate dicte, quia pluralitatem latine grece *apolpoeton* [ἀπὸ πλείων?]
dicitur. Pliades sunt multi vage stelle quas etiam Botrum apellant. Pliades vii
fuerunt quorum nomina sunt Terope, Meropios, Cileno, Maia, Altione, Tagete,
Electra. Dicte autem pliades *apo tu plictos* [cf. Isidore, *Etymologiae*, 3, 70, 13: ἀπὸ
τοῦ πλεῖστον], id est a pluralitate, sive a pluvia vel a mare, ut sint filie Athlantis et
Pliadis.'

[39] The accounts of the constellations in the Venetian MS., though based upon
the Greek catalogues, are not directly translated. E. g. (f. 33 v), 'equus qui et
bellorum fons ' [i. e., Bellerophon]; 'navis que apud Argivos Argo vocatur' (f. 35).

the style is not that of a direct translation, and the quotations from Augustine and Bede show that the matter was worked over in the West.

The dialogue bears clear traces of Syrian origin, for the disciple Ioathon or Ioanton [40] can be none other than the fourth son of Noah who appears as Ionton, Ionaton, Ionites, Ἰώνητος, Ὑιώνητος, Μονήτων, and Munt in Christian writers of the Middle Ages. Unknown to the Hebrew tradition, he is found in works of Syrian origin and in these only,[41] and is there brought into direct relation to Nimrod. Thus in the *Cave of Treasure*, which in its Syrian form is probably of the sixth century, Ionton is visited by Nimrod in the land of Nod and teaches him that wisdom and learning of the stars which the Persians call the oracle and the Romans astronomy.[42] Similar and apparently related is the account which appears toward the close of the seventh century in the *Apocalypse* of the Pseudo-Methodius,[43] where we read that Noah sent his son Ionitus to the east, to the land of the sea and the sunrise, where God granted him the gift of wisdom so that he became the discoverer of astronomy and the teacher of Nimrod. Their relations continued friendly, and Ionitus wrote a letter to Nimrod prophesying the destruction of the dominion of the sons of Ham.[44] The astronomical attainments of Ionithon are described in greater detail in a third and considerably later Syrian source, the so-called *Causa causarum*,[45] but it was through the Pseudo-Metho-

---

[40] The *Catalogus codicum astrologicorum*, v. 86, cannot identify him.

[41] So Sackur, who has collected the material relating to him in his *Sibyllinische Texte und Forschungen* (Halle, 1898), pp. 15, 54, 64.

[42] Bezold, *Die Schatzhöhle* (Leipzig, 1883–88), i. 33 f. and notes; Götze, "Die Schatzhöhle," in Heidelberg *Sitzungsberichte*, 1922, no. 4, pp. 57 f.

[43] A critical edition of the Greek text, with studies of Latin and Slavic versions, is given by Istrin, *Otkrovenie Methodiya Patarskogo* in the *Čteniya* of the Historical and Archaeological Society of the University of Moscow, 1897, parts 2 and 4. The Latin version is edited by Sackur, *Sibyllinische Texte*, pp. 59–96. Cf. Gervase of Tilbury, ed. Leibnitz, p. 899.

[44] Οὗτος δὲ ὁ Μονήτων (al. Ἰώνητος, Ὑιώνητος) ἔλαβε παρὰ τοῦ θεοῦ χάρισμα σοφίας, ὥστε πρῶτος ἀστρονομίας τέχνην ἐφεῦρε. Πρὸς τοῦτον κατῆλθε Νεβρὼδ καὶ παιδευθεὶς παρ' αὐτοῦ εἴληφε βουλὴν ἐφ' ᾧ βασιλεῦσαι αὐτόν. Istrin, text, p. 9 f.; cf. pp. 52, 77, and Sackur, pp. 63 f.

[45] Kayser, *Das Buch von der Erkenntniss der Wahrheit* (Strasbourg, 1893), pp. 259 f.

dius that he passed into the West and found mention in a number of chroniclers and other writers of the Middle Ages.[46]   In all these sources Ionitus is the master and Nimrod the pupil, but the reversal of the relation might easily arise under the influence of the tradition which we find in Malalas and others that Nimrod was the founder of astronomy.

As regards the date of Nimrod and Ioathon our text stands in general agreement with the chronology of the Pseudo-Methodius, who mentions Ionites in A.M. 2799 and Nimrod in 3008:

> Et ab initio seculi usque ad tempus Nemroth fortissimi et Ioanton discipuli sui in quo anno circumivit Mercurius per omnia signa circulum .i., qui sunt .xxii. circuli et anni .iii. clxxxiiii. et ab ipso anno usque ad finem mundi currit.[47]

This is the only indication on this point, and unfortunately the similar cycles given for each planet [48] throw no light on the date of the treatise itself, the years being in each case carried out to the close of the cycle next preceding A.M. 7000, doubtless on the theory which we find in the Pseudo-Methodius, that the end of the world will coincide with the close of the seventh millenary period.   The same theory appears in the table of solar eclipses,[49] which is carried to the year 6995:

> Si vis scire in quo anno fit eclipsis, sume annos ab origine mundi, scito quot sunt, et subtrahe ex ipsis vi cc xc viiii, et quot remanent divide eos per decem et novem, et sicut scriptum est in rota ita invenies eclypsis solis in tempore ipsius.

There follows a table, but no *rota*, beginning, *In vi anno non erit eclypsis, in xxiiii anno erit eclypsis,* and so on at intervals of twenty-four years to *in dcxcvi anno erit eclypsis.* Here, however, the year 6299 is evidently chosen because it is the date of writing

---

[46] To the passages collected by Sackur, p. 64, should be added the *Summa philosophie* of Grosseteste, in Baur, *Die philosophischen Werke des Robert Grosseteste* (*Beiträge*, ix), p. 275; and the Slavic material collected by Istrin and by Veselovsky in his *Razyskaniya* (St. Petersburg, 1880–91), no. x.

[47] MS. *d'rt*, apparently corrupted from *c'rit*, which appears constantly in this part of the text.

[48] MS. Venice, ff. 17–19 v. Mars is carried to the year 6990, Mercury to 6936, Jupiter to 6912, Venus to 6922, and Saturn to 6800. The text of the numbers is quite corrupt.

[49] MS. Vat., f. 9; MS. Venice, f. 11 v.

or at least of the beginning of the current nineteen-year period, which would bring the treatise between A.D. 791 and 810 according to the Byzantine era or between 807 and 826 according to the era of Antioch. With the ninth century the style and manner of treatment in general correspond. The home of the work should probably be sought in Gaul, where throughout the early Middle Ages relations were maintained with Syria [50] which have left literary monuments in the Latin version of the Pseudo-Methodius and in the translation of the legend of the Seven Sleepers by Gregory of Tours.

The various astronomical questions involved in Nimrod's treatise I cannot pretend to discuss, still less can I enter into the problem of its sources and its affinities with other works. My purpose has been merely to bring to light an unused source for the study of Byzantine and Syrian astronomy and for the astronomical and cosmological ideas current in western Europe in the early Middle Ages.

---

[50] See particularly Scheffer-Boichorst, "Zur Geschichte der Syrer im Abendlande," in *Mitteilungen des Instituts für österreichische Geschichtsforschung*, vi. 535 ff. (1885); L. Bréhier, "Les colonies d'Orientaux en Occident au commencement du moyen-âge," in *B. Z.*, xii. 1–39 (1903).

CHAPTER XVII

SOME EARLY TREATISES ON FALCONRY[1]

Works on falconry occupy a not inconsiderable place in the literature of the later Middle Ages, whether in Latin or in the various vernaculars. Interesting as a phase of the court life and manners of the period, these are also significant in the history of mediaeval science, not only as illustrating the current medical notions, but also as marking the growth of knowledge based upon detailed personal observation. For the most part these treatises consist of collections of remedies for diseases, in which traditional lore, superstition, and practical experience are curiously mingled. Many of them describe with some fulness various species of birds of prey and their uses, and in the later period the actual practice of falconry receives minute attention. There is much translation and much borrowing back and forth, and the interrelations of the several works constitute an exceedingly intricate subject. As no survey of this literature has been attempted since the study of Werth in 1888,[2] it may not be out of place to call attention to certain unknown or little known manuals, chiefly of the twelfth and thirteenth centuries, which have come to my notice in the course of a study of the most famous of such treatises, the *De arte venandi cum avibus* of the Emperor Frederick II.[3]

1. Adelard of Bath

The earliest treatise on hawking so far identified in western Europe was written in England in the time of Henry I. Its author, Adelard of Bath, was not only attached in some fashion to

---

[1] Reprinted from the *Romanic Review*, xiii. 18–27 (1922).

[2] H. Werth, "Altfranzösiche Jagdlehrbücher, nebst Handschriftenbibliographie der abendländischen Jagdlitteratur überhaupt," in *Zeitschrift für romanische Philologie*, xii. 146–191, 381–415; xiii. 1–34 (1888–89). Cf. Biedermann's supplementary notes, *ibid.*, xxi. 529–540; and J. E. Harting, *Bibliotheca accipitraria* (London, 1891).

[3] See Chapter XIV.

the English court, but had studied in France, southern Italy, and the Mohammedan East, and was one of the pioneers in introducing Arabic learning into western Europe. Yet his little work on falconry ignores eastern experience and concerns itself chiefly with old English recipes for the diseases of hawks. Moreover, it refers specifically to earlier writings on the subject, the *libri Haroldi regis*, probably books once in the possession of the last Anglo-Saxon king.[4] The beginning of Adelard's treatise indicates that it was an interlude in the more serious studies represented by the author's *Questiones naturales*, also in the form of a dialogue with his nephew. The nephew begins: [5]

Quoniam in causis disserendis rerum animus noster admodum fatigatus [6] est, ad eiusdem relevationem id magis delectabile quam grave interponendum est. Intellectus enim similiter ut arcus si nunquam cessas tendere mollis erit. Quare in eo iudicio tale ad quod et iocundum et utile sit eligendum est. Id autem recte fieri spero si de accipitrum natura et usu [7] elegantius aperias, precipue cum et nos Angli sumus genere et eorum inde scientia pre ceteris gentibus probata sit et ea deinde scientie qualitas constat [8] ut [9] quanto pluribus dividitur tanto magis efflorescet. *Adel[ardus]*. Sit sane ne aut inscientia aut invidia [10] arguamus. Ea igitur disseremus que et modernorum magistrorum usu didicimus et non minus que Haraoldi [11] regis libris reperimus scripta, ut quicunque his intentus disputatione[m] habeat si negotium exercuit paratus [12] esse possit. Tuum itaque sit inquirere, meum explicare.

It ends:

Hec habui que de cura accipitrum dicerem. Ceterum si tibi vel alicui alii suam addere sententi[am] placet, non invideo.

Adelard's little work does not seem to have been widely used. The only complete copy I have found is in MS. 2504 of the Nationalbibliotek at Vienna (ff. 49–51). The greater part is incorporated into a compilation of the thirteenth century to which we shall come below (Clare College, Cambridge, MS. 15, ff. 185–187). The earlier portion at least is used by the author of an

---

[4] See my note on "King Harold's Books," in *E. H. R.*, xxxvii. 398–400 (1922); and for Adelard, supra, Chapter II.

[5] Vienna, MS. 2504, f. 49 (ca. 1200). [8] MS. *et stat*.

[6] MS. *fatigatitus*. [9] MS. *ett* (?).

[7] Corrected from *usque ad*. [10] MS. *individua*.

[11] The scribe may have tried to correct the *a* into an *o* or vice versa.

[12] MS. *paritus*.

Anglo-Norman poem in the British Museum (Harleian MS. 978).[13]

No other treatise connected with the Anglo-Norman court is known to have survived. Daude de Pradas, writing his *Romans dels auzels cassadors* early in the thirteenth century,[14] cites:

> En un libre del rei Enric
> d'Anclaterra lo pros el ric,
> que amet plus ausels e cas
> que non fes anc nuill crestias,
> trobei d'azautz esperimens
> on no coue far argumens.[15]

Whether the reference is to Henry I or Henry II it is impossible to say, though the latter is more likely. This would be a particularly interesting treatise to recover.

2. WILLIAM THE FALCONER

Like the Norman kings of England, the Norman rulers of Sicily were mighty hunters and hawkers, and the first who bore the royal title, Roger II (1130–54), is said to have had a falconer, William, whose precepts are frequently cited. Thus Albertus Magnus, in the chapters of his *De animalibus* devoted to falcons,[16] cites in three passages William the falconer, in one instance specifically as King Roger's falconer, followed as an authority by Frederick II: [17]

[13] Compare the extract given by Paul Meyer in *Romania*, xv. 278 f., with the passage from Adelard printed below, note 36.

[14] The biographical data on Daude given in the standard works are very meagre. He dedicates his poem on the cardinal virtues to Stephen, bishop of Le Puy (1220–31); and Torraca has found him attesting as canon of Rodez in 1214–18: *Studi su la lirica italiana del duecento* (Bologna, 1902), pp. 244 f.

[15] Ed. Monaci (in *Studî di filologia romanza*, v. 65–192), lines 1930–35; ed. Sachs (Brandenburg, 1865), lines 1905–10. Werth (xii. 154 f., 166–171) thinks he can identify other passages in Daude derived from the *libre del rei Enric*. The incantations of lines 1937 ff. reappear in Albertus Magnus, c. 19.

[16] Bk. xxiii, c. 40. Ed. Stadler (*Beiträge*, xvi), pp. 1453–93; *Opera* (Paris, 1891), xii. 451–487. These chapters often appear in the manuscripts as a separate work on falconry, e. g., Bodleian, MS. Rawlinson D. 483, ff. 1–47 v, from Bologna.

[17] C. 10, ed. Stadler, p. 1465; not in the known text of Frederick's *De arte*. Cf. c. 20 below.

Hunc falconem [*i. e.*, nigrum] Federicus imperator sequens dicta Guilelmi, regis Rogerii falconarii, dixit primum visum esse in montanis quarti climatis quae Gelboe vocantur, et deinde iuvenes expulsos a parentibus venisse in Salaminae Asiae montana, et iterum expulsos nepotes primorum devenisse ad Siciliae montana et sic derivata esse per Ytaliam.

These citations can be identified in a brief treatise which in several manuscripts [18] follows the Latin text of the so-called 'Dancus.' [19] The last chapter of 'Dancus' runs:

Iste magister non fuit mendax sed verax, iste medicine sunt bone et perfecte et multum probate. Guilielmus falconerius qui fuit nutritus in curia regis Rogerii qui postea multum moratus fuit cum filio suo et habuit quendam magistrum qui vocatus fuit Martinus qui fuit sapiens et doctus in arte falconum, et iste discipulus suus Guilielmus scivit omnia que ipse scivit et tanto plus quod ipse composuit libellum unum de arte ista cuius principium tale est. Nolite dubitare sed firmiter sciatis quod nullus talis magister vivit modo in mundo.

Explicit liber Galacianus rex [*sic*] de avibus.

[*Chapter headings, then*] Incipit tractatus Guilielmi de avibus et eorum medicamine, et primo capitulo incipit de dolore capitis qui dicitur furtinum [*or* siurtinum].

Quando vides quod habet furtinum accipe mumiam et da ei comedere cum carne porcina et alio die da ei carnem gatti et tene eum donec liberabitur. . . .

Seventeen chapters contain brief remedies of this sort; the remaining chapters, 18–24, treat briefly of the training and species of falcons. In the midst of chapter 20 we read:

Nullus magister scit ita de naturis falconum unde sunt et unde exierunt sicut iste magister Guillelmus filius Malgerii Neapoletani scivit et ideo tractat de naturis falconum quia plus scivit quam aliquis homo. Falcones qui prius apparuerunt in mundo ipse bene agnovit. Falcones nigri prius apparuerunt.

---

[18] I have used in the Vatican MSS. Vat. lat. 5366, ff. 40 v–44 v (saec. xiii); Ott. lat. 1811, ff. 37–40 (saec. xiv); Reg. lat. 1227, ff. 51–56 (saec. xv); Reg. lat. 1446, ff. 74–76 (saec. xiv); and in the Bibliothèque Nationale, MS. 7020, ff. 45 v–49 (saec. xv). The text of the extracts printed follows MS. Vat. lat. 5366, with some obvious corrections from the others. See also the French version of Dancus, anterior to 1284, ed. Martin-Dairvault (Paris, 1883), pp. 19–29, and its notes; and the Italian version in *Il propugnatore*, ii, part 2, pp. 221 ff. (1869). An Italian version of William, now in MS. Ashburnham 1249 of the Laurentian, is cited by G. Mazzatinti, *La biblioteca dei Re d'Aragona* (Rocca S. Casciano, 1897), p. 172.

[19] On which see Werth, xii. 148–160. There is a series of extracts from Dancus and others at the University of Bologna, MS. 1462 (2764), saec. xiv.; and a copy of the Latin Dancus at Modena, Estense MS. 15, followed by an anonymous *Liber curarum avium*, beginning, 'Notandum est quod meliores aves viventes de rapina . . .'

Venerunt a Babilonia in Montem Gebeel et deinde venerunt in Sclavoniam et deinde venerunt ad Palunudum [20] quod est in pertinentiis Policastri.

Magister Guillelmus is again quoted in chapter 22:

Propter carnem non perdet voluntatem venandi set propter sanguinem tantum, et hoc probavit magister Guillelmus qui plus modo fecit quam aliquis qui vivat.

The treatise ends with the chapter on *ysmerli* cited from William by Albertus Magnus: [21]

Sed tamen si bonus est magister potest eos facere capere grues tali dieta et tali custodia ut alii falcones, et si vult capere grues oportet habere duodecim ysmerlos.

Apparently we have not William's manual in its original form, but extracts from it, which, however, have something of the brevity to be expected from a practical falconer of the early period. The connection with Sicily is clear, not only in the statements respecting the king and the Neapolitan falconer Malgerio, but, more certainly, in the reference to the region of Policastro. If the treatise in its original form should be discovered, we should probably have one of the important sources for later writers.

3. THE COURT OF FREDERICK II AND HIS SONS

In the thirteenth century the chief centre of literary activity on subjects of falconry was the court of the Emperor Frederick II. A tireless sportsman from his youth, the emperor called in expert falconers from many lands and devoted long years to the observation of birds and the practice of the art. He had the treatise of Moamyn, and probably that of Yatrib, translated from the Arabic under his personal supervision, and appears in general to have systematically collected the authorities on the subject. After thirty years of preparation he dedicated to his son Manfred the *De arte venandi cum avibus*, which is the most noteworthy mediaeval work on the subject, noteworthy for its independent and scientific spirit even more than for the eminence of its author. In the form known to us the *De arte* consists of a systematic account

---

[20] Lat. 7020 has 'Palumbidum'; Reg. lat. 1446 interlines in a later hand 'Paludinum.' The place is evidently Monte Palladino on the gulf of Policastro.

[21] Ed. Stadler, p. 1468.

of birds in general and falcons in particular, followed by a detailed examination of lures and the methods of hunting with the several types of falcons. There is reason for thinking that the emperor also discussed hawks and the diseases of falcons, but this part of his work has not been recovered.[22] Besides half a dozen manuscripts of the Latin original, in a six-book edition and a two-book recension by Manfred, we have two different French versions made before the end of the thirteenth century.[23]

Frederick's favorite son Manfred inherited in large measure the intellectual interests of his father. We learn from the preface that Frederick's *De arte* was finally put into form at Manfred's request, and it was he who later searched out the notes and loose sheets of the author which are incorporated in his recension.[24]

Another son, Enzio, well known in the literary circle of the *Magna Curia*, was likewise a patron of writers on falconry. His "servenz et hom de lige," Daniel of Cremona, dedicates to him French versions of Moamyn and Yatrib which afford interesting evidence of the prevalence of French in North Italy;[25] while an anonymous young writer composed for him, as king of Torres and Gallura, a brief set of excerpts on the species of falcons and their diseases, which is preserved at Clare College, Cambridge (MS. 15, ff. 185–187). It begins:

Incipit practica avium. Ex primis legum cunabilis impericie mee solacium querens scemam virorum honestatisque sigillum mente ne facto viri deinceps

---

[22] See the chapters on diseases in Albertus Magnus 'secundum falconarios Federici imperatoris' (c. 19) and 'secundum experta Federici' (c. 20). The greater part of chapter 19 appears in a treatise in the Vatican (MS. Reg. lat. 1446, ff. 76–77) headed 'Gerardus falconarius,' possibly one of the emperor's falconers.

[23] Supra, Chapter XIV.

[24] Supra, Chapters XII, XIV. The treatise of Adam des Esgles, "falconer of the prince of Tarento," dates doubtless not from Manfred's time but from one of the later bearers of this title. It is found in a manuscript of the fifteenth century at Le Mans, MS. 79, ff. 116 v–128 v, beginning:

'Aultres medicines pour faulcons fait par Adam des Esgles chevalier faulconnier du prince de Tarente, et premierement faulconnerie veult que soyes doulx et courtoys et debonnaire. Se ung faulcon aver qui soit blanc et blont et de gros plumage . . .'

[25] Ciampoli, *I codici francesi della R. Biblioteca di S. Marco* (Venice, 1897), pp. 112–114; cf. Paul Meyer, in *Atti* of the Roman Congress of History, iv. 78 (1904); supra, Chapter XIV, nn. 128–130.

videar contrarius set honeste pretendi pocius condescendens, igitur ut principi nostro excellentissimo, .E. Turrensi principi, qui causa aucupantium delectat precipue ceterisque eiusdem generis [26] satisfactioni[bus], utiliora ex libris antiquorum collecta in huius libelli compendium de natura avium breviter enodavi, opus hoc meum esse non affirmans nisi per compilationem. Eius seriem in .v. particulas divisi quarum prima continetur qualiter Aquila et Simachus et Theodosion Tholomeo imperatori Egipti scripserunt et quid de avibus senserunt et eorum accidentibus, variis enim subiacent periculis ut corpus humanum et variis succurritur medicinis. Et nota quod unus pro omnibus rationari intelligitur. Secunda continet quid [27] Alexander grecus medicus Cosme de vario casu ancipitrum et eorum medela [28] scripsit. Tercia quid Girosius [29] hyspanus Theodosio imperatori. Quarta quid Alardus anglicus nepoti suo interroganti responderit. Quinta quid M. G. de Monte P. expertus sit, et sic liber terminatur.

The nature of the work is indicated by this preface: the species of hawks and falcons, and their diseases. Of our author's sources, the letters of Ptolemy and Theodosius are well known,[30] and Adelard's treatise has just been described. The supposed letter of the Greek physician Alexander, I have not identified.[31] Master G. of Montpellier may be Gilbert the Englishman, chancellor of Montpellier, well known as a medical writer about 1250;[32] his contribution deals entirely with diseases.

4. ARCHIBERNARDUS

Among the Rossi manuscripts recently returned from Vienna to Rome and now on deposit in the Vatican [33] there is found a codex of the thirteenth century containing a Latin poem of 324 hexameter lines entitled *Liber falconum*.[34] The author, who calls himself Archibernardus, is evidently an Italian, using such expressions as *pulzinus*, *buzza*, *pollastra*, and twice having the line,

Ars mea sanari docet hunc Italis medicari.

[26] MS. *genera*.

[27] MS. *grecus*.

[28] MS. *ex medelo*.

[29] As later. MS. here *Gñosius*.

[30] Werth, xii. 160–165.

[31] Alexander is cited by Daude de Pradas, line 2319; cf. Werth, xii. 165.

[32] *Histoire littéraire*, xxi. 393–403; cf. Duhem, iii. 291; Thorndike, ii, ch. 57. There is an early copy of his *Liber morborum* at the University of Madrid, MS. 120, f. 20.

[33] On this collection see Bethmann, in Pertz's *Archiv*, xii. 409–415; [Silva-Tarouca], in *Civiltà cattolica*, 18 February 1922, pp. 320–335; *Neues Archiv*, xlv. 102.

[34] MS. VII. 58, ff. 85–87 v.

The subject matter is of the usual kind, the species, food, and diseases of falcons:

A nostra prohemaria ductris sit virgo Maria!
Archibernardi per carmen disce mederi
Leso falconi nec dedignere doceri
Miles mille valens si vis urbanus haberi.

. . . .

Sit hic locus mete musarum avete cetus
Egregios iuvenes equites peditesque docetis.
Explicit liber falconum.

5. Egidio di Aquino

Friar Egidius de Aquino is given as the author of a brief treatise preserved in a manuscript of the fifteenth century in Corpus Christi College, Oxford (MS. 287, ff. 74 v–78 v). It covers the training, diseases, and species of birds of prey, beginning with falcons and ending with hawks, and is particularly full in distinguishing the varieties used in Italy. Thus the species of hawks include those of Ventimiglia, Slavonia, Calabria (*calavresi*), Istria, Sardinia, Germany, and the Alps (*alpisiani*);[35] while among *astures* we find those of Tuscany, Lombardy, the March, Apulia, Germany, and Sicily:

Incipit liber avium viventium de rapina et [de] morbis et curis et generationibus eorum.

Quoniam vidimus et experimento cognovimus morbos doctrinas naturas et generationes avium et plures de nobilioribus, scilicet viventibus de rapina et eorum generationibus documentis infirmitatibus curis et naturis, omnibus aliis generationibus pretermissis ad presens tractatulum intendimus inchoare. . . . Quoniam inhonestum est retinere ancipitrem in manu cum pennis fractis sive tortis.

Explicit liber de naturis morbis et generationibus omnium avium viventium de rapina. Compositus est a fratre Egidio de Aquino.

Laus tibi sit, Christe, quoniam liber explicit iste.
Et facto fine pia laudetur virgo Maria.
Amen.

[35] The manual of Egidio is followed quite closely in the anonymous Italian treatise published by A. Mortara, *Scritture antiche toscane di falconeria* (Prato, 1851), pp. 1–21. Chapter 6 of this appears as a fragment in MS. Rawlinson D, 483, ff. 47 v–48 v, following the Latin text of Albertus Magnus.

This is followed in the manuscript (ff. 78 v–84) by an anonymous *Liber de ancipitribus et falconibus et curis eorum*, beginning:

Nimis sumit precipue volucres sparvarius et pre cunctis passeres . . .

It makes use of personal experience, but at the end incorporates a condensed version of William the falconer.

6. PETRUS FALCONERIUS

Of uncertain date is the brief Italian tract of a certain Peter on the care of falcons, preserved in a manuscript of the fifteenth century in the Vatican (MS. Urb. lat. 1014, ff. 53 v–56), in the midst of a copy of Moamyn:

Petrus falconerius aliter dictus Petrus de la stōr composuit ista. Qui fuit et est si vivit de melioribus falconeriis totius mundi et magister magistrorum imprimis.

Chi vol fare uno falcone ramage saur sitost come preso e vol mangiare su lopugno hoiuli [*sic*] de dar mangiare .viii. grani gorge entre lagente apresso si de hom quattro giorni carne lassativa lavata e apresso ledevo lomo dar uno membro de gallina. . . . e poi lo mecti su la pertica e lassalo stare che non de multo gettara lapiumata e quello sella se non la gettara quello pure. Allo sparvieri smeriglio daneli promicta.

7. ANONYMOUS WORKS

The care and cure of falcons is the subject of an anonymous treatise of the late thirteenth century preserved in a manuscript in the library of the University of Cambridge. At the beginning there is a suggestion of the earlier portion of Adelard of Bath,[36] while the remedies often coincide with those of the falconer of Frederick II quoted by Albertus Magnus. The beginning of the treatise has been printed by Paul Meyer;[37] it ends:

Aneti et piperis grana sex insimul tere et cum pullina carne sibi tribue.

---

[36] Adelard has: 'Inde audire desidero quales esse velis qui huic studio conveniant. Sobrios, pacientes, castos, bene hanhelantes, necessitatibus expeditos. Quare? Ebrietas enim oblivionis mater est. Ira lesiones generat. Meretricum frequentatio tineosos ex tactu accipitris facit.' MS. Vienna 2504, f. 49; MS. Clare 15, f. 186.

[37] MS. Ff. vi. 13, ff. 69 v–73; *Romania*, xv. 279 (1886).

Two French treatises, likewise anterior to 1300, have been noted by Paul Meyer in the same manuscript.[38]

Another French treatise of the same period is noted by Meyer in a manuscript at Lyons; as a different French version is found at Cheltenham, it is likely that both go back to a Latin original.[39]

[38] *Ibid.*, pp. 279–281.

[39] *Romania*, xiii. 506 (1884); *Bulletin de la Société des anciens textes français*, xi. 75–77 (1885).   Not in Werth.

CHAPTER XVIII

A LIST OF TEXT-BOOKS FROM THE CLOSE OF THE
TWELFTH CENTURY [1]

To the historian of the influence of classical antiquity upon the
civilization of the Middle Ages the study of mediaeval text-books
yields information of the first importance. It was almost wholly
as formulated in a few standard texts that the learning of the
ancient world was transmitted to mediaeval times, and the authority of these manuals was so great that a list of those in use in
any period affords an accurate index of the extent of its knowledge and the nature of its instruction. For the later Middle Ages
the names of the text-books in use are known to us chiefly from
the statutes prescribing the course of study in the several faculties of the various universities, but, unfortunately, the documents of this sort which have reached us do not belong to the
earlier period of university history. If we except the brief list of
books in logic, grammar, and rhetoric drawn up by the papal
legate in 1215,[2] our earliest information respecting the arts course
at the University of Paris comes from 1255 [3] and at Oxford from
1267; [4] the first medical statutes, those of Paris, Naples, and
Salerno, belong to the decade following 1270; [5] while the oldest
extant statutes of Bologna [6] and Montpellier [7] date from the
fourteenth century. By this time, however, important changes
had taken place in the subject-matter of both liberal and profes-

---

[1] Revised from *Harvard Studies in Classical Philology*, xx. 75–94 (1909). For the
results cf. Baeumker, in *Philosophisches Jahrbuch*, xxvii. 478–487 (1914); Grabmann, *Aristotelesübersetzungen*, pp. 22–24; L. J. Paetow, *The Arts Course at Mediaeval Universities* (Urbana, 1910), pp. 15 f.

[2] Denifle and Chatelain, *Chartularium Universitatis Parisiensis*, i. 78.

[3] *Ibid.*, i. 277. There is a compendious account of the principal text-books in
arts in Paul Abelson, *The Seven Liberal Arts* (Columbia thesis, New York, 1906).

[4] *Munimenta academica*, pp. 34–36.

[5] *Chart. Univ. Par.*, i. 517; de Renzi, *Collectio Salernitana* (Naples, 1852), i. 361.

[6] Malagola, *Statuti delle università e dei collegi dello studio bolognese*, pp. 3–44.

[7] Germain, *Cartulaire de l'Université de Montpellier*, i, nos. 25, 65, 68, 75.

sional study. The decline of the classics before the triumph of
the scholastic logic, the diffusion of the Aristotelian metaphysics
and natural philosophy, the introduction of new texts in grammar
and mathematics, the rise of Arabian medicine — these are some
of the changes which made the curriculum of the fourteenth century a very different thing from that of the twelfth. Special interest, accordingly, attaches to an anonymous list of text-books
in arts and in the various professional studies which was composed toward the end of the twelfth century and is for the first
time printed below. The list, it is true, contains no mention of
university organization, still less of any particular institution, but
the arrangement of books in order under the seven liberal arts and
the professional studies of medicine, civil and canon law, and
theology, presupposes something like the university organization
of the four faculties; and as reason will be shown for ascribing the
list to Alexander Neckam, who studied and taught at Paris in the
last quarter of the twelfth century, we may fairly regard it as an
unofficial enumeration of the books then in use in the schools of
Paris. The importance of Paris as an intellectual centre and of
this period as an age of transition gives this text a certain significance in the history of mediaeval education.

The list in question forms part of a descriptive vocabulary of
terms relating to ecclesiastical matters, court life, and learning,
which is preserved in a manuscript in the library of Gonville and
Caius College, Cambridge.[8] This portion of the volume was
written in England by an unlearned copyist in the latter half of
the thirteenth century, and is accompanied by an elaborate gloss
which is quite full but has an almost exclusively lexicographical
interest. As the vocabulary has no title or indication of authorship, we shall cite it by the opening words, *Sacerdos ad altare*

---

[8] MS. 385 (605), pp. 7–61, for the repeated use of which I am greatly indebted
to the Master and Fellows of the college. The vocabulary is preceded by a brief
table of contents, as follows: 'De vestimentis sacerdotalibus. De ornamentis
altaris. De officiis cenobii. De ornatu regio. De tyrannorum excerticiis. De
oblectamentis curialium. De erudicione scolarium. De notario. De gramatica.
De logica. De arsmetrica et musica. De geometria. De astronomia. De phisica.
De iure ecclesiastico. De iure civili. De celesti pagina. De librario.' The
rubric 'De notario' is here misplaced; in the text it comes after 'De celesti
pagina.'

accessurus. Most of the other tracts in the volume are from the
pen of John of Garland, and as this vocabulary is likewise ascribed to him in the table of contents inserted at the beginning of
the volume,[9] it has been treated as one of Garland's works by all
who have had occasion to mention it.[10] This table of contents,
however, was written in the fifteenth century by the donor of the
manuscript, Roger Marchall, and as its statements cannot be
shown to rest on anything better than Marchall's own opinion,
we are obliged, in default of any contemporary authority, to treat
the matter of authorship as an open question to be determined, if
possible, by internal evidence.

Even a cursory examination proves fatal to the hypothesis that
Garland was the author. The simple and direct style is in striking contrast with the overloaded pedantry of Garland's writings,[11]
as seen, for example, in the well known *Dictionarius*[12] which he
prepared for the students of Paris, or in the unpublished *Com-
mentarius curialium*[13] designed for the instruction of courtiers;
nor does the subject-matter show parallels to these or to his other

[9] 'Diccionarius Mᵣⁱ Iohannis de Garlandia cum commento.' In his description
of the MS. James inserts 'Dictionarius Joh. de Garlandia' as if this occurred on
p. 7 of the text, but there is nothing of the sort in the MS.

[10] Bernard, *Catalogi librorum MSS. Angliae et Hiberniae* (Oxford, 1697), no. 1045
of the Cambridge MSS.; Tanner, *Bibliotheca Britannico-Hibernica* (London, 1748),
p. 310; Way, *Promptorium parvulorum* (Camden Society), iii, pp. xxviii, note, xxx;
Smith, *Catalogue of MSS. in the Library of Gonville and Caius College*, p. 179; *Dictionary of National Biography*, under "Garland," no. 13; Sandys, *History of Classical Scholarship*[3], i. 550; Abelson, *The Seven Liberal Arts*, p. 28; James, *Descriptive
Catalogue*, ii. 441.

[11] On Garland's writings see Hauréau, *Notices sur les oeuvres authentiques ou
supposées de Jean de Garlande*, in the *Notices et extraits des MSS.*, xxvii, 2, pp. 1–86
(1877); the article in the *Dictionary of National Biography*; E. Habel, in *Mitteilungen der Gesellschaft für deutsche Erziehungsgeschichte*, xix. 1–34, 119–130 (1909);
and E. Faral, *Les arts poétiques du XIIᵉ et du XIIIᵉ siècle* (Paris, 1923), pp. 40–46.
None of these mentions the grammatical exercises at Basel, MS. B. viii. 4, ff. 47–76.
Cf. also Paetow, *The Arts Course*, pp. 16–18, 40–44.

[12] Edited by Géraud, *Paris sous Philippe-le-Bel*, pp. 585–612; T. Wright, *A
Volume of Vocabularies* (London, 1857), pp. 120–138; Scheler, in the *Jahrbuch für
englische und romanische Litteratur*, vi. Cf. the 'Dictionarius versificatus' at
Douai, MS. 438.

[13] Caius College, MS. 385, pp. 199–211; Bruges, MS. 546, ff. 77–83 v; Rome,
Biblioteca Casanatense, MS. 2052, ff. 64–72 (also dated 1246). For specimens see
Scheler, *o. c.*, vi. 52; Way, *Promptorium parvulorum*, iii, p. xxix.

works. Moreover, we shall shortly see reasons for assigning the *Sacerdos ad altare* to the close of the twelfth century, while Garland's earliest datable work, the *Dictionarius*, is subsequent to 1218 [14], his *De triumphis ecclesie* was written as late as 1252, and his *Exempla honeste vite* after 1257.[15] Garland and the author of our vocabulary were plainly a full generation apart.[16]

There is, on the other hand, enough resemblance of style and matter to suggest some connection between the author of the *Sacerdos ad altare* and an older lexicographer of considerable repute, Alexander Neckam. Neckam was born at St. Albans in 1157,[17] taught for some years at Dunstable in the time of Warin, abbot of St. Albans [18] (1183–95), and later became a canon of

[14] It contains a reference to the siege of Toulouse in this year and was written after the close of the Albigensian war ('sedato tumultu belli'): ed. Scheler, *Jahrbuch*, vi. 153; Hauréau, *Notice*, pp. 45–46.

[15] *Joannis de Garlandia De triumphis ecclesiae libri octo*, ed. Wright (London, Roxburgh Club, 1856), pp. ix, 139, where there is a reference to the crusade projected by Ferdinand III for 1252; E. Habel, "Die *Exempla honestae vitae*," in *Romanische Forschungen*, xxix. 131–154 (1910). The *Poetria* (ed. Mari, *I trattati medievali di ritmica latina*, Milan, 1899, pp. 35–80; and *Romanische Forschungen*, xiii. 883–965) is assigned to ca. 1260 by Hauréau, *Notice*, p. 82. Cf. Mari, *I trattati*, p. 7; and Rockinger, in *Quellen und Erörterungen zur bayerischen und deutschen Geschichte*, ix. 489.

[16] It is usually stated by the biographers of John of Garland that he studied at Paris under Alain de Lille, who died in 1202, but the passage in the *De triumphis ecclesie* (p. 74) which is cited in support of this view affords no evidence that John was Alain's pupil. As Alain entered the Cistercian order some time before his death (Hauréau, in *Mémoires de l'Académie des Inscriptions*, xxxii, 1, p. 27), it is exceedingly unlikely that he was the master of a man who was writing in 1257 or later. In his introduction to the *De triumphis* (p. vi) Wright argues that John was at the University of Paris as early as 1204, but he reaches this conclusion by translating *quater* "four" in a line of the *De mysteriis ecclesie* which will not scan as he prints it (*delegat* instead of *decem ligat* in the following line). In the text given by Otto, *Commentarii critici in codices bibliothecae Academicae Gissensis* (Giessen, 1842), p. 147, line 644, this line reads:

Mille ducentenis quater inde decem ligat annis.

Unless we emend the next line in some way so as to read *quinque annos* or something of the sort for *qui nos* (cf. *De triumphis*, p. 127), there is some difficulty in reconciling this with the year 1245 of which Garland is writing, but the reference to the council of Lyons and the death of Alexander of Hales is too plain to admit of any other year. In any case 1204 is quite out of the question.

[17] See the extract printed in Tanner, *Bibliotheca*, p. 539, note d.

[18] *Gesta abbatum S. Albani* (Rolls Series), i. 196.

Cirencester, where he was made abbot in 1213 and died in 1217.[19] He studied and taught at Paris, where he became a pillar of the school of the Petit-Pont, the range of his studies covering not only the liberal arts but also theology, medicine, and civil and canon law.[20] The exact time of his sojourn at Paris cannot be determined, the date of 1180 given by modern writers resting, like more than one supposed fact of mediaeval literary history, upon an unsupported statement of Du Bóulay;[21] but for reasons of age he can hardly have begun his studies there before 1175, and he must have returned some years before the death of Abbot Warin in 1195. Neckam was a man of much learning and a prolific author, his writings comprising fables, books on natural history, theological commentaries, and grammatical and lexicographical treatises; and while a comprehensive and critical study of his unpublished works is still lacking, enough is available to permit of satisfactory comparison with the Caius College vocabulary.[22]

We naturally take up first the *De nominibus utensilium*, written, like Garland's *Dictionarius*, to illustrate in descriptive form the meanings of as many words as possible, but comparison with the

[19] *Annales monastici* (Rolls Series), i. 63; ii. 289; iii. 40; iv. 409.

[20] See the *De laudibus*, ed. Wright, p. 503, and cf. in the same volume pp. 311, 414, 453.

[21] *Historia Universitatis Parisiensis*, ii. 725: 'Alexander Nekamus natione Anglus circa an. 1180 Lutetiae legebat adhuc publice.'

[22] The list of Neckam's works given by Bishop Tanner in his *Bibliotheca Britannico-Hibernica*, pp. 539–541, needs sifting and supplementing. Contributions have been made especially by Hauréau, in the *Nouvelle biographie générale*, xxxvii. 569, and in his study of the *De motu cordis*, *Mémoires de l'Académie des Inscriptions*, xxviii, 2, pp. 317–334; and by Paul Meyer, *Notice sur les Corrogationes Promethei d'Alexandre Neckam*, in the *Notices et extraits des MSS.*, xxxv, 2, pp. 641–682; and now by the elaborate bibliographical study of M. Esposito, *E. H. R.*, xxx. 450–471 (1915), who has further work in preparation. While citing this chapter in its original form (1909), Esposito fails to discuss the *Sacerdos ad altare*. The printed works comprise the *Fables*, published by Hervieux, *Fabulistes latins*[2], ii. 392–416; the *De naturis rerum* and its metrical paraphrase, the *De laudibus divine sapientie*, edited by Wright in the Rolls Series (1863); and the *De nominibus utensilium*, edited, without sufficient study of the glosses, by Wright, *A Volume of Vocabularies*, pp. 96–119, and by Scheler in the *Jahrbuch für englische und romanische Literatur*, vii. 58–74, 155–173. The memoir of Meyer gives extracts from the *Corrogationes*. The poem *De vita monachorum* attributed to Neckam by Wright, *Anglo-Latin Satirical Poets*, ii. 175–200, has been shown by Hauréau to be the work of another (*Notices et extraits de quelques MSS.*, i. 79). Cf. Thorndike, ii. ch. 43.

Sacerdos ad altare is rendered difficult by the fact that the two do not cover the same ground, the *De nominibus* dealing with the vocabulary of the household and of everyday life, while the *Sacerdos ad altare* is confined to court life, learning, and ecclesiastical terms. The Caius College vocabulary is also briefer and more elementary, being evidently designed for a lower stage of instruction. At one of the few points where the two treatises overlap, namely in dealing with the implements of the *scriptorium*, they show some things in common:

Caius College, MS. 385, p. 58: Librarius vero, qui vulgo scriptor dicitur, cathedram habeat cum ansis porrectis ad sustinendum asserem cui quaternus superponendus est. Asser autem centone operiatur cui pellis cervina maritetur ut pargameni vel menbrane superfluitates rasorio seu novacula queant apcius eradi. Dehinc pellicula ex qua (p. 59) formabitur quaternus pumice mordaci purgetur et planula leni adequetur superficies. Folia iungantur tam in superiori [quam in inferiori] parte quaterni appendicis officio circumvolute. Quaterni margines altrinsecus punctorio distinguantur proporcionaliter ut certius usu [23] regule lineetur quaternus errore sublato. Si vero in scribendo liture occurrunt aut obliteracio, non cancelletur scriptura sed abradatur. Opus est autem ut dente apri poliatur locus abrasionis aut panniculo lineo complicito frequenter superinducto confricetur. Sicut vero rubrica est obnoxia minio, sic etiam littere capitales nunc minio, nunc viridi colore, nunc [24] veneto se debent (?), nunc atiro [25] superbire videntur.

De nominibus utensilium, ed. Scheler, pp. 167–169: Scriptor rasorium vel novaculum ad abradendum sordes pergameni sive membrane. Pumicem habeat mordacem et planulam ad purgandum et equandum superficiem pergameni; plumbum etiam habeat et lineam quibus linetur pagina. . . . Cidula sive appendice tam in superiori quam inferiori parte folia habeat coniuncta. . . . Scripturus etiam in cathedra sedeat ansis utrimque elevatis pluteum sive asserem sustinentibus. . . . Habeat etiam dentem verris sive apri sive liofe ad polliendum percamenum cum liquescat litera (non dico elementum), sive litura facta sit, sive literas ascriptas cancellaverit. . . . Habeat et minium ad formandum literas rubeas vel puniceas vel feniceas sive capitales. Habeat etiam fuscum pulverem et azarram.

These resemblances are not conclusive, but when we turn to Neckam's principal printed work, the *De naturis rerum,* the agree-

[23] MS. *usus.*   [24] MS. *nuc.*   [25] I. e., *azuro.*

ment is very close. We find not only characteristic turns of phrase, like *filii Ade*,[26] *celestis pagina*,[27] *vir maturi pectoris*,[28] *civilis iuris peritia*,[29] and other similarities to which attention is called in the notes, but some passages have been taken over bodily from one work into the other. The following is a good illustration of such borrowing:

MS. 385, p. 39: Admirationem item pariat oculis intuencium [30] psitacus, qui vulgo dicitur papagabio, cuius forma corporis aliquantisper falconem vel hobelum representat sed plumis intensissimi viroris decoratur. Pectore rotundo et rostro adunco munitur, tante virtutis ut cum in cavea recluditur, effectus etiam domesticus, ex virgis ferreis domuncula eius contexatur. Duris enim ictibus et corrosioni rostri non possent resistere [31] virge lignee. Linguam habet spissam et formacioni soni vocis humane ydoneam. Mire caliditatis et adulacionis est, in eccitando risu preferendus histrionibus.

Miraberis [32] etiam et ciconiam, que et crotolistria dicitur, que rostris crepitantibus crotolans horas diei distinguere perhibetur crepitacione sua. In yeme autem latet in aquis sed verno tempore Naiadum regna relinquens sub divo degit clemencioris aure leta salutatrix.

De naturis rerum, pp. 87–88: Psittacus, qui vulgo dicitur papagabio, id est principalis seu nobilis gabio, eoas inhabitat oras. . . . Forma corporis aliquantisper falconem vel hobelum representat, sed plumis intentissimi viroris decoratur. Pectore rotundo et rostro adunco munitur, tante virtutis ut cum in cavea recluditur, effectus etiam domesticus, ex virgis ferreis domuncula eius contexatur. Duris enim ictibus et corrosioni rostri non possent resistere virge lignee. Linguam habet spissam et formationi soni vocis humane idoneam. Mire calliditatis est et in excitando risu preferendus histrionibus.

P. 112: Ciconia, que et crotalistria, rostris crepitantibus crotolans, horas diei distinguere perhibetur crepitatione sua. In hieme autem latet in aquis, sed verno tempore Naiadum regna linquens, sub divo degit clementioris aure leta salutatrix.

The *Sacerdos ad altare* stands in close relation with still another of Neckam's works, the so-called *Corrogationes Promethei*, a treatise in two parts comprising a brief summary of Latin grammar and an elaborate verbal commentary on the Bible. The

[26] Ed. Wright, pp. 81, 83, 241, 333. Cf. pp. 119, 241: 'posteritas Ade.' *De laudibus*, p. 463: 'natis Ade'; p. 499: 'Ade successio.'

[27] Pp. 3, 185, 257; *De laudibus*, pp. 414, 453, 500.

[28] P. 255.

[29] P. 311. Cf. Meyer, *Corrogationes Promethei*, p. 658.

[30] MS. *intuecium*. Cf. *De naturis rerum*, p. 94.

[31] MS. *risistere*.   [32] MS. *mirabilis*.

following passage from the first part of the *Corrogationes* can be paralleled in almost every phrase by the text of the *Sacerdos*: [33]

Habet igitur gramatica suas regulas, dialetica maximas, rethorica locos communes, arismetica aporismata, musica anxiomata, geometria theoremata, astronomia continet canones sicut et decretorum volumen, medicina aphorismos, civilis iuris peritia regulas iuris, theologia regulas sicut et gramatica, unde etiam regulas Ticonii dicimus in celesti pagina.[34]

Still more striking are the parallels between both parts of the *Corrogationes* and the gloss in the Caius College manuscript, which, being essentially lexicographical, follows the same method in illustrating the use of words and explaining their meaning and etymology. French equivalents are freely given in the gloss,[35] as in the *Corrogationes*, and the two works are usually in close verbal agreement. Examples are: [36]

Quoniam igitur effluentia tempora cicius effectum suum apparere faciant in illa regione capitis que gall. dicitur *temples* (p. 8; Meyer, p. 664). Equi fortes emissarii dicuntur gall. *estaluns* (p. 11; Meyer, p. 674). Commissa sunt pignora, gall. *encuru* (p. 12; Meyer, p. 677). Pincerne debet dici, Re-

[33] See especially lines 51–56.

[34] MS. 72 of the library of Evreux, f. 3; and in the British Museum, Harl. MS. 6, f. 150; Royal MSS. 2, D, VIII, f. 17, and 5, C, V, f. 2 v; *Notices et extraits*, xxxv, 2, p. 660. For other MSS. see *E. H. R.*, xxx. 463.

[35] There are many French words in the gloss which are not in the *Corrogationes*. Examples are: nastilus, *butun* (p. 8); manipulum, *fanun* (p. 9); calx, *chauz* (p. 11); antidonum, *werdun* (p. 12); abdicare, *desavoer* (p. 13); lavatorium, *lavur* (p. 14); capus, avis, *muschet*; cippus, *cep*; acceptifero, *clamer quite*; accipiter, *ostur*, ab australi parte veniens (p. 17); munium, *forcele*; matricuria, *custerere*; subula, *aleyne* (p. 18); catovolatilibus, *cheysil*; apote et antapote, *taile et contretaile*; instauramenta, *les estors de la mesun*; statera, *balance* (p. 20); locusta, *languste* (p. 21); classicum, *glas*; testudines, *voutes*, et dicuntur a testudine, gall. *limazun* (p. 25); serum, *mege*; sero, *enter* (p. 30); manutergium, *tuayle* (p. 33); musca, *musche*; rancor, *rancun*; sompnus, *dormir*; sompnia, *sunges*; catalaunensia, *chaluns* (p. 34); obses, *ostage*; superest, *remeynt* (p. 35); odorinsecos, *brachez* (p. 36); pilus, *pestel* (p. 37); palestris, *lute* (p. 38); municipium, *forcele*; munusculum, *benbelet* (p. 39); pedagium, *page*; larva, *visere* (p. 43); rostrum, *bec* (p. 44); cavea, *cage*; alvearia, *rusches* (p. 45); lurtisca, *lure* (p. 47); volumen, *parchemin* (p. 49); legare, *deviser*; satirici, quidam dii rurales, gall. *saleceus* (p. 50); fragum, *frese* (p. 51); operam, *entente* (p. 52); primum pilum, *baneur* (p. 55); cancellus, *chancel*. . . . item cancellus, *kenil* (p. 60). In some cases the scribe has left a blank space for the French word. An instructive study could be made of the French glosses to Neckam's works, especially those in the commentary on the *De nominibus utensilium*, where a collation of the MSS. has not yet been made. Cf. P. Meyer in the *Revue critique*, 1868, ii. 295 ff., and in *Romania*, xxxvi. 483–485; and for the MSS., *E. H. R.*, xxx. 461.

[36] See also below, nn. 40, 42, and note 2 to the text.

cense ciphum, gall. *Reschet cest hanap* (p. 13; Meyer, p. 666). Botrus est congregatio racemorum, racemus congregatio uvarum; botrus, gall. *muissine*, racemus *grape* (p. 15; Meyer, p. 674). Scorpio, *escurge* (pp. 16, 49; Meyer, p. 677). Examitus, gall. *samite* (p. 19; Meyer, p. 666). Criptas, gall. *crute* (p. 25; Meyer, p. 678). The gloss on Martial's murrina pocula (p. 28; Meyer, p. 667; cf. the use of the phrase in *De naturis rerum*, i). Protectum = *apentiz* (p. 30; Meyer, p. 679). Taxare iudicis est, *amesurer* gall. (p. 36; Meyer, p. 674). Taxus pro arbore que gal. dicitur *yf* (*ibid.; Revue Critique*, 1868, ii. 295). Macula est in oculo meo, g. *mayle est en le oyl* . . . Macula corporis est lesura, gall. *mayme* (p. 38; Meyer, pp. 673–674).

Examination of earlier lexicons would doubtless reveal the origin of the Latin portion of the greater part of these glosses, indeed the correspondence between the *Sacerdos ad altare* and any one of Neckam's writings might be explained on the ground of copying or the use of a common source; but such considerations are not sufficient to destroy the cumulative force of the argument. The close agreement of the text with the *De nominibus utensilium* and the *De naturis rerum*, and the exact correspondence of the gloss, in both Latin and French, with the *Corrogationes*, taken with the general similarities of style, point clearly to the conclusion that text and gloss are the work of one writer and that this writer is Alexander Neckam. This view is strengthened by considerations which show that both text and gloss were composed toward the close of the twelfth century [37] by one familiar with the schools of Paris, and that the gloss, at least, was written in England.

Let us begin with the gloss. Its author had studied at Paris, for he cites the *magistri Parisienses* on a question of etymology,[38] and knows the city even to its stenches,[39] and he gives as an ex-

---

[37] Only further critical study can determine its chronological place among Neckam's works, whose dates have so far been but little investigated. In general it would seem that the grammatical works belong to the earlier period of his literary activity; the *Corrogationes* are certainly anterior to the *De naturis rerum*, in which they are cited (p. 16), and this is plainly earlier than its metrical paraphrase, the *De laudibus* (cf. Wright's introduction, p. lxxiv), to which he later composed a supplement (*E. H. R.*, xxx. 460).

[38] He says (p. 15) apropos of the word *cassilide* in certain MSS. of the Book of Tobit: 'Quidam autem qui in oculis suis scioli sunt capsilide dicunt; dicunt enim quod est dictio composita ex capsa et sedile. Magistri autem Parisienses dicunt cassilide a casse, quod est rethe.'

[39] P. 22: 'Unde, "Adveniente rota fetet Babilonia tota." Item dicitur (?) bene, Parisius Babilonia vult imitari in fetore suo.'

ample of a two days' journey the distance from Paris to a place which in the original was doubtless Orleans, as in the *Corrogationes*, but which the copyist, with strange disregard of space, has made into England.[40] Yet our glossator is no Frenchman; he speaks of tournaments as the "sport of French knights," [41] and he lives near enough to Wales — Cirencester was in a border county — to use the Welsh wars as an illustration of fighting.[42] As he cites the decree of the Third Lateran Council forbidding tournaments as "detestable fairs," [43] he must have written after 1179, and as they are still a French custom to him, he probably wrote before their introduction into England by Richard I, in 1194.[44]

[40] P. 38: 'Sunt enim ab Anglicanis due diete Parisius.' Cf. Meyer, *Corrogationes*, p. 667.

[41] P. 38: 'Troiana agmina a vulgo tormenta dicuntur ad differentiam hastiludiorum, que Alexander papa tercius detestabiles vocat nundinas. Item dici solent ab exercicio francorum militum.' On the French origin of tournaments and the mediaeval opinion which derived them from the games described in the Aeneid, see Du Cange, *Glossarium*, under *torneamentum*, and his sixth dissertation on Joinville. Neckam also refers to the *Troiana agmina* in the *De nominibus*, ed. Scheler, p. 70.

[42] P. 38, where after the passage concerning *oploma* printed by Meyer (*Corrogationes*, p. 667) he says: 'Unde Seneca in declamationibus [3, praef., 10], "Quidam cum oplomatis, quidam cum Tracibus bene pugnant" . . sed pugna cum Tracibus vel cum Wallensibus non est imaginaria pugna sed vera, sicut illa que cum viciis fit.' This passage is also in the *Corrogationes* (Royal MS. 2. D. viii, f. 43), and the same idea appears in a brief poem of Neckam addressed to Thomas, abbot of Gloucester (1179–1205), and preserved in a volume of extracts from Neckam's works, now in the library of the University of Cambridge (Gg. VI, 42, f. 223):

MAGISTER ALEXANDER DOMINO T. ABBATI CLAUDIE

> Munus sed munusculum tibi mitto, Thoma,
> Optans ut nec videas Romam nec te Roma,
> Nec Romanum audias rursus ydioma.
> Vix minus displiceat tibi vile scoma;
> Romanorum oculos excecet glaucoma.
> Revertentes felix vos reduxit duploma.
> Claudie te teneat sancti claustri doma;
> Ibi corpus maceres, ibi carnem doma;
> Pugnantem cum viciis te tegat opploma.

. . . .

[43] C. 20, Mansi, *Concilia*, xxii. 229.

[44] Rymer, *Foedera* (Record edition), i. 65; Roger of Hoveden, iii. 268. Cf. Ralph de Diceto, ii, pp. lxxx–lxxxi, 120; William of Newburgh, in Howlett, *Chronicles of Stephen*, ii. 422–423.

The text is, of course, not later than the gloss, and internal evidence assigns it to the same period. The most specific indices of date are afforded by the books enumerated under canon law and logic. The absence of any canonical works more recent than the decretals of Alexander III not only carries us back of the *Decretals* of Gregory IX (1234), but makes it improbable that the author wrote long after 1191, the latest date for the publication of the so-called *Compilatio prima* of Bernard of Pavia, the earliest of the collections of decretals known as the *Quinque compilationes*.[45] 'Decretales Alexandri tertii' may have meant either some collection of that Pope's decretals made in his lifetime,[46] or the canons of the Lateran Council of 1179, or one of the collections composed under his immediate successors in which his letters still formed the dominant element;[47] but in any case the expression would not have been used more than a very few years after Alexander's time, inasmuch as the grouping of decretals by Popes very soon gave way to the arrangement by subjects which was universally followed from Bernard of Pavia on. Not earlier than Alexander III, the list of books on canon law cannot be much later than 1191.[48]

[45] The limits for the *Compilatio prima* are 1187 and 1191: Schulte, *Geschichte der Quellen des canonischen Rechts*, i. 82.

[46] Such as the collection in the British Museum described by Seckel, *Neues Archiv*, xxv. 527 (1899).

[47] The so-called *Collectio Casselana* (in Böhmer, *Corpus juris canonici*, Halle, 1747, ii, appendix, pp. 180 ff.) is entitled 'Decretales Alexandri III in concilio Lateranensi tertio generali anno MCLXXIX celebrato editae,' a title which fits only the first part of the compilation.

On the whole subject of the collections of this period see Schulte, *Beiträge zur Geschichte des canonischen Rechts von Gratian bis auf Bernhard von Pavia*, in Vienna *Sitzungsberichte* (1873), phil.-hist. Kl., lxxii. 481 ff.; Friedberg, *Die Canones-sammlungen zwischen Gratian und Bernhard von Pavia*, Leipzig, 1897 (with Seckel's review in the *Deutsche Litteraturzeitung*, 1897, coll. 658 ff.); Seckel, "Ueber drei Kanonessammlungen des ausgehenden 12. Jahrhunderts," in *Neues Archiv*, xxv. 521–537; H. Singer, *Neue Beiträge über die Dekretalensammlungen vor und nach Bernhard von Pavia*, in Vienna *Sitzungsberichte*, clxxi (1913).

[48] The line cannot be drawn sharply, for some time must be allowed for the spread of the newer collections. Stephen of Tournai, writing between 1192 and 1203, speaks of the 'inextricabilis silva decretalium epistolarum' sold under the name of Alexander III, but he does not say that the 'novum volumen,' of which he complains, composed of papal letters and read in the schools of Paris, bore this Pope's name. *Chartularium Universitatis Parisiensis*, i. 47, no. 48. Seckel thinks

This conclusion is confirmed by the list of books given under logic, where besides the familiar apparatus of the twelfth century — the *Old* and *New Logic* and the lesser treatises which regularly accompanied them — we find the *Metaphysics* of Aristotle, the *De generatione et corruptione*, and the *De anima*. Although the channels through which the *Metaphysics* and natural philosophy of Aristotle passed into western Europe are now fairly well understood,[49] the exact dates of their introduction have not been determined further than that they reached Paris, then the centre of philosophical and theological speculation, about the year 1200. Denifle pointed out that the *Metaphysics* is cited at second-hand by Peter of Poitiers, chancellor of the University of Paris, who died in 1205,[50] and by Simon of Tournai, who seems to have written before 1201, while he also maintained that the *De anima* was known to Simon [51] and is quoted by Absalom of St. Victor, who died in 1203; [52] but none of these instances has withstood successfully the attacks of subsequent critics,[53] though these and other works of Aristotle were certainly used by Neckam's friend, Alfred of 'Sareshel,' before 1217.[54] Indeed the whole trend of recent inquiry points in the direction of an early date for the translations of the *Metaphysics* and the physical works, very possibly anterior to 1200. On the other hand, the public and private reading of Aristotle's books on natural philosophy and the com-

this reference is most probably to the *Compilatio* of Bernard of Pavia (Hauck-Herzog, *Realencyklopädie³*, xvi. 292).

[49] Cf. Chapter XI, n. 2.

[50] *Chartularium Universitatis Parisiensis*, i. 61, 71.

[51] *Chartularium*, i. 71; Hauréau, *Histoire de la philosophie scholastique*, part 2, i. 59; idem, *Notices et extraits de quelques MSS. de la Bibliothèque Nationale*, iii. 256. On Simon's date see *Chartularium*, i. 45; Hauréau, *Notices et extraits*, i. 179. Matthew Paris narrates as of 1201 the story of the miracle which is said to have ended his studies (*M. G. H., Scriptores*, xxviii. 116).

[52] *Chartularium*, i. 71. For the date of the abbot's death see *Gallia Christiana*, vii. 673. According to Hauréau, *Histoire de la philolosophie scholastique*, part 2, i. 63, Neckam's *De nominibus utensilium* has a reference to the *De anima*. See also Thorndike, ii. 194 f.

[53] Baeumker, *Die Stellung Alfreds von Sareshel*, especially pp. 35 f., 44–46; and in *Philosophisches Jahrbuch*, xxvii. 479; Grabmann, *Aristotelesübersetzungen*, pp. 19–21, 190 ff.; Minges, in *Archivum Franciscanum historicum*, vi. 17 (1913).

[54] Supra, Chapter VI, end. For citations of the *De anima* in 1143, see Chapter III, n. 151.

mentaries upon them at Paris was forbidden by a provincial
council in 1210,[55] and the prohibition was repeated and extended
to the *Metaphysics* by the statutes of the papal legate in 1215.[56]
They were still under the ban in 1231, when Gregory IX decreed
that they should not be used until they had been examined and
purged from error;[57] but they are found in general use shortly
afterward,[58] and the whole of the new Aristotle appears in the arts
course of 1255.[59] The meagreness of the list in the *Sacerdos ad
altare* as compared with the large number of Aristotelian and
Pseudo-Aristotelian treatises prescribed in 1255 points to a much
earlier date, while the prohibitions of 1210 and 1215 make it likewise probably anterior to 1210. Indeed, so far as the chronological considerations already urged carry weight, it would seem that
the *Sacerdos ad altare* contains one of the earliest mentions of the
Metaphysics and the *De generatione* in Latin Europe. If this
mention is an addition to the original list of the *Sacerdos ad altare*,
the original list is still earlier.

The texts enumerated in other subjects do not yield chronological information of quite so definite a character, but they abundantly confirm the general conclusion that the list represents the
learning of the twelfth century and not of the thirteenth. In
medicine the author is familiar with the early translations from
the Arabic, but not with Avicenna, whose influence dates from
the thirteenth century; the omission of the *Versus Egidii*, composed by Giles of Corbeil, contemporary of Philip Augustus, likewise points to an early date.[60] As compared with the texts pre-

[55] *Chartularium*, i. 70.

[56] *Ibid.*, i. 78. The *Metaphysics* may have been included in the *libri naturales*
condemned in 1210: Luquet, *Aristote et l'Université de Paris* (Paris, 1904), pp. 20–27.

[57] *Chartularium*, i. 136.

[58] Notably in the works of William of Auxerre, Philip de Grève, and William of
Auvergne: Jourdain, pp. 288–299; Valois, *Guillaume d'Auvergne*, p. 200; Minges,
in *Philosophisches Jahrbuch*, xxvii. 21–32 (1914); Grabmann, *Aristotelesübersetzungen*, pp. 28–38. See also Hauréau, in *Notices et extraits des MSS.*, xxxi, 2, p. 288;
and Roger Bacon, in Rashdall, *Universities*, ii. 754.

[59] *Chartularium*, i. 277. The *De anima* appears in 1252 in the statutes of the
English Nation (*ibid.*, i. 227).

[60] On Egidius see the note in the Paris *Chartularium*, i. 517; the introduction to
V. Rose, *Egidii Corboliensis Viaticus* (Leipzig, 1907); and C. Vieillard, *Gilles de
Corbeil* (Paris, 1909).

scribed in the earliest medical statutes, those of Paris between
1270 and 1274,[61] Naples in 1278, and Salerno in 1280,[62] the most
important difference is the inclusion of Alexander of Tralles and
of *materia medica* as represented in the works of Dioscorides and
the so-called Macer. *Iohannicius*, Hippocrates, Galen, and the
Pantegni are also mentioned in our list and not in these statutes,
but no inference can be drawn from the absence of these names
from the statutes, where they may have been included under the
ars medicinae, a phrase which apparently designated a well known
series of treatises rather than any particular work.[63]

In mathematics and astronomy the author of the *Sacerdos ad
altare* knows only Euclid and the astronomical compendium of
Alfraganus, which were put into Latin in the earlier part of the
twelfth century,[64] and Ptolemy's *Canons*; he does not mention
the *Almagest*, of which translations were made in Sicily ca. 1160
and in Spain in 1175,[65] or any of the mathematical works of the
early thirteenth century.

In grammar we find only the well known texts of the earlier

---

[61] *Chartularium, l. c.*

[62] De Renzi, *Collectio Salernitana* (Naples, 1852), i. 361; Haeser, *Geschichte der
Medizin* (Jena, 1875), i. 829, where the date is wrongly given as 1276.

[63] *Chartularium*, i. 517: 'Debet audivisse bis artem medicine ordinarie et semel
cursorie, exceptis urinis Theophili, quas sufficit semel audivisse ordinarie vel cursorie.' Rashdall, *Universities*, i. 429, identifies this *Ars medicine* with the *Ars parva*
or *Tegni* of Galen. But it plainly includes the *De urinis* of Theophilus and seems to
denote a regular set of treatises which students were in the habit of using. The
language of the Naples and Salerno statute is still clearer in support of this view:
'Teneatur baccalarius audivisse bis ordinarie ad minus omnes libros artis medice,
exceptis urinis Theofili et libro pulsuum Filareti, quos sufficit audivisse semel ordinarie vel cursorie' (de Renzi, i. 362). The title *Ars medicine* occurs in various library catalogues (e. g. Delisle, *Cabinet des MSS.*, iii. 66), and the Erfurt library likewise has examples of an *Ars commentata*, copied in 1260 and 1288, which contains
the treatises of Philaretus and Theophilus, the *Iohannicius*, the *Tegni*, and the
Aphorismi and *Pronostica* of Hippocrates (MSS. F 264 and F 285: Schum, *Beschreibendes Verzeichnis der Amplonianischen Handschriften-Sammlung*, pp. 172, 192).

[64] On the translations of Euclid see Weissenborn, *Z. M. Ph.*, hist.-litt. Abth., xxv;
and Steinschneider, *ibid.*, xxxi; supra, Chapter II, n. 26. On Alfraganus (al-
Fargani) and his translators see Mädler, *Geschichte der Himmelskunde* (Braun-
schweig, 1873), i. 91–93; Wüstenfeld, pp. 26, 63; Suter, p. 18; Steinschneider, in
B. M., 1892, pp. 55–56, and his *H. U.*, pp. 554–556.

[65] Supra, Chapter V, n. 53, where it is noted that the translation of the *Canons*
still requires investigation.

Middle Ages, Donatus and Priscian and Remigius of Auxerre, with no mention of the popular works of the thirteenth century, the *Doctrinale* of Alexander of Villedieu, composed in 1199, or the *Grecismus* of Evrard of Béthune, which appeared in 1212.[66]

But if our list represents in general the learning of the twelfth century and not that of the thirteenth, it still belongs to the last quarter of its century and not to an earlier age. Apart from the decisive indications afforded by the mention of the *Decretals* of Alexander III and the *Metaphysics* and natural philosophy of Aristotle, it is plainly subsequent to the *Eptatheuchon* of Thierry of Chartres, composed before 1155 and itself in many respects far advanced for its time.[67] In the studies of the trivium there is substantial agreement, although Thierry does not have Remigius, Apuleius, or the 'Apodoxim';[68] but when we come to geometry, we find that Thierry knows only the Pseudo-Boethius and the *agrimensores*, and in astronomy he is restricted to the *Canons* of Ptolemy and certain tables.[69]

The respectable list of classical authors which our text contains also points to the twelfth century rather than the thirteenth, when dialectic had driven the poets, historians, and moralists of ancient Rome from the curriculum in arts.[70] In the contest between the humanists and the logicians, Neckam is on the whole to be reckoned on the side of the humanists, not only by reason of his familiarity with the Roman poets but also because of the contempt he expresses for the subtleties of scholastic reasoning.[71] In the *De naturis rerum* and the *Corrogationes* he quotes frequently and often at some length from Lucan, Ovid, Virgil, Claudian, Juve-

---

[66] See Reichling's introduction to his edition of the *Doctrinale* (Berlin, 1893).

[67] Supra, Chapter V, n. 51.

[68] On the *Posterior Analytics* (*Apodoxim*) see Chapter XI. Neckam, *De naturis rerum*, p. 293, speaks of the period before it was known at Paris.

[69] He knows, but does not here use, the *Planisphere*.

[70] This is seen in the earliest university curriculum in arts, the Paris course of 1215(*Chartularium*, i. 78). Cf. Denifle, *Universitäten*, i. 758; Rashdall, *Universities*, i. 71, 433; Norden, *Die antike Kunstprosa*, ii. 725–726; Paetow, "The Neglect of the Ancient Classics at the Early Mediaeval Universities," in *Transactions of the Wisconsin Academy*, xvi. 311–319 (1908); the same, *The Arts Course at Medieval Universities*; the same, *The Battle of the Seven Liberal Arts* (Berkeley, 1914).

[71] *De naturis rerum*, pp. 302 ff. Cf. *E. H. R.*, xxx. 451, n. 10.

nal, Martial, Statius, and Horace. He also draws largely from Solinus, and cites Pliny, Cicero, and Macrobius. How much further his classical knowledge went, cannot be determined without a study of his unprinted works, and even then we cannot be sure to what extent he relied upon collections of extracts [72] or upon citations in Priscian and similar works.[73] For the same reason we cannot be certain how many of the writers mentioned in the *Sacerdos ad altare* were really known to its author, and we must be careful not to take the list too literally as representing what was actually read in the schools of Neckam's day. The number of authors is naturally less than the number of those cited by the most learned classical scholar of the preceding generation, John of Salisbury,[74] who is particularly full on the side of the historians; but save for the mention of Martial and the omission of Persius, the list of poets stands in substantial agreement with the more ambitious attempts of Conrad of Hirschau [75] and Hugh of Trimberg.[76] Of the ancient writers not mentioned in the text the gloss cites Persius, Claudian, Plautus,[77] Terence,[78] Valerius Maximus, Josephus, Macrobius, Prudentius, Fulgentius (*Mythologiae*), Chrysostom, and Martianus Capella.

As I have not been able to find another copy of the *Sacerdos ad altare*, the portion printed below is a faithful reproduction of the Caius College MS. Occasionally an obvious slip of the scribe has been corrected in the text, but in all such cases the MS. reading is given in a note.

[72] Such as the Paris collection described by Wölfflin, *Philologus*, xxvii. 153; cf. Norden, *o. c.*, ii. 720; and the doctoral dissertation of Miss Eva M. Sanford on mediaeval *florilegia, Harvard Studies in Classical Philology*, xxxiv. 195–197 (1923). For such a set of extracts see MS. Vat. Pal. lat. 957, f. 97 (saec. xiii).

[73] Cf. Abelson, *Seven Liberal Arts*, pp. 23, 39, note 2.

[74] Schaarschmidt, *Johannes Saresberiensis*, pp. 81–125; Webb's edition of the *Policraticus*, i. pp. xxi–xlviii; A. C. Krey, in *Transactions of the Wisconsin Academy*, xvi, 2, pp. 948–987 (1910). The list of historians which John's pupil, Peter of Blois, says he has read (*Chartularium Univ. Par.*, i. 29) has a suspicious resemblance to that given by his master in the *Policraticus*, 8, 18. Cf. Rashdall, *Universities*, i. 65; Norden, *Kunstprosa*, ii. 719.

[75] *Conradi Hirsaugiensis Dialogus super auctores*, ed. Schepss, Würzburg, 1889.

[76] Huemer, *Das Registrum multorum auctorum des Hugo von Trimberg*, in Vienna *Sitzungsberichte*, phil.-hist. Kl., cxvi. 145–190.

[77] *Aulularia*, 400 (p. 41), and one or two doubtful citations.

[78] P. 24: 'lacrime pro gaudio' (*Adelphoe*, 536–537).

(P. 47.) Scolaris liberalibus educandus artibus dipticas gerat quibus
scitu digna scribantur. Ferat palmatoriam sive volariam vel ferulam
qua manus puerilis leniter feriatur ob minores excessus, virgis vero
cedatur cum res id fieri desideraverit. Absint flagella et scorpiones, ne
5 modum excedat castigando. Postquam alphabetum didicerit et ceteris
puerilibus rudimentis imbutus fuerit, Donatum et illud utile mora- (p.
48) litatis compendium quod Catonis esse vulgus opinatur addiscat et
ab egloga Theodoli ¹ transeat ad egglogas bucolicorum, prelectis tamen
quibusdam libellis informacioni rudium necessariis. Deinde satiricos et
10 ystoriographos legat, ut vicia etiam in minori etate addiscat esse fu-
gienda et nobilia gesta eroum desideret imitari. A thebaide iocunda
transeat ad divinam eneida, nec neggligat vatem quem Corduba genuit ²
qui non solum civilia bella describit sed et intestina. Iuvenalis moralia
dicta in archano pectoris reservet, et Flacium nature summopere vitare
15 studeat. Sermones Oracii et epistolas legat et poetriam et odas cum
libro epodon. Elegias Nasonis et Ovidium metamorfoseos audiat ³ sed
et precipue libellum de remedio amoris familiarem habeat. Placuit
tamen viris autenticis carmina amatoria cum satiris subducenda esse a
manibus adolescencium, ac si eis dicatur,

20   Qui legitis flores et humi nascencia fraga,⁴
           Frigidus, o pueri, fugite hinc, latet anguis in herba.⁵

Librum fastorum non esse legendum nonnullis placet. Stacius Achil-
leidos etiam a viris multe gravitatis probatur. Bucolica Maronis et
georgica multe sunt utilitatis. Salustius et Tullius de oratore et thus-
25 canarum et de amicicia et de senectute et de fato multa commendacione
digni sunt et paradoxe. Liber inscriptus de multitudine deorum ⁶ a
quibusdam reprobatur. Tullius de officiis utilissimus est. Martialis
totus et Petronius ⁷ multa continent in se utilia sed multa auditu in-

¹ On the popularity of the Eclogues of Theodulus in the Middle Ages, when they
were closely associated with the *Disticha Catonis* and Avianus, see Manitius, in the
Mitteilungen der Gesellschaft für deutsche Erziehungs- und Schulgeschichte, xvi. 38–39,
233–235 (1906); and *Lateinische Litteratur*, i. 570–574, ii. 811; G. L. Hamilton, in
Modern Philology, vii. 169–185 (1909); Osternacher, in *Neues Archiv*, xl. 331–376
(1915). The *Disticha Catonis* is now conveniently edited, with an English translation, by Wayland J. Chase in the University of Wisconsin *Studies* (Madison, 1922).
² Here the gloss says (p. 50): 'Corduba est nomen civitatis de qua oriundus est
Seneca, et inde Lucanus Cordubanus nomen accepit. Et nota quod Lucanus non
ponitur in numero poetarum quia historiam composuit et non poema.' Cf. *De naturis rerum*, pp. 309, 337. Sandys, *History of Classical Scholarship*³, i. 550, note.
omits Lucan from his list of the authors mentioned in this text, which he still (1921)
ascribes to Garland.
³ MS. *audeat*.
⁴ MS. *fragra*, but the gloss has *fraga*.
⁵ Virgil, *Bucol.*, 3, 92–93.
⁶ I. e., *De natura deorum*.
⁷ According to Manitius, *Rheinisches Museum*, xlvii, Erg.-Heft, p. 57, citations
of Petronius are rare in France in the Middle Ages.

digna. Simachi breve genus dicendi admiracionem [8] parit. Soliqum [9]
30 de mirabilibus mundi et Sydonium et Suetonium et Quintum Curcium
et Trogium Pompeium [10] et Crisippum [11] et Titum Liphium commendo,
sed Senecam ad Lucillum (p. 49) et de questionibus phisicis et de beneficiis relegere tibi utile censeas. Tragediam ipsius et declamaciones
legere non erit inutile.

35   (P. 52.) Gramatice daturus operam audiat et legat barbarismum
Donati et Prisciani maius volumen cum libro constructionum [12]   et
Remigium et Priscianum de metris et de ponderibus et duodecim versibus Virgilii et Priscianum de accentibus, quem tamen multi negant
editum esse a Prisciano, inspiciat diligenter.

40   Secundo inter liberales artes invigilare desiderans audiat librum
cathegoricorum sillogismorum editum a Boecio et thopica eiusdem
et librum divisionum et ysagogas Porphiri et cathegorias Aristotilis et
librum periarmenias [13] et librum elenchorum et priores analetichos et
apodoxim [14] eiusdem et topica et topica Ciceronis et librum periar-
45 menias Apuleii. Inspiciat etiam methafisicam Aristotilis et librum
eiusdem de generacione et corrupcione et librum de anima.[15]

---

[8] MS. *admiracioni*. Cf. the passage printed above, p. 362; and the *De naturis
rerum*, p. 94.

[9] Solinus is freely used in the *De naturis rerum*. On his popularity in the Middle
Ages see Manitius, *loc. cit.*, pp. 78 f., and in *Philologus*, xlvii. 562–565, li. 191 f.

[10] Justin is generally so styled in mediaeval catalogues. Manitius, in *Rheinisches
Museum*, xlvii, Erg.-Heft, p. 38.

[11] This name presents a problem, since, even if the author could have known of
the philosopher Chrysippus, he would have had no reason for inserting his name
among the historians of his list. Sandys conjectures Hegesippus, a plausible emendation in view of his appearance among the historians enumerated by John of Salisbury (*Policraticus*, 8, 18) and Peter of Blois (*Chart. Univ. Par.*, i. 29). I am inclined,
however, to read 'Crispum,' under which name Sallust is cited by John of Salisbury
(*Pol.*, 3, 12). This might easily have been changed to 'Crisippum' by a scribe who
knew the name from the Roman satirists. Our author may have thought Sallust
and Crispus distinct persons, which would not be surprising in view of a similar error
on the part of the best classicist of the age, John of Salisbury, who makes two historians out of Suetonius Tranquillus; or he may have used the two words merely
for variety, as in the case of Ovid and Naso. The repetition of Sallust's name is
natural here, since it is obviously as an orator and moralist that he is mentioned
with Cicero above.

[12] Here a space of six letters is left blank.

[13] A common mediaeval form for the *De interpretatione*.

[14] Sandys, in *Hermathena*, xii. 440, takes some pains to show that *apodoxium*, as
he reads the word, is a corruption of ἀποδείξεων and denotes the *Posterior Analytics*.
The matter is perfectly plain from the *De naturis rerum*, p. 293, where *apodixis* is
used as a synonym for the *Posterior Analytics*, if not from the gloss (p. 53): 'Apodiptica apellatur res demonstrativa que tractatur in libro priorum [*i. e.*, posteriorum]
analeticorum ab Aristotile.' See above, Chapter XI.

[15] Baeumker (*Philosophisches Jahrbuch*, xxvii. 485 f.) points out that this sen-

(P. 53.) In rethorica educandus legat primam Tullii rethoricam et librum ad Herrennium et Tullium de oratore et causas Quintiliani et Quintilianum de oratoris institucione.

50   Institutis arsmetice informandus arsmeticam Boecii et Euclidis [16] legat. Postea musicam Boecii legat. Sic a regulis gramatice transeat quis ad maximas dialetice, dehinc ad communes locos rethorice, post-modum ad aporismata arismetice, postea ad axiomata musice.

(P. 54.) Deinde ad theoremata geometrie que ordine artificiosissimo
55   disponit Euclides in suo libro.[17]

Demum ad canones Tholomei accedat astronomie secretis daturus operam. In artem vero quam subtilissime ediscerit Tholomeus ysa-gogas scripsit compendiosas Alfraganus.

Studium medicine usibus filiorum Ade perutile subire quis desiderans
60   audiat Ihohannicium [18] et tam aphorismos quam pronostica Ypocratis et tegni [19] Galieni et pantegni. Huius operis auctor est Galienus sed trans-lator Constantinus.[20] Legat etiam tam particulares quam universales dietas Ysaac et librum urinarum [21] et viaticum Constantini [22] cum libro

tence is plainly a later addition to the original list, as these works do not belong here under dialectic and probably represent a later phase of the curriculum here described; but the addition may very well be by Neckam himself while these treatises were still a novelty and before the prohibitions of 1210 and 1215.

[16] It is not clear why Euclid is mentioned here. In the next line *sic* is repeated.

[17] Cf. *De naturis rerum*, p. 299; 'secundum artificiosam Euclidis dispositionem.'

[18] The Latin name of the *Isagoge in artem parvam Galeni* of Honein ben Ishak, probably one of the earliest works translated into Latin from the Arabic. Cf. Stein-schneider, *H. U.*, pp. 709 ff.; and *E. U.*, no. 81; Rose, *Hermes*, viii. 338; Neu-burger, *Geschichte der Medizin*, ii, 1, pp. 166 f.

[19] I. e., τέχνη. The *Tegni* is cited in the *De naturis rerum*, p. 267. On mediaeval versions of Galen and Hippocrates, see the MSS. listed by Diels, in *Abhandlungen* of the Berlin Academy, 1905.

[20] The real author of the general text-book of theoretical and practical medicine known under the Latin title of *Pantegni* was Ali ben el-Abbas, an Arabic physician of the tenth century. See Wüstenfeld, pp. 12–16; Haeser, *Geschichte der Medizin*, i. 576; Steinschneider, *H. U.*, p. 669. On the translations of Constantinus Afri-canus see the elaborate monograph of Steinschneider, in Virchow's *Archiv für patho-logische Anatomie*, xxxvii. 351–410 (1866); and cf. Pagel, in Puschmann's *Hand-buch der Geschichte der Medizin* (Jena, 1902), i. 643 ff.; Thorndike, i, ch. 32; supra, Chapter VII; and on the use of his works in the twelfth century, Sudhoff, in *Archiv für die Geschichte der Medizin*, ix. 348 (1916), who discusses the contents of a medi-cal library ca. 1160.

[21] Of the four treatises of the Jewish physician Isaac translated by Constantinus, the *Liber febrium* is here omitted. On Isaac's works cf. Steinschneider, *H. U.*, pp. 755 ff.

[22] The original of the *Viaticum* was the work of ibn el-Jezzar, a pupil of Isaac: Steinschneider, in Virchow's *Archiv*, xxxvii. 363 ff., and *H. U.*, p. 703; Dugat, in *Journal Asiatique* (1855), 5, i. 289 ff.

urinarum et libro pulsuum [23] et Diascoriden et Macrum in quibus de
65 naturis herbarum agitur [24] et libros Alexandri.[25]

In ecclesiastico iure informandus legat Burcardum et canones seu
decreta Graciani [26] et decreta Yvonis et decretales Alexandri tertii.

(P. 55.) Iuris civilis periciam volens quis addiscere primo institutis
institucionum informetur, apices vero iuris intelligere volens audiat
70 codicem Iustiniani et utrumque digestorum volumen et tres partes et
forzatum.[27] Decimum autem librum codicis et undecimum cum duodecimo vix presumit quis legere pre nimia sui difficultate.[28]

(P. 56.) Celestem paginam audire volens, vir maturi pectoris, audiat
tam vetus instrumentum quam novum testamentum. Non solum
75 penthateuchum audiat set etiam eptatheucum, scilicet librum geneseos
et exodum, leviticum, numeros et deuteronomium, Iosue et iudicum.
Audiat postea Ruth et librum regum et librum paralipomenon qui et
liber dierum dicitur ab Ebreis. Audiat Hesdram et Neemiam et
Tobyam, Iudith et Hester. Felix erit si in noticiam venerit prophetice
80 doctrine que in Ethe,[29] Ysaya, Ieremya et Daniele et in libro duodecim
prophetarum continetur. Pascet pias meditaciones mentis liber Iob.
Accedat etiam ad librum parabolarum Salomonis et ad ecclesiastem et
ad cantica canticorum. Utiles etiam erunt auditu tam liber sapientie qui
Philonis dicitur quam ecclesiasticus quem conditum esse a Iesu filio

[23] Probably the works of Theophilus are meant.

[24] Macer is the second title of a work *De naturis herbarum* probably written by
Odo of Meung-sur-Loire in the eleventh century. See Rose, in *Hermes*, viii. 63;
Manitius, in *Philologus*, li. 171 (= lii. 545), and in *Mitt. Gesells. Erziehungsgeschichte*,
xvi. 251–253; H. Stadler, in *Archiv für die Geschichte der Naturwissenschaften*, i. 52–
65 (1909); C. Resak, *Odo Magdunensis* (Leipzig diss., 1917); Manitius, *Lateinische
Litteratur*, ii. 539–547; Thorndike, i. 612–615, following his account of the Latin
Dioscorides (pp. 608–611, with references). See also above, Chapter VII, n. 15.
Macer and Dioscorides are mentioned in the *De naturis rerum*, p. 275.

[25] Alexander of Tralles. On his writings see Bloch in Puschmann's *Handbuch*, i.
535–544; Thorndike, ch. 25.

[26] Here the gloss reads: 'Decreta Gratiani dicuntur decreta que tantum modernis sunt in usu, que ultimo composita sunt a Grationo et autenticata (?) a sede
Romana ita quod alia ab aliis composita publice legerentur, ut cum dicitur, Iste
legit decreta, semper intelligendum Gratiani que sola approbata sunt a sede apos-
tolica. . . . Sed decreta que Yvo composuit et Burcardus omnino recesserunt ab
aula nisi ea que inde sumuntur a Gratiano in suis decretis.'

[27] I. e., *infortiatum*, the mediaeval name for the portion of the Digest extending
from 24, 3, to 35, 2, 82, where the *Tres partes* begins.

[28] The last three books of the Code, treating of the administrative law of the later
empire, were naturally less important and less intelligible in the Middle Ages than
the other books. Under the title of *Tres libri* they were commonly grouped with the
treatises which made up the *Volumen parvum*, and occupied a subordinate place in
the course of legal instruction.

[29] So in MS.

85   Sirach perhibent. Liber [30] Machabeorum prelia Iude et Ionathe fratris
      eius et Symonis explicabit. Quam vero sit utilis liber (p. 57) psalmorum
      nemo satis fideliter verbis posset explicare. Novum autem testamentum
      audire quis desiderans audiat Matheum cum Marco, Lucam et Iohan-
      nem, epistolas Pauli cum canonicis epistolis, actus apostolorum, et
90   apochalipsim Iohannis.

[30] MS. *leber*.

INDEXES

INDEX OF MANUSCRIPTS
AND LIBRARIES

Unless otherwise indicated, the library is in each case the public library of the town.

SUBJECT INDEX

INDEX OF PROPER NAMES

Henricus Aristippus, 53, 142 f., 150, 152,
159–163, 165–172, 179, 181, 182, 183,
190, 191, 225, 236.
Henry of Avranches, 276, 316.
Henry Bate, 111.
Henry, archbishop of Benevento, 195 f.
Henry of Blois, 29.
Henry of Cologne, 279.
Henry VI, emperor, 208.
Henry I, king of England, 20, 26, 27,
119, 328, 332, 333, 346, 348.
Henry II, king of England, 28 f., 34, 35,
41, 169, 170, 188, 189, 317, 328, 332, 348.
Henry III, king of England, 255, 274.
Henry, patriarch of Grado, 196.
Henry, C., 35.
Heraclides of Pontus, 88, 89.
Herbert of Braose, 187.
Herbert of Middlesex, 187.
Hercules, cave of, 19.
Hereford, 124 f.
Hergenröther, J., 195, 196, 213, 214.
Heriger of Lobbes, 334.
Hermann of Carinthia, 9, 11 f., 30, 33,
35, 43–66, 67, 68, 82, 89, 90, 91, 96,
104, 120, 121, 122.
Hermann the Dalmatian, *see* Hermann of
Carinthia.
Hermann the German, 15, 16, 43.
'Hermannus,' 53 f., 159, 161.
Hermannus Alemannus, *see* Hermann the
German.
Hermannus Contractus, 43, 52, 53, 115,
162.
Hermes Trismegistus, 30, 51, 57, 58, 61,
66, 79, 80, 220, 270, 288.
Hero of Alexandria, 39, 143, 153, 160,
181 ff., 189, 244.
Hertfordshire, 329.
Hervieux, L., 360.
Hesiod, 65.
Heyd, W. von, 198.
Hierocles, 269.
Hilarius Pictaviensis, 211, 212.
Hilka, A., 147.
Hipparchus, 62, 66, 88, 89, 109, 164.
Hippocrates, 3, 5, 15, 18, 66, 67, 88, 94,
98, 145, 151, 208, 218, 316, 369, 374.

Hocedez, E., 152.
Hönger, F., 250.
Hofmeister, A., 55, 142, 213, 223, 226,
228, 230, 231, 236.
Holder-Egger, O., 173, 174, 276.
Homerocentones, 200.
Honein ben Ishak, 374.
Honorius of Autun, 338.
Honorius III, pope, 138, 274, 282.
Horace, 28, 38, 41, 200, 371, 372.
Horna, K., 173.
Huart, M. d', 202.
Huber, M., 142.
Huemer, J., 371.
Huesca, 118.
Hugh de Bocland, 328 f., 332.
Hugh, Master, 239.
Hugh, cardinal priest of St. Sabina, 139.
Hugh of St. Victor, 338.
Hugh of Santalla, *see* Hugo Sanctallensis.
Hugh of Trimberg, 371.
Hugo Eterianus, 131, 146, 196, 197, 210,
213–218.
Hugo Falcandus, 156, 160, 188.
Hugo Sanctallensis, 9, 12, 33, 67–81.
Hugutio, 150, 251.
Huillard-Bréholles, J. L. A., 242, 247,
248, 249, 250, 252, 253, 255, 256, 257,
261, 268, 269, 270, 279, 284, 299, 300,
324.
Hultsch, F., 178.
Hungary, 146.
Hurter, H., 195.
Hyginus, 83, 91, 338, 341.
Hypsikles, 91.

Iberi, 205.
Iceland, 316. See *Ysland*.
Illyricum, 210.
India, 63, 97, 267, 290, 306.
Innocent II, pope, 169.
Innocent III, pope, 276.
Innocent IV, pope, 139, 276.
'Ioannes Ocreatus,' *see* Ocreatus.
Ioanton, Ioathon, 286, 341–344.
Iohannicius, see Johannitius.
'Iorma Babilonicus,' 66.
Irak, 264.